Atmosphere and Climate

The Handbook of Natural Resources, Second Edition

Series Editor:
Yeqiao Wang

Volume 1
Terrestrial Ecosystems and Biodiversity

Volume 2
Landscape and Land Capacity

Volume 3
Wetlands and Habitats

Volume 4
Fresh Water and Watersheds

Volume 5
Coastal and Marine Environments

Volume 6
Atmosphere and Climate

Atmosphere and Climate

Edited by
Yeqiao Wang

CRC Press
Taylor & Francis Group
Boca Raton London New York

CRC Press is an imprint of the
Taylor & Francis Group, an **informa** business

CRC Press
Taylor & Francis Group
6000 Broken Sound Parkway NW, Suite 300
Boca Raton, FL 33487-2742

First issued in paperback 2022

ISBN 13: 978-1-03-247441-0 (pbk)
ISBN 13: 978-1-138-33967-5 (hbk)

DOI: 10.1201/9780429440984

Library of Congress Cataloging-in-Publication Data

Names: Wang, Yeqiao, editor.
Title: Handbook of natural resources / edited by Yeqiao Wang.
Other titles: Encyclopedia of natural resources.
Description: Second edition. | Boca Raton: CRC Press, [2020] | Revised edition of: Encyclopedia of natural resources. [2014]. | Includes bibliographical references and index. | Contents: volume 1. Ecosystems and biodiversity — volume 2. Landscape and land capacity — volume 3. Wetland and habitats — volume 4. Fresh water and watersheds — volume 5. Coastal and marine environments — volume 6. Atmosphere and climate. | Summary: "This volume covers topical areas of terrestrial ecosystems, their biodiversity, services, and ecosystem management. Organized for ease of reference, the handbook provides fundamental information on terrestrial systems and a complete overview on the impacts of climate change on natural vegetation and forests. New to this edition are discussions on decision support systems, biodiversity conservation, gross and net primary production, soil microbiology, and land surface phenology. The book demonstrates the key processes, methods, and models used through several practical case studies from around the world" — Provided by publisher.
Identifiers: LCCN 2019051202 | ISBN 9781138333918 (volume 1 ; hardback) | ISBN 9780429445651 (volume 1 ; ebook)
Subjects: LCSH: Natural resources. | Land use. | Climatic changes.
Classification: LCC HC85 .E493 2020 | DDC 333.95—dc23
LC record available at https://lccn.loc.gov/2019051202

Visit the Taylor & Francis Web site at
http://www.taylorandfrancis.com

and the CRC Press Web site at
http://www.crcpress.com

Contents

SECTION I Atmosphere

SECTION II   Weather and Climate

SECTION III   Climate Change

Preface

Atmosphere and Climate is the sixth volume of The *Handbook of Natural Resources*. This volume consists of 40 chapters authored by 68 contributors from 7 countries. The contents are organized in three sections: *Atmosphere* (16 chapters); *Weather and Climate* (16 chapters); and *Climate Change* (8 chapters).

Land and atmosphere interactions are driven by continuous exchange of heat, moisture, momentum, and various gases. Land surface heterogeneity leads to spatial variation of land–atmosphere exchanges resulting in various natural phenomena, such as sea breezes, valley winds, and monsoons, at wide ranges of spatial and temporal scales. Human induced land-use and land-cover change from deforestation, agriculture, and urbanization modify such exchanges and affect weather, climate, water, and carbon cycle at multiple scales. Growing demands for food, energy, and natural resources imply that land-use and land-cover change will continue to intensify in the future. Understanding the underlying process of land–atmosphere interactions is critical for predicting the impact of future change on the environment and natural resource base.

Climate change studies reveal the effects of global warming and provide understanding of the changing environment, the impacts and threats caused by changes, and the likely trends in the future for natural resources and associated ecosystems. Adaptive management and mitigation strategies require data and indicators to science-based assessment and effective implementation. This is especially important in view of a changing climate, with direct and indirect consequences on the natural resources.

With the challenges and concerns, the 40 chapters in this volume cover topics in *Atmosphere*, including acid rain and precipitation chemistry, acid rain and nitrogen deposition, air pollutants, albedo, Asian monsoon, atmospheric acid deposition, general atmospheric circulation, atmospheric and oceanic circulation patterns, dew point temperature, fronts, land-atmosphere interactions, ozone and ozone depilation, transpiration and physical evaporation, atmospheric water storage, arctic oscillation, and estimation of Arctic sea-ice shortwave Albedo; in *Weather and Climate*, including agroclimatology, climatology, climate classification, extreme events, moist enthalpy and long-term anomaly trends, trace gas exchange between crops and atmosphere, precipitation, evapotranspiration and soil moisture, drought resistance, drought management, El Niño, La Niña, and the Southern Oscillation, tropical meteorology, ocean-atmosphere interactions, ocean observation and prediction, tropical cyclones, urban heat islands, and wind speed probability distribution; and in *Climate Change*, including climate change, climate change and boreal forests, climate change in coastal marine environment, climate change in polar regions, climate change effects on habitat suitability, climate change and ecosystem dynamics, spatial and temporal variation in global land surface phenology, and in precipitation and temperature.

The chapters provide updated knowledge and information in general environmental and natural science education and serve as a value-added collection of references for scientific research and management practices.

Yeqiao Wang
University of Rhode Island

About The Handbook of Natural Resources

With unprecedented attentions to the changing environment on the planet Earth, one of the central focuses is about the availability and sustainability of natural resources and the native biodiversity. It is critical to gain a full understanding about the consequences of the changing natural resources to the degradation of ecological integrity and the sustainability of life. Natural resources represent such a broad scope of complex and challenging topics.

The *Handbook of Natural Resources, Second Edition (HNR)*, is a restructured and retitled book series based on the 2014 publication of the *Encyclopedia of Natural Resources (ENR)*. The *ENR* was reviewed favorably in February 2015 by CHOICE and commented as *highly recommended for lower-division undergraduates through professionals and practitioners*. This *HNR* is a continuation of the theme reference with restructured sectional design and extended topical coverage. The chapters included in the *HNR* provide authoritative references under the systematic relevance to the subject of the volumes. The case studies presented in the chapters cover diversified examples from local to global scales, and from addressing fundamental science questions to the needs in management practices.

The *Handbook of Natural Resources* consists of six volumes with 241 chapters organized by topical sections as summarized below.

Volume 1. Terrestrial Ecosystems and Biodiversity
Section I. Biodiversity and Conservation (15 Chapters)
Section II. Ecosystem Type, Function and Service (13 Chapters)
Section III. Ecological Processes (12 Chapters)
Section IV. Ecosystem Monitoring (6 Chapters)

Volume 2. Landscape and Land Capacity
Section I. Landscape Composition, Configuration and Change (10 Chapters)
Section II. Genetic Resource and Land Capability (13 Chapters)
Section III. Soil (15 Chapters)
Section IV. Landscape Change and Ecological Security (11 Chapters)

Volume 3. Wetlands and Habitats
Section I. Riparian Zone and Management (13 Chapters)
Section II. Wetland Ecosystem (8 Chapters)
Section III. Wetland Assessment and Monitoring (9 Chapters)

Volume 4. Fresh Water and Watersheds
Section I. Fresh Water and Hydrology (16 Chapters)
Section II. Water Management (16 Chapters)
Section III. Water and Watershed Monitoring (8 Chapters)

Volume 5. Coastal and Marine Environments
Section I. Terrestrial Coastal Environment (14 Chapters)
Section II. Marine Environment (13 Chapters)
Section III. Coastal Change and Monitoring (9 Chapters)

Volume 6. Atmosphere and Climate
Section I. Atmosphere (16 Chapters)
Section II. Weather and Climate (16 Chapters)
Section III. Climate Change (8 Chapters)

With the challenges and uncertainties ahead, I hope that the collective wisdom, the improved science, technology and awareness and willingness of the people could lead us toward the right direction and decision in governance of natural resources and make responsible collaborative efforts in balancing the equilibrium between societal demands and the capacity of natural resources base. I hope that this *HNR* series can help facilitate the understanding about the consequences of changing resource base to the ecological integrity and the sustainability of life on the planet Earth.

Yeqiao Wang
University of Rhode Island

Acknowledgments

I am honored to have this opportunity and privilege to work on *The Handbook of Natural Resources, Second Edition (HNR)*. It would be impossible to complete such a task without the tremendous amount of support from so many individuals and groups during the process. First and foremost, I thank the 342 contributors from 28 countries around the world, namely, Australia, Austria, Brazil, China, Cameroon, Canada, Czech Republic, Finland, France, Germany, Hungary, India, Israel, Japan, Nepal, New Zealand, Norway, Puerto Rico, Spain, Sweden, Switzerland, Syria, Turkey, Uganda, the United Kingdom, the United States, Uzbekistan, and Venezuela. Their expertise, insights, dedication, hard work, and professionalism ensure the quality of this important publication. I wish to express my gratitude in particular to those contributors who authored chapters for this HNR and those who provided revisions from their original articles published in the *Encyclopedia of Natural Resources*.

The preparation for the development of this HNR started in 2017. I appreciate the visionary initiation of the restructure idea and the guidance throughout the preparation of this HNR from Irma Shagla Britton, Senior Editor for Environmental Sciences and Engineering of the Taylor & Francis Group/ CRC Press. I appreciate the professional assistance and support from Claudia Kisielewicz and Rebecca Pringle of the Taylor & Francis Group/CRC Press, which are vital toward the completion of this important reference series.

The inspiration for working on this reference series came from my over 30 years of research and teaching experiences in different stages of my professional career. I am grateful for the opportunities to work with many top-notch scholars, colleagues, staff members, administrators, and enthusiastic students, domestic and international, throughout the time. Many of my former graduate students are among and/or becoming world-class scholars, scientists, educators, resource managers, and administrators, and they are playing leadership roles in scientific exploration and in management practice. I appreciate their dedication toward the advancement of science and technology for governing the precious natural resources. I am thankful for their contributions in HNR chapters.

As always, the most special appreciation is due to my wife and daughters for their love, patience, understanding, and encouragement during the preparation of this publication. I wish my late parents, who were past professors of soil ecology and of climatology from the School of Geographical Sciences, Northeast Normal University, could see this set of publications.

Yeqiao Wang
University of Rhode Island

Aims and Scope

Land, water, and air are the most precious natural resources that sustain life and civilization. Maintenance of clean air and water and preservation of land resources and native biological diversity are among the challenges that we are facing for the sustainability and well-being of all on the planet Earth. Natural and anthropogenic forces have affected constantly land, water, and air resources through interactive processes such as shifting climate patterns, disturbing hydrological regimes, and alternating landscape configurations and compositions. Improvements in understanding of the complexity of land, water, and air systems and their interactions with human activities and disturbances represent priorities in scientific research, technology development, education programs, and administrative actions for conservation and management of natural resources.

The chapters of The *Handbook of Natural Resources, Second Edition (HNR)*, are authored by world-class scientists and scholars. The theme topics of the chapters reflects the state-of-the-art science and technology, and management practices and understanding. The chapters are written at the level that allows a broad scope of audience to understand. The graphical and photographic support and list of references provide the helpful information for extended understanding.

Public and private libraries, educational and research institutions, scientists, scholars, resource managers, and graduate and undergraduate students will be the primary audience of this set of reference series. The full set of the HNR and individual volumes and chapters can be used as the references in general environmental science and natural science courses at different levels and disciplines, such as biology, geography, Earth system science, environmental and life sciences, ecology, and natural resources science. The chapters can be a value-added collection of references for scientific research and management practices.

Editor

Yeqiao Wang, PhD, is a professor at the Department of Natural Resources Science, College of the Environment and Life Sciences, University of Rhode Island. He earned his BS from the Northeast Normal University in 1982 and his MS degree from the Chinese Academy of Sciences in 1987. He earned the MS and PhD degrees in natural resources management and engineering from the University of Connecticut in 1992 and 1995, respectively. From 1995 to 1999, he held the position of assistant professor in the Department of Geography and the Department of Anthropology, University of Illinois at Chicago. He has been on the faculty of the University of Rhode Island since 1999. Among his awards and recognitions, Dr. Wang was a recipient of the prestigious Presidential Early Career Award for Scientists and Engineers (PECASE) in 2000 by former U.S. President William J. Clinton, for his outstanding research in the area of land cover and land use in the Greater Chicago area in connection with the Chicago Wilderness Program.

Dr. Wang's specialties and research interests are in terrestrial remote sensing and the applications in natural resources analysis and mapping. One of his primary interests is the land change science, which includes performing repeated inventories of landscape dynamics and land-use and land-cover change from space, developing scientific understanding and models necessary to simulate the processes taking place, evaluating consequences of observed and predicted changes, and understanding the consequences of change on environmental goods and services and management of natural resources. His research and scholarships are aimed to provide scientific foundations in understanding of the sustainability, vulnerability and resilience of land and water systems, and the management and governance of their uses. His study areas include various regions in the United States, East and West Africa, and China.

Dr. Wang published over 170 refereed articles, edited *Remote Sensing of Coastal Environments* and *Remote Sensing of Protected Lands*, published by CRC Press in 2009 and 2011, respectively. He served as the editor-in-chief for the *Encyclopedia of Natural Resources* published by CRC Press in 2014, which was the first edition of *The Handbook of Natural Resources*.

Contributors

Kathryn M. Anderson
Department of Zoology
University of British Columbia
British Columbia, Vancouver, Canada

Konstantinos M. Andreadis
Jet Propulsion Laboratory
California Institute of Technology
Pasadena, California

Roger G. Barry
Cooperative Institute for Research in
    Environmental Sciences
National Snow and Ice Data Center
Boulder, Colorado

Jürgen Bender
Federal Research Institute for Rural Areas,
    Forestry and Fisheries
Thünen Institute of Biodiversity
Braunschweig, Germany

Jaclyn N. Brown
Wealth from Oceans National Research Flagship
CSIRO Marine and Atmospheric Research
Hobart, Tasmania, Australia

Thomas J. Butler
Cary Institute of Ecosystem Studies
Millbrook, New York

and

Cornell University
Ithaca, New York

Thomas N. Chase
Department of Civil, Environmental and
    Architectural Engineering
and
Cooperative Institute for Research in the
    Environmental Sciences (CIRES)
University of Colorado at Boulder
Boulder, Colorado

Long S. Chiu
Department of Atmospheric, Oceanic and Earth
    Sciences
George Mason University
Fairfax, Virginia

John Clark
Woods Hole Research Center
Falmouth, Massachusetts

Jill S. M. Coleman
Department of Geography
Ball State University
Muncie, Indiana

Endre Dobos
Department of Physical Geography and
    Environmental Sciences
University of Miskolc
Miskolc-Egyetemváros, Hungary

Haibo Du
Key Laboratory of Geographical Processes and
    Ecological Security in Changbai Mountains
Ministry of Education
School of Geographical Sciences
Northeast Normal University
Changchun, China

Souleymane Fall
College of Agriculture Environment and
    Nutrition Science
and
College of Engineering
Tuskegee University
Tuskegee, Alabama

Joshua B. Fisher
Jet Propulsion Laboratory
California Institute of Technology
Pasadena, California

Patrick J. Fitzpatrick
Stennis Space Center
Mississippi State University
Starkville, Mississippi

Congbin Fu
Institute of Atmospheric Physics
Chinese Academy of Sciences
Beijing, China

Scott Glenn
Coastal Ocean Observation Laboratory
Institute of Marine and Coastal Sciences
School of Environmental and Biological Sciences
Rutgers University
New Brunswick, New Jersey

Felicity S. Graham
Wealth from Oceans National Research
    Flagship
CSIRO Marine and Atmospheric Research
and
Institute for Marine and Antarctic Studies
University of Tasmania
Hobart, Tasmania

Xiaoyi Guo
Key Laboratory of Geographical Processes and
    Ecological Security in Changbai Mountains
Ministry of Education
School of Geographical Sciences
Northeast Normal University
Changchun, China

Pertti Hari
Department of Forestry
University of Helsinki
Helsinki, Finland

Christopher D. G. Harley
Department of Zoology
University of British Columbia
British Columbia, Vancouver, Canada

Michael J. Hayes
National Drought Mitigation Center
Lincoln, Nebraska

Jennifer P. Jorve
Department of Zoology
University of British Columbia
British Columbia, Vancouver, Canada

Olivia Kellner
Department of Earth, Atmospheric, and
    Planetary Sciences
Purdue University
West Lafayette, Indiana

Josh Kohut
Coastal Ocean Observation Laboratory
Institute of Marine and Coastal Sciences
School of Environmental and Biological Sciences
Rutgers University
New Brunswick, New Jersey

Rebecca L. Kordas
Department of Zoology
University of British Columbia
British Columbia, Vancouver, Canada

Jürgen Kreuzwieser
Institute of Forest Botany and Tree Physiology
University of Freiburg
Freiburg, Germany

Gene E. Likens
Cary Institute of Ecosystem Studies
Millbrook, New York

and

University of Connecticut
Storrs, Connecticut

Toshihisa Matsui
NASA Goddard Space Flight Center
National Aeronautics and Space
    Administration (NASA)
Greenbelt
and
ESSIC
University of Maryland
College Park, Maryland

Forrest M. Mims III
Geronimo Creek Observatory
Seguin, Texas

Vasubandhu Misra
Department of Earth, Ocean and
    Atmospheric Science
and
Center for Ocean-Atmospheric Prediction
    Studies
Florida State University
Tallahassee, Florida

David M. Mocko
NASA Goddard Space Flight Center
National Aeronautics and Space
    Administration (NASA)
Greenbelt
and
SAIC
Beltsville, Maryland

Adam H. Monahan
School of Earth and Ocean Sciences
University of Victoria
Victoria, British Columbia, Canada

Jocelyn C. Nelson
Department of Zoology
University of British Columbia
British Columbia, Vancouver, Canada

Dev Niyogi
Department of Agronomy
and
Department of Earth, Atmospheric, and
    Planetary Sciences
Purdue University
West Lafayette, Indiana

David Noone
Department of Civil, Environmental and
    Architectural Engineering
and
Cooperative Institute for Research in the
    Environmental Sciences (CIRES)
University of Colorado at Boulder
Boulder, Colorado

Jesse Norris
Centre for Atmospheric Science
School of Earth, Atmospheric and
    Environmental Sciences
University of Manchester
Manchester, United Kingdom

Manon Picard
Department of Zoology
University of British Columbia
British Columbia, Vancouver, Canada

Roger A. Pielke, Sr.
Cooperative Institute for Research in
    Environmental Sciences (CIRES)
University of Colorado at Boulder
Boulder, Colorado

Ying Qu
Key Laboratory of Geographical Processes
    and Ecological Security in Changbai
    Mountains
Ministry of Education
School of Geographical Sciences
Northeast Normal University
Changchun, China

Nageswararao C. Rachaputi
Queensland Department of Primary Industries
Kingaroy, Queensland, Australia

Deeksha Rastogi
Department of Atmospheric Sciences
University of Illinois
Urbana, Illinois

Heinz Rennenberg
Intitute of Forest Botany and Tree Physiology
University of Freiburg
Freiburg, Germany

John S. Roberts
New Technology Department
Research and Development
Rich Products Corporation
Buffalo, New York

Gilbert L. Rochon
Tuskegee University
Tuskegee, Alabama

Somnath Baidya Roy
Department of Atmospheric Sciences
University of Illinois
Urbana, Illinois

Grace Saba
Coastal Ocean Observation Laboratory
Institute of Marine and Coastal Sciences
School of Environmental and Biological Sciences
Rutgers University
New Brunswick, New Jersey

Oscar Schofield
Coastal Ocean Observation Laboratory
Institute of Marine and Coastal Sciences
School of Environmental and Biological Sciences
Rutgers University
New Brunswick, New Jersey

David M. Schultz
Centre for Atmospheric Science
School of Earth, Atmospheric and Environmental
   Sciences
University of Manchester
Manchester, United Kingdom

Philip Sura
Department of Earth, Ocean and Atmospheric
   Science
and
Center for Ocean-Atmospheric Prediction
   Studies
Florida State University
Tallahassee, Florida

George F. Vance
Department of Ecosystem Sciences and
   Management
University of Wyoming
Laramie, Wyoming

James A. Voogt
Department of Geography
University of Western Ontario
London, Ontario, Canada

Pao K. Wang
Atmospheric and Oceanic Sciences
University of Wisconsin
Madison, Wisconsin

Yeqiao Wang
Department of Natural Resources Science
University of Rhode Island
Kingston, Rhode Island

Zhuo Wang
Department of Atmospheric Sciences
University of Illinois at Urbana- Champaign
Urbana, Illinois

Hans-Joachim Weigel
Federal Research Institute for Rural Areas,
   Forestry and Fisheries
Thünen Institute of Biodiversity
Braunschweig, Germany

Robert L. Wilby
Department of Geography
University of Loughborough
Loughborough, United Kingdom

Donald A. Wilhite
National Drought Mitigation Center
Lincoln, Nebraska

John Wilkin
Coastal Ocean Observation Laboratory
Institute of Marine and Coastal Sciences
School of Environmental and Biological
   Sciences
Rutgers University
New Brunswick, New Jersey

Graeme C. Wright
Department of Primary Industries
Queensland Department of Primary
   Industries
Kingaroy, Queensland, Australia

Zhengfang Wu
Key Laboratory of Geographical Processes and
    Ecological Security in Changbai Mountains
Ministry of Education
School of Geographical Sciences
Northeast Normal University
Changchun, China

Marcia Glaze Wyatt
Department of Geology
University of Colorado at Boulder
Boulder, Colorado

Jason Yang
Department of Geography
Ball State University
Muncie, Indiana

Xu Yi
Coastal Ocean Observation Laboratory
Institute of Marine and Coastal Sciences
School of Environmental and Biological
    Sciences
Rutgers University
New Brunswick, New Jersey

Hongyan Zhang
Key Laboratory of Geographical Processes and
    Ecological Security in Changbai Mountains
Ministry of Education
School of Geographical Sciences
and
Urban Remote Sensing Application Innovation
    Center
School of Geographical Sciences
Northeast Normal University
Changchun, China

Ying Zhang
Key Laboratory of Geographical Processes and
    Ecological Security in Changbai Mountains
Ministry of Education
School of Geographical Sciences
Northeast Normal University
Changchun, China

Jianjun Zhao
Key Laboratory of Geographical Processes and
    Ecological Security in Changbai Mountains
Ministry of Education
School of Geographical Sciences
and
Urban Remote Sensing Application Innovation
    Center
School of Geographical Sciences
Northeast Normal University
Changchun, China

Jinping Zhao
Ocean University of China
Qingdao, China

Atmosphere

I

1

Acid Rain and Nitrogen Deposition

George F. Vance
University of Wyoming

Introduction

Air pollution has occurred naturally since the formation of the Earth's atmosphere; however, the industrial era has resulted in human activities greatly contributing to global atmospheric pollution.[1,2] One of the more highly publicized and controversial aspects of atmospheric pollution is that of acidic deposition. Acidic deposition includes rainfall, acidic fogs, mists, snowmelt, gases, and dry particulate matter.[3] The primary origin of acidic deposition is the emission of sulfur dioxide (SO_2) and nitrogen oxides (NO_x) from fossil fuel combustion; electric power generating plants contribute approximately two-thirds of the SO_2 emissions and one-third of the NO_x emissions.[4]

Acidic materials can be transported long distances, some as much as hundreds of kilometers. For example, 30–40% of the S deposition in the northeastern U.S. originates in industrial midwestern U.S. states.[5] After years of debate, U.S. and Canada have agreed to develop strategies that reduce acidic compounds originating from their countries.[5,6] In Europe, the small size of many countries means that emissions in one industrialized area can readily affect forests, lakes, and cities in another country. For example, approximately 17% of the acidic deposition falling on Norway originated in Britain and 20% in Sweden came from eastern Europe.[5]

The U.S. EPA National Acid Precipitation Assessment Program (NAPAP) conducted intensive research during the 1980s and 1990s that resulted in the "Acidic Deposition: State of the Science and Technology" that was mandated by the Acid Precipitation Act of 1980.[6] NAPAP Reports to Congress have been developed in accordance with the 1990 amendment to the 1970 Clean Air Act and present the expected benefits of the Acid Deposition Control Program[6,7] http://www.nnic.noaa.gov/CENR/NAPAP/. Mandates include an annual 10 million ton or approximately 40% reduction in point-source SO_2 emissions below 1980 levels, with national emissions limit caps of 8.95 million tons from electric utility and 5.6 million tons from pointsource industrial emissions. A reduction in NO_x of about 2 million tons from 1980 levels has also been set as a goal; however, while NO_x has been on the decline since 1980, projections estimate a rise in NO$_x$ emissions after the year 2000. In 1980, the U.S. levels of SO$_2$ and NO$_x$ emissions were 25.7 and 23.0 million tons, respectively.

Acidic deposition can impact buildings, sculptures, and monuments that are constructed using weatherable materials like limestone, marble, bronze, and galvanized steel,[7,8] http://www.nnic.noaa. gov/CENR/NAPAP/. While acid soil conditions are known to influence the growth of plants, agricultural impacts related to acidic deposition are of less concern due to the buffering capacity of these types of ecosystems.[2,5] When acidic substances are deposited in natural ecosystems, a number of adverse environmental effects are believed to occur, including damage to vegetation, particularly forests, and changes in soil and surface water chemistry.[9,10]

Sources and Distribution

Typical sources of acidic deposition include coal- and oil-burning electric power plants, automobiles, and large industrial operations (e.g., smelters). Once S and N gases enter the Earth's atmosphere they react very rapidly with moisture in the air to form sulfuric (H$_2$SO$_4$) and nitric (HNO$_3$) acids.[2,3] The pH of natural rainfall in equilibrium with atmospheric CO$_2$ is about 5.6; however, the pH of rainfall is less than 4.5 in many industrialized areas. The nature of acidic deposition is controlled largely by the geographic distribution of the sources of SO$_2$ and NO$_x$ (Figure 1.1). In the midwestern and northeastern U.S., H$_2$SO$_4$ is the main source of acidity in precipitation because of the coal-burning electric utilities.[2] In the western U.S., HNO$_3$ is of more concern because utilities and industry burn coal with low S contents and populated areas are high sources of NO$_x$.[2]

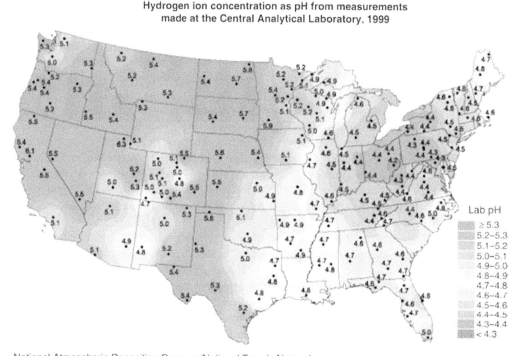

National Atmospheric Deposition Program/National Trends Network
http://nadp.sws.uiuc.edu

FIGURE 1.1   Acidic deposition across the U.S. during 1999.

Emissions of SO_2 and NO_x increased in the 20th century due to the accelerated industrialization in developed countries and antiquated processing practices in some undeveloped countries. However, there is some uncertainty as to the actual means by which acidic deposition affects our environment,[11,12] http://nadp.sws.uiuc.edu/isopleths/ maps 1999/. Chemical and biological evidence, however, indicates that atmospheric deposition of H_2SO_4 caused some New England lakes to decrease in alkalinity.[13,14] Many scientists are reluctant to over-generalize cause and effect relationships in an extremely complex environmental problem. Although, the National Acid Deposition Assessment Program has concluded there were definite consequences due to acidic deposition that warrant remediation[6,7] http://www.nnic.noaa.gov/CENR/NAPAP/. Since 1995, when the 1990 Clean Air Act Amendment's Title IV reduction in acidic deposition was implemented, SO_2 and NO_x emissions have, respectively, decreased and remained constant during the late 1990s.[4]

Both H_2SO_4 and HNO_3 are important components of acidic deposition, with volatile organic compounds and inorganic carbon also components of acidic deposition-related emissions. Pure water has a pH of 7.0, natural rainfall about 5.6, and severely acidic deposition less than 4.0. Uncontaminated rainwater should be pH 5.6 due to CO_2 chemistry and the formation of carbonic acid. The pH of most soils ranges from 3.0 to 8.0.[2] When acids are added to soils or waters, the decrease in pH that occurs depends greatly on the system's buffering capacity, the ability of a system to maintain its present pH by neutralizing added acidity. Clays, organic matter, oxides of Al and Fe, and Ca and Mg carbonates (limestones) are the components responsible for pH buffering in most soils. Acidic deposition, therefore, will have a greater impact on sandy, low organic matter soils than those higher in clay, organic matter, and carbonates. In fresh waters, the primary buffering mechanism is the reaction of dissolved bicarbonate ions with H^+ according to the following equation:

$$H^+ + HCO_3^- = H_2O + CO_2 \qquad (1.1)$$

Human Health Effects

Few direct human health problems have been attributed to acidic deposition. Long-term exposure to acidic deposition precursor pollutants such as ozone (O_3) and NO_x, which are respiratory irritants, can cause pulmonary edema.[5,6] Sulfur dioxide (SO_2) is also a known respiratory irritant, but is generally absorbed high in the respiratory tract. Indirect human health effects due to acidic deposition are more important. Concerns center around contaminated drinking water supplies and consumption of fish that contain potential toxic metal levels. With increasing acidity (e.g., lower pH levels), metals such as mercury, aluminum, cadmium, lead, zinc, and copper become more bioavailable.[2] The greatest human health impact is due to the consumption of fish that bioaccumulate mercury; freshwater pike and trout have been shown to contain the highest average concentrations of mercury.[5,15] Therefore, the most susceptible individuals are those who live in an industrial area, have respiratory problems, drink water from a cistern, and consume a significant amount of freshwater fish.

A long-term urban concern is the possible impact of acidic deposition on surface-derived drinking water. Many municipalities make extensive use of lead and copper piping, which raises the question concerning human health effects related to the slow dissolution of some metals (lead, copper, zinc) from older plumbing materials when exposed to more acidic waters. Although metal toxicities due to acidic deposition impacts on drinking waters are rare, reductions in S and N fine particles based on Clean Air Act Amendments will result in annual public health benefits valued at $50 billion with reduced mortality, hospital admissions and emergency room visits.[16]

Structural Impacts

Different types of materials and cultural resources can be impacted by air pollutants. Although the actual corrosion rates for most metals have decreased since the 1930s, data from three U.S. sites indicate that acidic deposition may account for 31–78% of the dissolution of galvanized steel-and copper,[7,8]

http://www.nnic.noaa.gov/CENR/NAPAP/. In urban or industrial settings, increases in atmospheric acidity can dissolve carbonates (e.g., limestone, marble) in buildings and other structures. Deterioration of stone products by acidic deposition is caused by: 1) erosion and dissolution of materials and surface details; 2) alterations (blackening of stone surfaces); and 3) spalling (cracking and spalling of stone surfaces due to accumulations of alternation crusts.[8] Painted surfaces can be discolored or etched, and there may also be degradation of organic binders in paints.[8]

Ecosystem Impacts

It is important to examine the nature of acidity in soil, vegetation, and aquatic environments. Damage from acidification is often not directly due to the presence of excessive H^+, but is caused by changes in other elements. Examples include increased solubilization of metal ions such as Al^{3+} and some trace elements (e.g., Mn^{2+}, Pb^{2+}) that can be toxic to plants and animals, more rapid losses of basic cations (e.g., Ca^{2+}, Mg^{2+}), and the creation of unfavorable soil and aquatic environments for different fauna and flora.

Soils

Soil acidification is a natural process that occurs when precipitation exceeds evapotranspiration.[2] "Natural" rainfall is acidic (pH of ~ 5.6) and continuously adds a weak acid (H_2CO_3) to soils. This acidification results in a gradual leaching of basic cations (Ca^{2+} and Mg^{2+}) from the uppermost soil horizons, leaving Al^{3+} as the dominant cation that can react with water to produce H^+. Most of the acidity in soils between pH 4.0 and 7.5 is due to the hydrolysis of Al^{3+},[17,18] http://www.epa.gov/airmarkets/acidrain/effects/ index.html. Other acidifying processes include plant and microbial respiration that produces CO_2, mineralization and nitrification of organic N, and the oxidation of FeS_2 in soils disturbed by mining or drainage.[2] In extremely acidic soils (pH < 4.0), strong acids such as H_2SO_4 are a major component.

The degree of accelerated acidification depends both upon the buffering capacity of the soil and the use of the soil. Many of the areas subjected to the greatest amount of acidic deposition are also areas where considerable natural acidification occurs.[19] Forested soils in the northeastern U.S. are developed on highly acidic, sandy parent materials that have undergone tremendous changes in land use in the past 200 years. However, clear-cutting and burning by the first European settlers have been almost completely reversed and many areas are now totally reforested.[5] Soil organic matter that accumulated over time represents a natural source of acidity and buffering. Similarly, greater leaching or depletion of basic cations by plant uptake in increasingly reforested areas balances the significant inputs of these same cations in precipitation.[20,21] Acidic deposition affects forest soils more than agricultural or urban soils because the latter are routinely limed to neutralize acidity. Although it is possible to lime forest soils, which is done frequently in some European countries, the logistics and cost often preclude this except in areas severely impacted by acidic deposition.[5]

Excessively acidic soils are undesirable for several reasons. Direct phytotoxicity from soluble Al^{3+} or Mn^{2+} can occur and seriously injure plant roots, reduce plant growth, and increase plant susceptibility to pathogens.[21] The relationship between Al^{3+} toxicity and soil pH is complicated by the fact that in certain situations organic matter can form complexes with Al^{3+} that reduce its harmful effects on plants.[18] Acid soils are usually less fertile because of a lack of important basic cations such as K^+, Ca^{2+}, and Mg^{2+}. Leguminous plants may fix less N_2 under very acidic conditions due to reduced rhizobial activity and greater soil adsorption of Mo by clays and Al and Fe oxides.[2] Mineralization of N, P, and S can also be reduced because of the lower metabolic activity of bacteria. Many plants and microorganisms have adapted to very acidic conditions (e.g., pH < 5.0). Examples include ornamentals such as azaleas and rhododendrons and food crops such as cassava, tea, blueberries, and potatoes[5,22] In fact, considerable efforts in plant breeding and biotechnology are directed towards developing Al-and Mn-tolerant plants that can survive in highly acidic soils.

Agricultural Ecosystems

Acidic deposition contains N and S that are important plant nutrients. Therefore, foliar applications of acidic deposition at critical growth stages can be beneficial to plant development and reproduction. Generally, controlled experiments require the simulated acid rain to be pH 3.5 or less in order to produce injury to certain plants.[22] The amount of acidity needed to damage some plants is 100 times greater than natural rainfall. Crops that respond negatively in simulated acid rain studies include garden beets, broccoli, carrots, mustard greens, radishes, and pinto beans, with different effects for some cultivars. Positive responses to acid rain have been identified with alfalfa, tomato, green pepper, strawberry, corn, lettuce, and some pasture grass crops.

Agricultural lands are maintained at pH levels that are optimal for crop production. In most cases the ideal pH is around pH 6.0–7.0; however, pH levels of organic soils are usually maintained at closer to pH 5.0. Because agricultural soils are generally well buffered, the amount of acidity derived from atmospheric inputs is not sufficient to significantly alter the overall soil pH.[2] Nitrogen and S soil inputs from acidic deposition are beneficial, and with the reduction in S atmospheric levels mandated by 1990 amendments to the Clean Air Act, the S fertilizer market has grown. The amount of N added to agricultural ecosystems as acidic deposition is rather insignificant in relation to the 100–300kg N/ha/yr required of most agricultural crops.

Forest Ecosystems

Perhaps the most publicized issue related to acidic deposition has been widespread forest decline. For example, in Europe estimates suggest that as much as 35% of all forests have been affected.[23] Similarly, in the U.S. many important forest ranges such as the Adirondacks of New York, the Green Mountains of Vermont, and the Great Smoky Mountains in North Carolina have experienced sustained decreases in tree growth for several decades.[6] Conclusive evidence that forest decline or dieback is caused solely be acidic deposition is lacking and complicated by interactions with other environmental or biotic factors. However, NAPAP research[6] has confirmed that acidic deposition has contributed to a decline in high-elevation red spruce in the northeastern U.S. In addition, nitrogen saturation of forest ecosystems from atmospheric N deposition is believed to result in increased plant growth, which in turn increases water and nutrient use followed by deficiencies that can cause chlorosis and premature needle-drop as well as increased leaching of base cations from the soil.[24]

Acidic deposition on leaves may enter directly through plant stomates.[1,22] If the deposition is sufficiently acidic (pH ~ 3.0), damage can also occur to the waxy cuticle, increasing the potential for direct injury of exposed leaf mesophyll cells. Foliar lesions are one of the most common symptoms. Gaseous compounds such as SO_2 and SO_3 present in acidic mists or fogs can also enter leaves through the stomata, form H_2SO_4 upon reaction with H_2O in the cytoplasm, and disrupt many metabolic processes. Leaf and needle necrosis occurs when plants are exposed to high levels of SO_2 gas, possibly due to collapsed epidermal cells, eroded cuticles, loss of chloroplast integrity and decreased chlorophyll content, loosening of fibers in cell walls and reduced cell membrane integrity, and changes in osmotic potential that cause a decrease in cell turgor.

Root diseases may also increase in excessively acidic soils. In addition to the damages caused by exposure to H_2SO_4 and HNO_3, roots can be directly injured or their growth rates impaired by increased concentrations of soluble Al^{3+} and Mn^{2+} in the rhizosphere[2,25] http://nadp.sws.uiuc.edu. Changes in the amount and composition of these exudates can then alter the activity and population diversity of soil-borne pathogens. The general tendency associated with increased root exudation is an enhancement in microbial populations due to an additional supply of carbon (energy). Chronic acidification can also alter nutrient availability and uptake patterns.[8,22]

Long-term studies in New England suggest acidic deposition has caused significant plant and soil leaching of base cations,[1,21] resulting in decreased growth of red spruce trees in the White Mountains.[6]

With reduction in about 80% of the airborne base cations, mainly Ca^{2+} but also Mg^{2+}, from 1950 levels, researchers suggest forest growth has slowed because soils are not capable of weathering at a rate that can replenish essential nutrients. In Germany, acidic deposition was implicated in the loss of soil Mg^{2+} as an accompanying cation associated with the downward leaching of SO_4^{2-}, which ultimately resulted in forest decline.[2] Several European countries have used helicopters to fertilize and lime forests.

Aquatic Ecosystems

Ecological damage to aquatic systems has occurred from acidic deposition. As with forests, a number of interrelated factors associated with acidic deposition are responsible for undesirable changes. Acidification of aquatic ecosystems is not new. Studies of lake sediments suggest that increased acidification began in the mid-1800s, although the process has clearly accelerated since the 1940s.[15] Current studies indicate there is significant S mineralization in forest soils impacted by acidic deposition and that the SO_4^{2-} levels in adjacent streams remain high, even though there has been a decrease in the amount of atmospheric-S deposition.[24]

Geology, soil properties, and land use are the main determinants of the effect of acidic deposition on aquatic chemistry and biota. Lakes and streams located in areas with calcareous geology resist acidification more than those in granitic and gneiss materials.[16] Soils developed from calcareous parent materials are generally deeper and more buffered than thin, acidic soils common to granitic areas.[2] Land management decisions also affect freshwater acidity. Forested watersheds tend to contribute more acidity than those dominated by meadows, pastures, and agronomic ecosystems.[8,14,20] Trees and other vegetation in forests are known to "scavenge" acidic compounds in fogs, mists, and atmospheric particulates. These acidic compounds are later deposited in forest soils when rainfall leaches forest vegetation surfaces. Rainfall below forest canopies (e.g., throughfall) is usually more acidic than ambient precipitation. Silvicultural operations that disturb soils in forests can increase acidity by stimulating the oxidization of organic N and S, and reduced S compounds such as FeS_2.[2]

A number of ecological problems arise when aquatic ecosystems are acidified below pH 5.0, and particularly below pH 4.0. Decreases in biodiversity and primary productivity of phytoplankton, zooplankton, and benthic invertebrates commonly occur.[15,16] Decreased rates of biological decomposition of organic matter have occasionally been reported, which can then lead to a reduced supply of nutrients.[20] Microbial communities may also change, with fungi predominating over bacteria. Proposed mechanisms to explain these ecological changes center around physiological stresses caused by exposure of biota to higher concentrations of Al^{3+}, Mn^{2+}, and H^+ and lower amounts of available Ca^{2+}.[15] One specific mechanism suggested involves the disruption of ion uptake and the ability of aquatic plants to regulate Na^+, K^+, and Ca^{2+} export and import from cells.

Acidic deposition is associated with declining aquatic vertebrate populations in acidified lakes and, under conditions of extreme acidity, of fish kills. In general, if the water pH remains above 5.0, few problems are observed; from pH 4.0 to 5.0 many fish are affected, and below pH 3.5 few fish can survive.[23] The major cause of fish kill is due to the direct toxic effect of Al^{3+}, which interferes with the role Ca^{2+} plays in maintaining gill permeability and respiration. Calcium has been shown to mitigate the effects of Al^{3+}, but in many acidic lakes the Ca^{2+} levels are inadequate to overcome Al^{3+} toxicity. Low pH values also disrupt the Na^+ status of blood plasma in fish. Under very acidic conditions, H^+ influx into gill membrane cells both stimulates excessive efflux of Na^+ and reduces influx of Na^+ into the cells. Excessive loss of Na^+ can cause mortality. Other indirect effects include reduced rates of reproduction, high rates of mortality early in life or in reproductive phases of adults, and migration of adults away from acidic areas.[16] Amphibians are affected in much the same manner as fish, although they are somewhat less sensitive to Al^{3+} toxicity. Birds and small mammals often have lower populations and lower reproductive rates in areas adjacent to acidified aquatic ecosystems. This may be due to a shortage of food due to smaller fish and insect populations or to physiological stresses caused by consuming organisms with high Al^{3+} concentrations.

Reducing Acidic Deposition Effects

Damage caused by acidic deposition will be difficult and extremely expensive to correct, which will depend on our ability to reduce S and N emissions. For example, society may have to burn less fossil fuel, use cleaner energy sources and/or design more efficient "scrubbers" to reduce S and N gas entering our atmosphere. Despite the firm conviction of most nations to reduce acidic deposition, it appears that the staggering costs of such actions will delay implementation of this approach for many years. The 1990 amendments to the Clean Air Act are expected to reduce acid-producing air pollutants from electric power plants. The 1990 amendments established emission allowances based on a utilities' historical fuel use and SO_2 emissions, with each allowance representing 1 ton of SO_2 that can be bought, sold or banked for future use,[4,6,7] http://www.nnic.noaa.gov/CENR/ NAPAP/. Short-term remedial actions for acidic deposition are available and have been successful in some ecosystems. Liming of lakes and some forests (also fertilization with trace elements and Mg^{2+}) has been practiced in European counties for over 50 years.[16,23] Hundreds of Swedish and Norwegian lakes have been successfully limed in the past 25 years. Lakes with short mean residence times for water retention may need annual or biannual liming; others may need to be limed every 5–10 years. Because vegetation in some forested ecosystems has adapted to acidic soils, liming (or over-liming) may result in an unpredictable and undesirable redistribution of plant species.

References

1. Smith, W.H. Acid rain *The Wiley Encyclopedia of Environmental Pollution and Cleanup*; Meyers, R.A., Dittrick, D.K. Eds.; Wiley: New York, 1999; 9–15.
2. Pierzynski, G.M.; Sims, J.T.; Vance, G.F. *Soils and Environmental Quality;* CRC Press: Boca Raton, FL, 2000; 459 pp.
3. Wolff, G.T. Air pollution. *The Wiley Encyclopedia of Environmental Pollution and Cleanup* Meyers, R.A., Dittrick, D.K. Eds.; Wiley: New York 1999; 48–65.
4. U.S. Environmental protection agency *Progress Report on the EPA Acid Rain Program* EPA-430-R-99-011 U.S. Government Printing Office: Washington, DC.
5. Forster, B.A. *The Acid Rain Debate: Science and Special Interests in Policy Formation;* Iowa State University Press: Ames, IA, 1993.
6. *National Acid Precipitation Assessment Program Task Force Report,* National Acid Precipitation Assessment Program 1992 Report to Congress; U.S. Government Printing Office: Pittsburgh, PA, 1992; 130 pp.
7. *National Science and Technology Council,* National Acid Precipitation Assessment Program Biennial Report to Congress: An Integrated Assessment; 1998 (accessed July 2001).
8. Charles, D.F., The acidic deposition phenomenon and its effects: critical assessment review papers. *Effects Sciences*; EPA-600/8-83-016B; U.S. Environmental Protection Agency: Washington, DC, 1984; Vol. 2.
9. McKinney, M.L.; Schoch, R.M. *Environmental Science: Systems and Solutions*; Jones and Bartlett Publishers: Sudbury, MA, 1998.
10. United Nations, World band and World Resources Institute-World Resources: *People and Ecosystems—The Fraying Web of Life;* Elsevier: New York, 2000.
11. *National Atmospheric Deposition Program (NRSP-3)/ National Trends Network. Isopleth Maps.* NADP Program Office, Illinois State Water Survey, Champaign, IL, 2000 (accessed July 2001).
12. Council on Environmental Quality. *Environmental Quality, 18th and 19th Annual Reports;* U.S. Government Printing Office: Washington, DC, 1989.
13. Charles, D.F., *Acid Rain Research: Do We Have Enough Answers?* Proceedings of a Speciality Conference. Studies in Environmental Science #64, Elsevier: New York, 1995.
14. Kamari, J. *Impact Models to Assess Regional Acidification;* Kluwer Academic Publishers: London, 1990.

15. Charles, D.F. *Acidic Deposition and Aquatic Ecosystems;* Springer-Verlag: New York, 1991.
16. Mason, B.J. *Acid Rain: Its Causes and Effects on Inland Waters;* Oxford University Press: New York, 1992.
17. U.S. environmental protection agency. Effects of acid rain: human health. EPA Environmental Issues Website. Update June 26, 2001 (accessed July 2001).
18. Marion G.M.; Hendricks, D.M.; Dutt, G.R.; Fuller, W.H. Aluminum and silica solubility in soils. *Soil Science* **1976**, *121*, 76–82.
19. Kennedy, I.R. *Acid Soil and Acid Rain;* Wiley: New York, 1992.
20. Reuss J.O.; Johnson, D.W. *Acid Deposition and the Acidification of Soils and Waters;* Springer-Verlag: New York, 1986.
21. Likens G.E.; Driscoll, C.T.; Buso, D.C. Long-term effects of acid rain: response and recovery of a forest ecosystem. *Science* **1996**, 272, 244–246.
22. Linthurst R.A. *Direct and Indirect Effects of Acidic Deposition on Vegetation;* Butterworth Publishers; Stoneham, MA, 1984.
23. Bush, M.B. *Ecology of a Changing Planet;* Prentice Hall: Upper Saddle River, NJ, 1997.
24. Alawell, C.; Mitchell, M.J.; Likens, G.E.; Krouse, H.R. Sources of stream sulfate at the Hubbard Brook experimental forest: long-term analyses using stable isotopes. *Biogeochemistry* **1999**, *44*, 281–299.
25. National Atmospheric Deposition Program. *Nitrogen in the Nation's Rain.* NADP Brochure 2000–2001a (accessed July 2001).

2

Acid Rain and Precipitation Chemistry

Pao K. Wang
University of Wisconsin

Introduction

Whereas an aqueous solution is acidic if its pH value is less than 7.0, acid rain refers to rainwater with pH less than 5.6. This is because, even without the presence of man-made pollutants, natural rainwater is already acidic as CO_2 in the atmosphere reacts with water to produce carbonic acid:

$$CO_2 + H_2O(1) \Leftrightarrow H_2CO_3(1) \qquad (2.1)$$

The pH value of this solution is around 5.6. Even though the carbonic acid in rain is fairly dilute, it is sufficient to dissolve minerals in the Earth's crust, making them available to plant and animal life, yet not acidic enough to cause damage. Other atmospheric substances from volcanic eruptions, forest fires, and similar natural phenomena also contribute to the natural sources of acidity in rain. Still, even with the enormous amounts of acids created annually by nature, normal rainfall is able to assimilate them to the point where they cause little, if any, known damage.

However, large-scale human industrial activities have the potential of throwing off this acid balance, and converting natural and mildly acidic rain into precipitation with stronger acidity and far-reaching environmental effects. This is the root of the acid rain problem, which is not only of national but also international concern. This problem may have existed for more than 300 years starting at the time when the industrial revolution demanded a large scale burning of coal in which sulfur was a natural contaminant. Several English scholars, such as Robert Boyle in the 17th century and Robert A. Smith of the 19th century, wrote about the acids in air and rain; though there was a lack of appreciation of the magnitude of the problem at that time. Individual studies of the acid rain phenomenon in North America started in the 1920s, but true appreciation of the problem came only in the 1970s.

To address this problem, the U.S. Congress established the National Acid Precipitation Assessment Program (NAPAP) to study the causes and impacts of the acid deposition. This research established that acid rain does cause broad environmental and health effects; the pollution causing acid deposition can travel hundreds of miles, and electric power generation is mainly responsible for SO_2 (~65%) and

NO_x emissions (~30%). Subsequently, Congress created the Acid Rain Program under Title IV (Acid Deposition Control) of the 1990 Clean Air Act Amendments. Electric utilities are required to reduce their emissions of SO_2 and NO_x significantly. It was expected that by 2010 they would lower their emissions by 8.5 million tons compared to their 1980 levels. They also need to reduce their NO_x emissions by 2 million tons each year compared to the levels before the Clean Air Act Amendments.

However, it may not be adequate to solve the acid emission merely at the national level. With increasing industrialization of the Third World countries in the twenty-first century, one can expect great increase of the atmospheric loading of SO_2 and NO_x because many of these countries will burn fossil fuels to satisfy their energy needs. Clearly, some form of international agreements need to be forged to prevent serious environmental degradation due to acid rain.

The Chemistry of Acid Rain

Sulfuric acid (H_2SO_4) and nitric acid (HNO_3) are the two main acid species in the rain. The partitioning of acids in rain may be different in different places. In the United States, the partitioning is H_2SO_4 (~65%), HNO_3 (~30%), and others (~5%). While there are many possible chemicals that may serve as the precursors of acid rain, the two main substances are SO_2 and NO_x (and NO_x consists of NO and NO_2), and both are released to the atmosphere via the industrial combustion process. While power generation is the predominant source of these precursors, industrial boilers and automobiles also contribute substantially. When these precursors enter the cloud and precipitation systems, acid rain occurs. Figure 2.1 shows a schematic of the acid rain formation process.

Once airborne, these chemicals can be involved in milliards of chemicals reactions. The main paths that lead to acid rain formation are described as follows.

Sulfuric Acid

SO_2 is believed to be the main precursor for the formation of sulfuric acid drops. Its main source in the atmosphere is the combustion of fossil fuels. This is because sulfur is a natural contaminant in coal (especially the low grade ones) and oil. The following reactions are thought to occur when SO_2 is absorbed by a water drop (see e.g., Pruppacher and Klett,[1] Seinfeld and Pandis[2]):

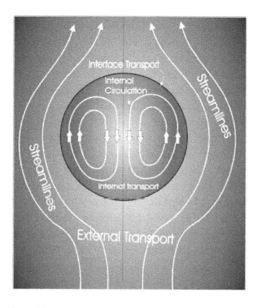

FIGURE 2.1   A schematic of the acid rain formation process.

$$SO_2(g) + H_2O(l) \Leftrightarrow SO_2 \bullet H_2O(l) \tag{2.2}$$

$$SO_2 \bullet H_2O \Leftrightarrow H^+ + HSO_3^- \tag{2.3}$$

$$HSO_3^- \Leftrightarrow H^+ + HSO_3^{-2} \tag{2.4}$$

$$SO_3^{-2} \xrightarrow{\text{oxidation}} SO_4^{-2} \tag{2.5}$$

The oxidant of the last step can be H_2O_2, O_3, OH, and others. There are still controversies about the identity of the oxidants.

Note that the equilibrium of the above reaction system is controlled by the pH values of the drop, and the presence of ammonia is often considered together with these reactions since it affects the pH of the drop. A detailed discussion of these reactions and their rates is given in Chapter 17 of Pruppacher and Klett.[1]

Nitric Acid

The main ingredients for the formation of nitric acid are NO and NO_2 (and are often combined into one category, NO_x). It is commonly thought that the main path of nitric acid found in clouds and raindrops is the formation of gas phase, HNO_3, followed by its uptake by liquid water. Although there are reactions of NO_x with liquid water that can lead to nitric acid, they are thought to be unimportant due to their slow reaction rates.

The main reaction for HNO_3 formation is

$$NO_2 + OH + M \rightarrow HNO_3 + M \tag{2.6}$$

where M can be any neutral molecule. NO can be converted to NO_2 by the following reaction:

$$2NO + O_2 \rightarrow 2NO_2 \tag{2.7}$$

Drop-Scale Transport Processes of Acid Rain

The chemical reactions described earlier must be considered together with the transport processes to obtain a quantitative picture of acid rain formation. This is especially true for SO_2 because absorption and reactions occur simultaneously. The convective transport influences the concentrations of different species and hence the reaction rates. Figure 2.2 illustrates these processes schematically. These include the following.

External Transport

This refers to the transport of SO_2 gas toward the surface of the drop. It is a convective diffusion process (both convective transport and diffusional transport occur) and is influenced by the flow fields created by the falling drop and atmospheric conditions (pressure and temperature).

Interfacial Transport

Once SO_2 is adsorbed on the surface of the drop, it must be transferred into the interior for further reactions to occur. The time for establishing phase equilibrium is controlled by Henry's law constant and mass accommodation coefficient of SO_2.

FIGURE 2.2   A schematic of the drop-scale transport process of sulfur species involved in the acid rain.

Internal Transport

In the interior of the drop, reactions[2] occur. At the same time, these species are transported by both diffusion and internal circulation. The latter is caused by the motion of the liquid drop in a viscous medium and can influence the production rates of these species (see Pruppacher and Klett[1]).

Environmental Factors Influencing the Acid Rain Formation and Impacts

Like many environmental hazards, the acid rain process is not driven by a few well-controlled physical and chemical processes, but involves complicated interactions between the chemicals and the environments they exist in. While the main ingredients of acid rain come from industrial activities, many other factors may influence the formation of acid rain and its impacts. The following are some of the most important.

Meteorological Factors

Acid rain occurs in the atmosphere and hence is greatly influenced by meteorological factors such as wind direction and speed, amount and frequency of precipitation, pressure patterns, and temperature. For example, in drier climates, such as the western United States, wind-blown alkaline dust is abundant and tends to neutralize the acidity in the rain. This is the *buffering effect* of the dust. In humid climates, like the Eastern Seaboard, less dust is in the air, and precipitation tends to be more acidic.

Seasonality may also influence acid precipitation. For example, while it is true that rain may be more acidic in summer (because of higher demands for energy and hence more fossil fuel used), the snow in winter can also pick up a substantial amount of acids. These snow-borne acids can accumulate throughout winter (if the weather is cold enough) and then are released in large doses during the spring thaw. These large doses of acid may have more significant effects during fish spawning or seed germination than the same doses at some less critical time.

Topography and Geology

The topography and geology of an area have marked influence on acid rain effects. Research from the U.S. EPA pointed out that areas most sensitive to acid precipitation are those with hard, crystalline bedrock and very thin surface soils. Here, in the absence of buffering properties of soil, acid rains will have direct access to surface waters and their delicate ecosystem. Areas with steep topography, such as mountainous areas, generally have thin surface soils and hence are very vulnerable to acid rain. In contrast, a thick soil mantle or one with high buffering capacity, such as most flatlands, helps keep acid rain damage down.

The location of water bodies is also important. Headwater lakes and streams are especially vulnerable to acidification. Lake depth, the ratio of water-shed area to lake area, and the residence time in lakes all play a part in determining the consequent threat posed by acids. The transport mode of the acid (rains or runoff) also influences the effects.

Biota

Acid rain may fall on trees causing damages. The kinds of trees and plants in an area, their heights, and whether they are deciduous or evergreen may all play a part in the potential effects of acid rain. Without a dense leaf canopy, more acid may reach the Earth to impact on soil and water chemistries. Stresses on the plants will also affect the balance of local ecosystem. Additionally, the rate at which different types of plants carry on their normal life processes influences an area's ratio of precipitation to evaporation. In locales with high evaporation rates, acids will concentrate on leaf surfaces. Another factor is that leaf litter decomposition may add to the acidity of the soil due to normal biological actions.

References

1. Pruppacher, H.R.; Klett, J.D. *Microphysics of Clouds and Precipitation*; Kluwer Academic Publishers: 1997; 954 pp.
2. Seinfeld, J.H.; Pandis, S.N. *Atmospheric Chemistry and Physics*; John Wiley and Sons: 1998.

3

Air Pollutants: Elevated Carbon Dioxide

Hans-Joachim
Weigel and
Jürgen Bender
*Thünen Institute
of Biodiversity*

Introduction

The concentrations of various compounds in the atmosphere have changed during the last century, and they continue to change. Most of these compounds interact with the terrestrial biosphere as they are part of the overall biogeochemical cycling of, e.g., carbon, oxygen, nitrogen, and sulfur.[1] For example, depending on their concentrations, gaseous compounds [sulfur dioxide (SO_2), nitrogen monoxide and dioxide (NO_2/NO)] may be beneficial to terrestrial ecosystems or remain inert (O_3) at low concentrations, whereas at higher levels, they may act as air pollutants affecting these systems in an adverse manner. Although atmospheric CO_2 is the basic plant resource for photosynthesis, its current concentration is still limiting C_3 plant growth. The rapid increase of the global atmospheric CO_2 concentration [CO_2], along with the overall changes in climate and atmospheric chemistry, require an assessment of the potential future interactive effects of air pollutants and elevated [CO_2] on terrestrial ecosystems.

Atmospheric Change: Concentrations and Trends

On a global scale, the concentrations of a variety of gaseous and particulate compounds in the atmosphere, including CO_2, NO/NO_2, SO_2, O_3, ammonia (NH_3), heavy metals, and volatile organic compounds (VOC) have undergone temporal and spatial changes during the last century.[2] After peak emissions in the 1960s to the 1980s in industrialized countries particularly, the concentrations of SO_2, and to a smaller extent of NO_x (NO/NO_2), VOCs, and particulate matter, have declined during the past decades in Europe and North America. NH_3, which is the most important reduced N species, is of importance as a direct air pollutant in the vicinity of local emitters. However, wet and dry N deposition from oxidized and reduced N species are predicted to increase in other regions of the world.[3] The occurrence and distribution of airborne VOCs are difficult to assess because there are both anthropogenic and biogenic sources. With respect to heavy metals such as lead (Pb), cadmium (Cd), nickel (Ni), mercury (Hg), and zinc (Zn), a decline in emission and subsequent deposition was observed in most of Europe since the late 1980s. Unlike the development in Europe and North America, emissions and consequently atmospheric concentrations of many of the above-mentioned compounds have been increasing over the last two decades particularly in the rapidly growing regions of Asia, Africa, and Latin America.[4] For example, China and India are now the leading emitters of SO_2 in the world. Also, the predicted further increase in global nitrogen oxide (NO_x) emissions may be attributed largely to these countries. On the other hand, concentrations of ground-level O_3 and atmospheric CO_2 have increased and continue to increase on a global scale. In most industrialized countries, O_3 concentration [O_3] has nearly doubled during the last 100 years. Current background [O_3] in the Northern hemisphere is within the range of 23–34 ppbv (parts-per-billion by volume). Although at least in most parts of Western Europe there is a clear trend of decreasing O_3 peak values ("photosmog episodes"), models predict that background [O_3] will continue to increase at a rate of 0.5% to 2% per year in the Northern Hemisphere during the next several decades, and that global surface [O_3] is expected to be in the range of 42–84 ppbv by 2100.[5] O_3 pollution has also become a major environmental problem in many of the countries with rapidly developing population and related economic growth, respectively. O_3 is currently considered the most important atmospheric pollutant that has direct negative effects on vegetation worldwide. Its concentrations vary considerably in time and space and show distinct annual and diurnal patterns.

Since the beginning of the 19th century, the CO_2 in the atmosphere has increased globally from approximately 280 ppmv (parts-per-million by volume) to current values of about 395 ppmv. It is expected that CO_2 will continue to increase even more rapidly and may reach about 550–650 ppmv between 2050 and 2070.[6] CO_2 is the substrate for plant photosynthesis, and its current atmospheric concentration is limiting for photosynthesis and growth of C_3 plants. It is expected that the increase in CO_2 will have far-reaching consequences for most types of vegetation.

Effects of O_3 and CO_2 Alone

Due to their global importance and their contrasting effects on vegetation, plant growth responses to either O_3 or CO_2 alone are briefly described. Primary O_3 effects include subtle biochemical and ultra-structural changes in the plant cell, which may result in impaired photosynthesis, alterations of carbon allocation patterns, symptoms of visible injury, enhanced senescence, reduced growth and economic yield, altered resistance to other abiotic and biotic stresses, or reduced flowering and seed production at the whole-plant level[7,8] At the ecosystem level, this may result in a loss of competitive abilities of plant species in communities along with shifts in biodiversity and impaired ecosystem functions and services like reduced carbon sequestration and altered hydrology.[9] For example, current ambient (O_3) in many industrialized areas has been shown to suppress crop yields of sensitive species and to retard growth and development of trees and other plant species of the non-woody (semi)natural vegetation. Overall quantification of O_3 effects on vegetation is complicated by large inter- and intra-specific variability in the O_3 susceptibility of plants.[10]

By contrast, plants of the C_3 type most frequently respond to elevated CO_2 with a stimulation of photosynthesis accompanied by a reduced stomatal conductance and transpiration rate, an enhanced concentrations of soluble carbohydrates, and a stimulation of biomass production and economic yield.[11] Similarly, in C_4 plants, higher CO_2 concentrations reduce stomatal conductance and transpiration, i.e., both C_3 and C_4 plants may benefit from elevated CO_2 by improved water-use efficiency and a reduced demand for water. Under well-watered conditions, no significant growth stimulation has been found so far in C_4 plants, because C_4 photosynthesis is saturated under ambient CO_2.[12,13] Growth and yield enhancements of up to 25–35% as compared to ambient CO_2 have been observed when crop plants were exposed to 550–750 ppmv CO_2. Experiments with tree species ranging from short-term studies with seedlings to long-term whole-stand manipulations have also shown that elevated CO_2 stimulated net photosynthesis and resulted in enhanced tree growth in almost all cases.[14,15] As with O_3, plant species differ widely in their response to high CO_2, which makes an overall assessment of its potential effects on vegetation difficult.

Interactive Effects of Air Pollutants and CO_2

Along with the ongoing and predicted further changes of global climate and atmospheric chemistry, there is considerable interest in how terrestrial ecosystems will respond to these multiple environmental changes and particularly how the individual changes in atmospheric constituents may interact with each other when they impact vegetation. The majority of studies dealing with this issue have addressed two-way interactions of O_3 and elevated CO_2, although there is much less information on how other air pollutants interact with high CO_2. There are no studies describing three-way interactions, i.e., in which two air pollution components and elevated CO_2 are combined together. In a biological sense, the combined action of multiple factors in comparison to single-factor effects can be described as additive (effect directly predictable from single-factor treatment) or as interactive. Interactive effects can be synergistic (effect > than expected from single-factor treatment) or antagonistic (effect < than expected from single-factor treatment).[16]

CO_2 and O_3

A great number of previous and more recent studies using different experimental approaches ranging from controlled environment to free-air O_3- and CO_2-enrichment systems have been carried out on the combined effects of the two gases. The bulk of these studies has shown that high CO_2 in the range of 200–400 ppmv above current ambient CO_2 levels either partially or totally compensates for adverse O_3 effects, whether these effects have been addressed at the biochemical and physiological level or at the whole-plant level including growth and yield. This has been demonstrated for crop (e.g., wheat, soybean, potato, rice) as well as for tree (e.g., trembling aspen, paper birch, sugar maple) species, although little information is available for grassland species.[17–20] For example, elevated CO_2 reduces O_3 effects, such as a loss in root and main stem biomass, a decrease in leaf area and mass, general foliar damage, lower growth and yield, lower starch levels, and an altered carbon balance. Results from recent free-air concentration enrichment (FACE) studies, however, have indicated that the mitigating effect of elevated CO_2 against O_3 damage might be less than predicted from earlier chamber studies.[12]

The proposed mechanisms to explain the protective effect of elevated CO_2 against the phytotoxic effects of O_3 include the following: i) reduced uptake or flux of O_3 through the stomata due to a CO_2-induced stomatal closure, ii) improved supply of carbon skeletons supporting the synthesis of antioxidants involved in the scavenging of O_3 and its toxic products, iii) protection of the RuBisCo protein from O_3-induced degradation, and iv) CO_2-induced changes in the cell surface/volume ratio.[9,21,22] However, it has been shown that in spite of decreased stomatal conductance under elevated CO_2, adverse effects of O_3 may still occur.[8,19] Additionally, elevated O_3 has been found to impair stomatal responsiveness to CO_2, i.e., O_3 causes less-sensitive ("sluggish") stomatal responses to elevated CO_2.[23] As CO_2 effects on stomatal conductance may be species specific, it is not yet possible to support a general concept of a CO_2-induced reduction in the flux of O_3 into the plant. Nevertheless, a reduction in stomatal conductance and thus in the O_3 uptake may increase atmospheric O_3 in the boundary layer.[24] Moreover, in a given plant species, protection by high CO_2 from a particular adverse effect is not necessarily associated with the protection against another adverse effect. For instance, in wheat plants, elevated CO_2 provided full protection from effects of O_3 on total plant biomass, but not on grain yield. From the available database of studies that have examined the interactive effects of O_3 and CO_2, the information is not entirely consistent, as several studies revealed that elevated CO_2 may not always protect plants from the adverse effects of O_3 (Table 3.1).

TABLE 3.1 Selected Examples of the Effects of Elevated O_3 and CO_2, Alone or in Combination, on Plant Responses (Examples with Significant Adverse Effects of O_3 on Visible Injury, Photosynthesis, Growth, and Yield)

Species	O_3 Effect	CO_2 Effect	O_3/CO_2 Effect
Potato	Decreased chlorophyll content; visible foliar leaf injury	n. e.	Adverse effect of O_3 on chlorophyll content unchanged; reduced degree of visible O_3-induced leaf injury
Wheat	Visible leaf injury; reduced photosynthesis; reduced growth; reduced yield	Increased photosynthesis; increased growth; increased yield	Reduced degree of visible O_3-induced leaf injury; amelioration of negative O_3 effects on photosynthesis, growth, and yield
Soybean	Reduced photosynthesis; reduced growth; reduced seed yield	Increased photosynthesis; increased growth; insignificant increase of seed yield	O_3 impact on photosynthesis lessened; amelioration of negative O_3 effects on photosynthesis, growth, and yield
Cotton	Reduced leaf area per mass; reduced starch contents	Increased leaf area per mass and starch contents	Prevention of adverse effects of O_3 by CO_2
Norway spruce	Visible leaf injury (chlorotic mottling)	n.e.	No effect of CO_2 on the degree of O_3-induced leaf injury
Trembling aspen (different O_3-sensitive and -tolerant clones)	Reduced tree growth parameters (height, diameter, volume)	Enhancement of growth parameters	No effect of CO_2 on the degree of O_3-induced growth reductions
Paper birch	Reduced photosynthesis; decreased dry matter production	Increased photosynthesis; increased dry matter production	Decrease in photosynthesis and dry matter production similar to O_3 alone
White clover (sensitive clone)	Visible leaf injury	n.e.	Little effect on the degree of O_3-induced foliar injury

Source: Adapted from Karnosky et al.,[19] Vandermeiren et al.,[20] Polle & Pell,[21] and Runeckles.[26] Abbreviations: CO_2, carbon dioxide; n.e., No effect; O_3, Ozone.

CO_2 and Other Air Pollutants

Very few studies have addressed the combined effects of elevated CO_2 and of other air pollutants. SO_2 has long been known to adversely affect agricultural crops and forest plants above a certain threshold concentration.[25] Reduced photosynthesis, altered water relations, growth retardations, yield losses, and altered susceptibilities to other stresses are common plant responses observed under SO_2 stress. Due to the diminishing importance of SO_2 as a widespread air pollutant, few studies have been conducted on the combined action of SO_2 and elevated CO_2. In earlier studies, it was shown for some crop species that elevated CO_2 reduced the sensitivity of the plants to SO_2 injury or protected them from the negative effects of SO_2 on growth and yield.[26] With the combined exposure of crop species to both gases, the yield increments were sometimes even larger when compared to the stimulation observed with exposure to elevated CO_2 alone, suggesting that the plants were able to use the airborne sulfur more effectively under the conditions of enhanced carbon availability. Low-to-moderate SO_2 concentrations may confer a nutritional benefit to plants, particularly under conditions of low sulfur availability in the soil.

Studies on the interactive effects of elevated CO_2 and nitrogen oxides (NO and NO_2) are confined to commercial greenhouses under conditions of horticultural crop production under very high CO_2 and are not considered here. However, it has been shown repeatedly that positive plant growth responses to elevated CO_2 are smaller at low relative to high soil N supply. This is related to the question on the role of future atmospheric N deposition and "aerial carbon fertilization" by elevated CO_2 in shaping the size of the terrestrial carbon sink and how plant biodiversity might be affected by these inputs. Assuming that aerial N supply via enhanced N deposition causes similar effects as soil N fertilization, a few experimental and modeling studies addressed the question of how elevated CO_2 interacts with N deposition. For example, it has been shown that N addition enhanced the CO_2 stimulation of plant productivity in the first phase of a multiyear CO_2–N manipulation study with a herbaceous wetland plant community. But in the longer term, the observed N-induced shift in the plant community composition suppressed the CO_2 stimulation of plant productivity, indicating that plant community shifts can act as a feedback effect that alters ecosystem responses to elevated CO_2.[27] In a long-term study with simulated grassland systems with 16 species, high N supply reduced species richness by 16% under ambient CO_2 but only by 8% under elevated CO_2, i.e., high CO_2 ameliorated negative N effects.[28,29] Elevated CO_2 and N addition have been found to affect above and belowground C allocation in temperate forest trees in an opposite way, i.e., elevated CO_2 increases belowground allocation, whereas N increases aboveground allocation; however, the ratio of above vs. belowground C flow does not change in the combination of both treatments.[30]

Conclusion

The assessment of the potential combined effects of air pollutants and elevated atmospheric concentrations of CO_2 on vegetation is of critical importance during the next decades. Interactive effects of these atmospheric compounds on crops, trees, and other types of vegetation have been shown. Existing evidence on such interactions is almost entirely restricted to CO_2 and O_3, the concentrations of which are increasing globally. Although rising CO_2 will be mostly beneficial to plants, current ambient O_3 are high enough to impair plants in many regions of the world. There is prevailing information that elevated CO_2 may protect plants from adverse effects of O_3, but this has not been demonstrated unequivocally. There is also some information that rising CO_2 may protect plants against phytotoxic SO_2 concentrations. The future interactions of elevated CO_2 and enhanced atmospheric N deposition are of concern in many ecosystem types with respect to carbon sequestration and biodiversity. Overall, our understanding has to be improved about how other growth variables, such as plant genotype, soil water deficit, nutrient availability, or temperature, may modify the interaction between air pollutants and elevated CO_2.

References

1. Dämmgen, U.; Weigel, H.J. Trends in atmospheric composition (nutrients and pollutants) and their interaction with agroecosystems. In *Sustainable Agriculture for Food, Energy and Industry*; El Bassam, N., Behl, R.K., Prochnow, B., Eds.; James & James (Science Publishers) Ltd.: London, 1998; 85–93.
2. Bender, J.; Weigel, H.J. Changes in atmospheric chemistry and crop health: A review. Agron. Sustain. Dev. **2011**, *31*, 81–89.
3. Grübler, A. Trends in global emissions: Carbon, sulfur, and nitrogen. In *Encyclopedia of Global Environmental Change. Causes and Consequences of Global Environmental Change*; Douglas, I., Ed.; John Wiley & Sons: Chichester, 2003; Vol. 3, 35–53.
4. Emberson, L.D., Ashmore, M.R., Murray, F., Eds.; *Air Pollution Impacts on Crops and Forests: a Global Assessment*; Imperial College Press: London, 2003.
5. Vingarzan R. A review of surface ozone background levels and trends, Atmos. Environ. **2004**, *38*, 3431–3442.
6. IPCC. *The 4th Assessment Report, Working Group I Report: The Physical Scientific Basis*, Report of the Intergovernmental Panel on Climate Change; USA Cambridge University Press: Cambridge, UK and New York, 2007.
7. Ashmore, M.R. Assessing the future global impacts of ozone on vegetation. Plant Cell Environ. **2005**, *28*, 949–964.
8. Fiscus, E.L.; Booker, F.L.; Burkey, K.O. Crop responses to ozone: uptake, modes of action, carbon assimilation and partitioning. *Plant Cell Environ.* **2005**, *28*, 997–1011.

9. Fuhrer, J. Ozone risk for crops and pastures in present and future climates. Naturwissenschaften **2009**, *96*, 173–194.
10. Bender, J.; Weigel, H.-J. Ozone stress impacts on plant life. In *Modern Trends in Applied Terrestrial Ecology*; Ambasht, R.S., Ambasht, N.K., Eds.; Kluwer Academic/Plenum Publishers: New York, 2003; 165–182.
11. Ainsworth, E.A.; McGrath, J.M. Direct effects of rising atmospheric carbon dioxide and ozone on crop yields. In *Climate Change and Food Security, Advances in Global Change Research 37*; Lobell, D., Burke, M., Eds.; Springer Science+Business Media: Dordrecht, 2010; 109–130.
12. Long, S.P.; Ainsworth, E.A.; Leakey, A.D.B.; Morgan, P.B. Global food insecurity. Treatment of major food crops with elevated carbon dioxide or ozone under large-scale fully open-air conditions suggests recent models may have overestimated future yields. Phil. Trans. Royal Soc. **2005**, *B 360*, 2011–2020.
13. Leakey, A.D.B. Rising atmospheric carbon dioxide concentration and the future of C_4 crops for food and fuel. Proc. R. Soc. Lond. B **2009**, *276*, 2333–2343.
14. Norby, R.J.; DeLucia, E.H.; Gielen, B.; Calfapietra, C.; Giardina, C.P.; King, J.S.; Ledford, J.; McCarthy, H.R.; Moore, D.J.P.; Ceulemans, R.; De Angelis, P.; Finzi, A.C.; Karnosky, D.F.; Kubiske, M. E.; Lukac, M.; Pregitzer, K.S.; Scarascia-Mugnozza, G.E.; Schlesinger, W.H.; Oren, R. Forest response to elevated CO_2 is conserved across a broad range of productivity. PNAS **2005**, *102*, 18052–18056
15. Wittig, V.E.; Ainsworth, E.A.; Naidu, S.L.; Karnosky, D.F.; Long, S.P. Quantifying the impact of current and future tropospheric ozone on tree biomass, growth, physiology and biochemistry: a quantitative meta-analysis. Glob. Change Biol. **2009**, *15*, 396–424.
16. Fangmeier, A.; Bender, J.; Weigel, H.J.; Jäger, H.J. Effects of pollutant mixtures. In *Air Pollution and Plant Life*; Bell, J.N.B., Treshow, M., Eds.; John Wiley & Sons: Chichester, 2002; 251–272.
17. Olszyk, D.M.; Tingey, D.T.; Watrud, R.; Seidler, R.; Andersen, C. Interactive effects of O_3 and CO_2: implications for terrestrial ecosystems. In *Trace Gas Emissions and Plants*; Singh, S.N., Ed.; Kluwer Academic Publishers: Netherlands, 2000; 97–136.
18. Feng, Z.; Kobayashi, K. Assessing the impacts of current and future concentrations of surface ozone on crop yield with meta-analysis. Atmos. Environ. **2009**, *43*, 1510–1519.
19. Karnosky, D.F.; Pregritzer, K.S.; Zak, D.R.; Kubiske, M.E.; Hendrey, G.R.; Weinstein, D.; Nosal, M.; Percy, K.E. Scaling ozone responses of forest trees to the ecosystem level in a changing climate. Plant Cell Environ. **2005**, *28*, 965–981.
20. Vandermeiren, K.; Harmens, H.; Mills, G.; De Temmerman, L. Impacts of ground level ozone on crop production in a changing climate. In *Climate Change and Crops, Environmental Science and Engineering*; Singh, S.N., Ed.; Springer: Berlin, 2009; 213–243.
21. Polle, A.; Pell, E.J. The role of carbon dioxide in modifying the plant response to ozone. In *Carbon Dioxide and Environmental Stress*; Luo, Y.; Mooney, H.A., Eds.; Academic Press: San Diego, 1999; 193–213.
22. Tausz, M.; Grulke, N.E.; Wieser, G. Defense and avoidance of ozone under global change. Environ. Pollut. **2007**, *147*, 525–531.
23. Onandia, G.; Olsson, A.-K.; Barth, S.; King, J.S.; Uddling, J. Exposure to moderate concentrations of tropospheric ozone impairs tree stomatal response to carbon dioxide. Environ. Pollut. **2011**, *159* 2350–2354.
24. Sanderson, M.G.; Collins, W.J.; Hemming, D.L.; Betts, R.A. Stomatal conductance changes due to increasing carbon dioxide levels: projected impacts on surface ozone levels. Tellus **2007**, *59b*, 404–411.
25. Legge, A.H., Krupa, S.V. Effects of sulphur dioxide. In *Air Pollution and Plant Life*; Bell, J.N.B., Treshow, M., Eds.; John Wiley & Sons: Chichester, 2002; 135–162
26. Runeckles, V. Air pollution and climate change. In *Air Pollution and Plant Life*; Bell, J.N.B., Treshow, M., Eds.; John Wiley & Sons Ltd.: Chichester, 2002; 431–454.

27. Langley, J.A.; Megonigal, J.P. Ecosystem response to elevated CO_2 levels limited by nitrogen-induced plant species shift. Nature **2010**, *466*, 96–99.
28. Reich, P.B., Hungate, B.A., Luo, Y. Carbon-nitrogen interactions in terrestrial ecosystems in response to rising atmospheric carbon dioxide. Annu. Rev. Ecol. Evol. Syst. **2006**, *37*, 611–36.
29. Reich, P.B. Elevated CO_2 reduces losses of plant diversity by nitrogen deposition. S²cience **2009**, *326*, 1399–1402.
30. Drake, J.E.; Oishi, A.C.; Giasson, M.-A.; Oren, R.; Finzi, A.C. Trenching reduces soil heterotrophic activity in a loblolly pine (Pinus taeda) forest exposed to elevated atmospheric CO_2 and N fertilization. Agric. Forest Meteorol. **2012**, *165*, 43–52.

4

Albedo

Endre Dobos
University of Miskolc

Introduction

The portion of solar radiation not reflected by the Earth's surface is absorbed by the soil or the vegetation, which interacts with the incident radiation. The absorbed energy can increase the soil temperature or the rate of evapotranspiration from the surface of the soil–vegetation system. Some of the energy that is absorbed and transformed into heat is reradiated at a longer wavelength than the incoming radiation. That is why the peak terrestrial radiation occurs in the infrared spectrum while the peak incident radiation occurs in the blue-green portion of the visible spectrum.

The albedo value ranges from 0 to 1. The value of 0 refers to a blackbody, a theoretical media that absorbs 100% of the incident radiation. Albedo ranging from 0.1–0.2 refers to dark-colored, rough soil surfaces, while the values around 0.4–0.5 represent smooth, light-colored soil surfaces. The albedo of snow cover, especially fresh, deep snow, can reach as high as 0.9. The value of 1 refers to an ideal reflector surface (an absolute white surface) in which all the energy falling on the surface is reflected. The mean albedo of the Earth system is 0.36 ± 0.06 (Table 4.1).[1]

TABLE 4.1 The Approximated Ranges of Albedo of Natural Surfaces

Natural Surface Types	Approximated Albedo
Blackbody	0
Forest	0.05–0.2
Grassland and cropland	0.1–0.25
Dark-colored soil surfaces	0.1–0.2
Dry sandy soil	0.25–0.45
Dry clay soil	0.15–0.35
Sand	0.2–0.4
Mean albedo of the Earth	0.36
Granite	0.3–0.35
Glacial ice	0.3–0.4
Light-colored soil surfaces	0.4–0.5
Dry salt cover	0.5
Fresh, deep snow	0.9
Water	0.1–1
Absolute white surface	1

Source: Adapted from Hoffer.[10]

Factors Affecting Albedo

Albedo varies diurnally and seasonally due to the changing sun angle.[2,3] In general, the lower the sun angle the higher the albedo. Besides the sun angle, many of the surface characteristics have large impact on the albedo. The most significant factors affecting the soil albedo are the type and condition of the vegetation covering the soil surface, soil moisture content, organic matter content, particle size, iron-oxides, mineral composition, soluble salts, and parent material)[4]

The type and the condition of the vegetation have a strong impact on the surface albedo. Forest vegetation with multilevel canopy has a low albedo because the incident radiation can penetrate deeply into the forest canopy where it bounces back and forth between the branches and leaves and get trapped by the canopy.[5] The albedo for grassland and cropland ranges between 0.1 and 0.25.[6-9]

Changes in soil moisture content change the absorbance and reflectance characteristics of the soil. Increase in soil moisture content increases the portion of the incident solar radiation absorbed by the soil system. This relationship is well known and used for soil color differentiation when the Munsell color chart is used. The colors of dry and moist soil samples are always different. The higher the soil moisture content, the darker the color and lower the albedo. However, this relationship is valid only for soil moisture contents up to the field capacity. Beyond field capacity, the increase in soil moisture content does not darken the color any more, but starts building up a water sheet on the aggregate surface, creating a shiny and better reflecting surface, which increases the reflectance and thus the albedo. This phenomenon is the major reason for differences in the albedo among soils of different textural classes. Clayey soils can maintain high moisture content in the presence of water supply, while the sandy textured soils drain and dry out much more rapidly. Due to the differences in the resulting soil moisture content between the texture classes, there are differences in the reflectance and absorbance characteristics and so in the albedo (Figure 4.1).

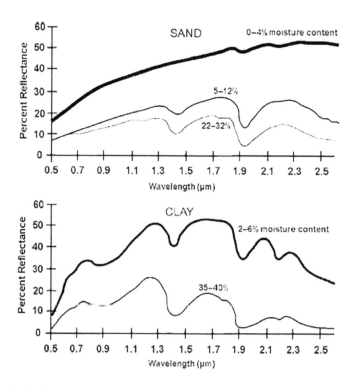

FIGURE 4.1 The higher the moisture content the lower the reflectance throughout the visible and near-infrared region, especially along the water absorption bands at $1.4\,\mu m$ and $1.7\,\mu m$. Notice the differences in the reflectance characteristics between the clayey and sandy soils.

Surface roughness defines the type of reflection. Shiny, smooth surfaces, like water body, plant leaves, or wet soil surfaces may be near-perfect, specular reflectors, which may reflect well and show relatively high albedo for lower sun angles. Rough surfaces represent lower albedo values, especially when sun angle is low and the shading effect lowers the reflection. There are measurable differences in the surface roughness among soil textural classes. Fine-textured, dry soils with small particle size produce high albedo due to relatively smooth surface. However, clayey soils are often wet, and soil moisture absorbs the incident radiation and decreases albedo. Conversely, dry, coarse textured soils with relatively large particles (sand grains) reflect larger portions of the incident radiation than clayey soils.

Surface color is determined by the interaction of the surface material with the visible spectra of the incident solar radiation. Soil color is a differentiating factor in all the soil classification systems. It reflects many of the most important soil physical and chemical characteristics. One of the most significant coloring agents of the soils is the soil organic matter content. Soil organic matter content increases the absorbance of the soil. Thus the higher the organic matter content, the lower the albedo. Iron oxides increase the reflectance in the red portion of the spectrum while causing a decrease in the blue-green and infrared portion. Salt crust on the surface increases the albedo dramatically. That is why mapping of salt-affected area with remotely sensed images is a very powerful tool for soil surveyors.

Measurement of Albedo

The theoretical concept of measuring albedo is simple. A radiation sensor (pyranometer) is pointed upwards to measure the incident radiation and then quickly flipped downwards to measure the reflected radiation. For deriving the albedo, the quantity of the reflected radiation has to be divided by the one for the incident radiation. In fact, the actual measurement of surface albedo under natural condition is rather complex. The problem is threefold. First, the incident radiation does not only come from the radiation source directly, but also from diffused light from other directions. Secondly, the reflector surfaces do not reflect equally in all directions and thirdly, the sensors gather light only from a small range of angles. Thus, our measurements of reflectance are only samples of the bidirectional reflectance distribution function (BRDF). Albedo is often defined as an overall average reflection coefficient of an object. More precisely the terms of spectral and total albedo are differentiated. The spectral albedo refers to the reflectance in a given wavelength, while the albedo is calculated as an integral of the spectral reflectivity times the radiation, over all wavelengths in the visible spectrum. A good estimation of the surface albedo can be done using clear-sky satellite measurements.[11]

Conclusions

Albedo measures the overall reflectance of the surface. It provides lots of useful information about the soil system and helps to better understand the soil energy balance. But different wavelengths of sunlight are normally not equally reflected, which gives rise to a variable color of surfaces and differences in reflectance of certain wavelengths due to differences in physical or chemical characteristics of the soil surface. Differences in soil albedo can be measured with radiometers.

References

1. Weast, R.C. *Handbook of Chemistry and Physics*; CRC Press: Boca Raton, FL, 1982.
2. Matthews, E. Vegetation, land-use and seasonal albedo data sets. In *Global Change Data Base Africa Documentation*; Appendix D; NOAA/NGDC, 1984.
3. Kotoda, K. *Estimation of River Basin Evapotranspiration. Environmental Research Center Papers*; University of Tsukuba: 1986; Vol. 8.
4. Baumgardner, M. F.; Sylva, L.F.; Biehl, L.L.; Stoner, E.R. Reflectance properties of soils. Adv. Agron. **1985**, 38, 1–44.

5. Geiger, R. *The Climate Near the Ground*; Harvard University Press: Cambridge, MA, 1965.

6. Jones, H.G. *Plants and Microclimate: A Quantitative Approach to Environmental Plant Physiology*, 2nd Ed.; The Cambridge University Press: Cambridge, U.K., 1992.

7. Jensen, M.E.; Burman, R.D.; Allen, R.G. Evapotranspiration and irrigation water requirements. In *ASCE Man. Rep. Pract. 70*; American Society of Civil Engineers: New York, NY, 1990.

8. Oke, T.R. *Boundary Layer Climates*; Methuen: New York, NY, 1978.

9. Van Wijk, W.R.; Scholte Ubing, D.W. Radiation. In *Physics of Plant Environment*; Van Wijk, W.R., Ed.; North-Holland Publishing Co.: Amsterdam, the Netherlands, 1963; 62–101.

10. Hoffer, R.M. Biological and physical considerations in applying computer-aided analysis techniques in remote sensing data. In *Remote Sensing: The Quantitative Approach*; Swain, P.H., Davis, S.M., Eds.; McGraw Hill: San Francisco, 1978; 227–289.

11. Li, Z.; Garand, L. Estimation of surface albedo from space: A parameterization for global application. J. Geophys. Res. **1994**, *99*, 8335–8350.

5

Arctic Oscillation

Jinping Zhao
Ocean University of China

Introduction

The Arctic Oscillation (AO) is a seesaw pattern in which sea level pressure (SLP) at the polar and middle latitudes in the Northern Hemisphere fluctuates between positive and negative phases (Figure 5.1). This pattern is obtained by the first mode of Empirical Orthogonal Function (EOF) of the SLP north of 20°N.[1] It is characterized by a pattern of one sign in the polar region with the opposite sign anomaly in the middle latitude centered about 37–45°N.[2]

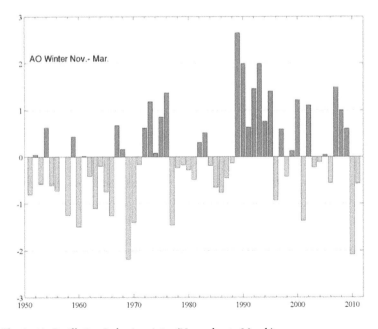

FIGURE 5.1   The Arctic Oscillation Index in winter (November to March).
Source: Data from NCEP.[3]

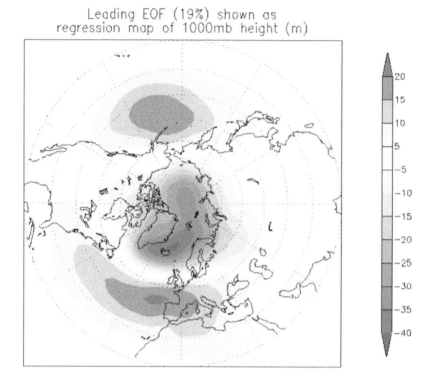

FIGURE 5.2   (See color insert.) The seesaw-like pattern of the Arctic Oscillation.
Source: Data adapted from http://www.cpc.ncep.noaa.gov/prod-ucts/precip/CWlink/daily_ao_index/ao.shtml.[4]

The standardized leading principal component (PC) time series is defined as the Arctic Oscillation Index (AOI, Figure 5.2). The monthly mean SLP anomaly fields are regressed upon the AOI. Based on this definition, the AO index is dimensionless and the regression maps have the unit of the SLP anomaly field. The amplitudes shown in the regression maps therefore correspond to anomaly values.

The polarity of the AO is defined by the AOI. If the AOI is in the positive phase, the AO in that time is called positive AO, otherwise, the AO is negative. The AO was found to exhibit high polarity or positive AOI in the early 1990s,[5] but has switched to a near-neutral or negative phase since 1996.[6] The year-to-year persistence of positive or negative values and the rapid transition from one to the other is often referred to as "regime-like." The AO presents a "positive phase" with relatively lower pressure over the polar region and higher pressure at mid-latitudes, and a "negative phase" with relatively higher pressure over the polar region and lower pressure at mid-latitudes.

AO and Atmospheric Circulation

The AOI in fact reflects the dominant mode of variation in SLP and zonal wind. Also, the time series of 50 hpa is strongly correlated with AOI. AO extends through the depth of the troposphere. During the "active season," January through March, it extends upward into the stratosphere where it modulates in the strength of the westerly vortex that encircles the Arctic polar cap region.[7] So AOI is an index to reflect the main characteristics of the overall atmospheric circulation in hemispheric scale.

In the negative phase of AO, the polar region is dominated by higher pressure, and the zonal wind becomes weaker (Figures 5.3 and 5.4). A very persistent, strong ridge of high pressure, or "blocking system," near Greenland allows polar cold air to easily extend to the south,[9] which causes a frigid winter in the eastern part of the U.S.A. and the Mediterranean. However, when the AO index is positive, surface pressure

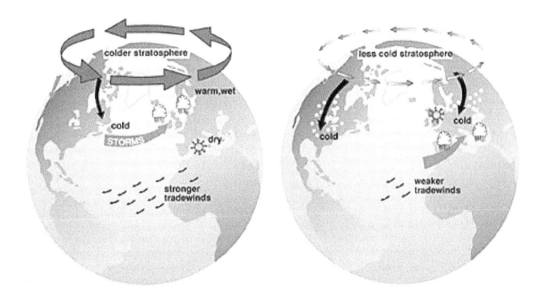

FIGURE 5.3 Positive and negative Arctic Oscillation.
Source: Courtesy of J. Wallace, University of Washington.

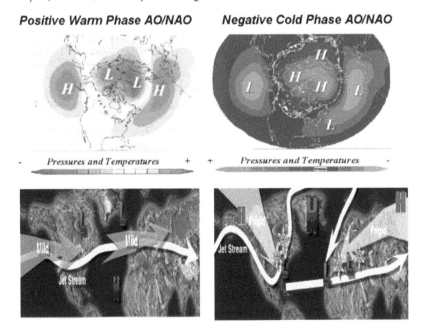

FIGURE 5.4 Atmospheric circulations in positive or negative AO.
Source: Data adapted from http://www.newx-forecasts.com/ao_2.html.[8]

is low in the polar region (Figure 5.4). This helps the middle latitude jet stream to blow strongly and consistently from west to east, thus keeping cold Arctic air locked in the polar region.

During the twentieth century, the AO mostly fluctuated between its positive and negative phases until the 1970s. Since then, the oscillation has stayed mostly in the positive phase for about 30 years, especially from 1988 to 1996. During this period, the U.S.A., China, and Europe kept at a lower air pressure and higher temperature. Since this century, the phase of the AO has tended to a neutral state.

Obvious negative AO appeared in the winters of 2010/2011 and 2011/2012, which is attributed to the causal factor of AO for the extremely cold winter in the U.S.A., Europe, and Eastern Asia.

In February 2010, the AO reached its most negative monthly mean value, – 4.266 (http://www.cpc.ncep.noaa.gov/products/precip/CWlink/daily_ao_index/monthly.ao.index.b50.current.ascii). The colder winter appeared in eastern U.S.A., such as New York, Washington D.C., and Baltimore. That month was characterized by three separate historic snowstorms that occurred in the mid-Atlantic region of the U.S.A.[8]

AO exerts strong influences on wintertime climate in the Northern Hemisphere.[10] However, the high correlation of negative AO and an excessive colder winter does not necessarily mean that extreme weather will surely occur. For a certain location, the climate is influenced not only by AO, but also by some other processes related with tropical or regional factors.

Global Effects of AO

As the AO reflects the variation of the atmospheric system, it will also embody a global effect. In the case with positive AO, four main phenomena are notable:

1. Surface air pressure. Positive AO produces a lower- than-normal pressure over the polar region. The westerly moves to the north and cold air is restricted in the polar region.
2. Air temperature. In the positive phase, frigid winter air does not extend as far into the middle of North America as it would during the negative phase of the oscillation. This keeps much of the U.S.A. east of the Rocky Mountains warmer than normal, but leaves Greenland and Newfoundland colder than usual (Figure 5.5a).
3. Flood and drought. With the positive AO, the atmospheric circulation brings wetter weather to Alaska, Scotland, and Scandinavia, as well as drier conditions to the western part of the U.S.A. and the Mediterranean, Spain, and the Middle East (Figure 5.5b).
4. Ocean storms. With the positive AO, higher pressure at mid-latitudes drives ocean storms farther north.

Weather patterns in the negative phase are in general "opposite" to those of the positive phase.

AO and Northern Hemisphere Annular Mode

In addition to defining AO, Thompson and Wallace[1] also pointed out a coupled signal of AO and a strong fluctuation at the 50 hPa level on the intraseasonal, interannual, and interdecadal time scales. The leading modes of variability of the extratropical circulation in both hemispheres are characterized by deep, zonally symmetric, or "annular" structures, with geopotential height perturbations of opposing signs in the polar cap region and in the surrounding zonal ring centered near 45° latitude.[5] Compared with the pattern of the annular circulation mode of the lower stratosphere in the Southern Hemisphere, AO is recognized in essence as the Northern Hemisphere annular mode (NAM)[4] and is interpreted as the surface signature of modulations in the strength of the polar vortex aloft (Figure 5.6).[4,10]

The structures of the Northern Hemisphere and Southern Hemisphere annular modes are shown to be remarkably similar, not only in the zonally averaged geopotential height and zonal wind fields, but in the mean meridional circulations as well. Both exist year-round in the troposphere, but they amplify with height upward into the stratosphere during midwinter in the Northern Hemisphere.[5]

The NAM is shown to exert a strong influence on wintertime climate, not only over the Euro-Atlantic half of the hemisphere, but over the Pacific half as well. It affects not only the mean conditions but also the day-to-day variability, modulating the intensity of mid-latitude storms and the frequency of occurrence of high-latitude blocking and cold air outbreaks throughout the hemisphere.[11]

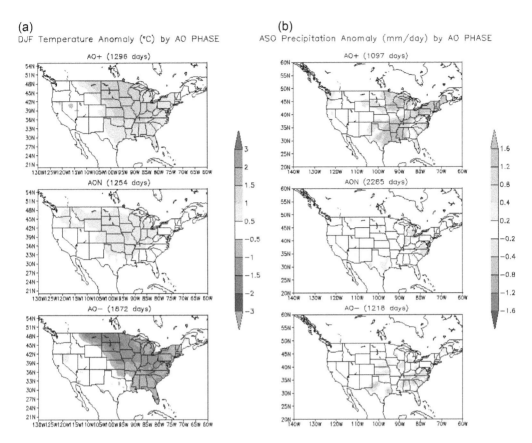

FIGURE 5.5   Temperature and precipitation anomalies by AO phase.
Source: Data adapted from http://www.cpc.ncep.noaa.gov/research_papers/ncep_cpc_atlas/8/figures/temp/, http://www.cpc.ncep.noaa.gov/research_papers/ncep_cpc_atlas/8/figures/precip/.[8]

AO and North Atlantic Oscillation

North Atlantic Oscillation (NAO) is a seesaw-like dipole oscillation along the Atlantic sector and defined by the difference of air pressure between Iceland and the Azores.[12,13] NAO was discovered in the 1920s by Walker[14] and has been studied extensively.[15] Since the NAO index correlates well with the AOI,[16] some studies focused on the physical reality of the AO and the connection between AO and NAO.

Although the definitions of AO and NAO were reached by quite different methods, the interesting thing is that both are closely correlated to each other with a correlation coefficient higher than 0.89. Some scientists believe that AO and NAO are the same process obtained from a different visual angle. They ignore the little differences between them and write it as AO/NAO to avoid the possible misunderstanding.

However, there is an argument about AO and NAO in regard to their scientific mechanism, whether one or the other is more fundamentally representative of the atmospheric dynamics. Some investigators argue that the NAO paradigm may be more physically relevant to and robust in the Northern Hemisphere than the AO paradigm.[17] Others proposed, based on the nonlinear principal component analysis (PCA), that the variability of the atmospheric system in the Northern Hemisphere is better characterized by an Arctic-Eurasian oscillation state. In the North Atlantic, such a state is replaced occasionally by a split-flow configuration state; AO represents a compromise between the two states.[18]

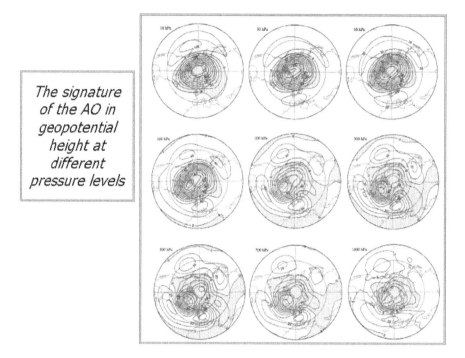

FIGURE 5.6 The signature of the AO in geopotential height at different pressure levels.
Source: Data adapted from http://www.newx-forecasts.com/ao_2.html.[8]

Deser[19] and Ambaum et al.[18] pointed out the inconsistency to demonstrate that the NAO is the more robust paradigm. Itoh[20] tried to explain the inconsistency by the concepts of true AO and apparent AO. However, AO is more strongly coupled to surface air temperature fluctuations over the Eurasian continent than the NAO. AO resembles the NAO in many respects, but its primary center of action covers more of the Arctic, giving it a more zonally symmetric appearance.[1] According to the National Snow and Ice Data Center, the AO and NAO are virtually "different ways of describing the same phenomenon."[9] Both are impossible to distinguish with purely statistical methods, as the two models are mathematically identical.[21]

AO and Arctic Warming

The Arctic region is experiencing obvious climatological warming at a rate that is nearly twice the global average during the past decades.[22] Most climatological indicators show a linear trend. The notable ones include the decline of the perennial ice coverage at a rate of 9% per decade,[23] and the thinning of the ice draft of more than a meter from two to four decades ago to the 1990s.[24] However, AO, which is believed to associate with the variation of atmospheric circulation in the Northern Hemisphere, behaves differently from such regularities. In AO, instead, a seesaw-like oscillation is still dominant.

In recent years, many studies have revealed the relationship between the AO/NAO and Arctic sea-ice condition. For example, when the AO/NAO turns to a high-index polarity, the coherent changes include the following: 1) a spin down of the Beaufort Gyre; 2) a slight increase in the ice advection out of the Arctic Basin through the Fram Strait; 3) an increase in ice import from the Barents/Kara Seas and ice advection away from the coast of the East Siberian and Laptev; 4) a decrease in ice advection from the western Arctic into the eastern Arctic; 5) a weakening of the Transpolar Drift Stream; and 6) rapid advection of ice out of the western Arctic with increased melting and inhibited accumulation of thicker ice.[25-30]

Spatial Variation of AO

Based on the introduction section, AO was derived by the EOF method, which is a popular method to analyze spatiotemporal variation[31] and gives a space-stationary and time-fluctuating characteristic of a variable.[18] However, as the AO is a spatially variable phenomenon, only the space stationary leading mode of EOF is not sufficient[32] to reflect the spatial variation of the AO.

By calculating the running correlation coefficients (RCCs) of AO index and gridded SLPs, a special region was identified by Zhao et al. called the Arctic Oscillation Core Region (AOCR),[33] in which SLP variation has never been influenced by a non-AO process as shown in Figure 5.7a. The normalized average SLP of AOCR is always consistent with the negative AOI since 1950 with a correlation coefficient of −0.949, both are nearly interchangeable each other (Figure 5.7b).[33] The AOCR includes the GIN Seas (Greenland Sea, Iceland Sea, and Norwegian Sea), part of the Barents Sea, and the area north of the Fram Strait. The high correlation of normalized averaged SLP and the AOI assigns the AOI a new physical significance, namely, the negative averaged SLP of the AOCR.

Based on the concept of AOCR, the high correlation of gridded SLP with AOI means the high negative correlation with the average SLP of AOCR. By calculating the RCC of gridded SLP with AOI, the ensemble of grids with high RCC in a certain time window could be identified, which is called the "AO-dominant region" as in this time span the gridded SLPs vary similar with AOI. By moving the time window with time, the AO-dominant region spatially varies by shape changing, area shrinking and extending, and boundary swinging, which is used to expresses the spatial variation of AO. The spatial variation of AO was then defined as "the temporal variation of the ensemble of grids with SLP varying consistent with the AO index in certain time span."[34] The difference of positive and negative relative area of the AO-dominant region is defined as a spatial variation index of AO (Figure 5.8). With this index, the three stages of long-term spatial variation of the AO were clearly identified. The positive SLP anomaly area dominated before 1970, showing the state before global warming. The negative SLP anomaly area dominated during 1971–1995, indicating the effect of global warming before the Arctic warming became apparent. Since 1996, both positive and negative SLP anomaly areas have been all small, possibly caused by the sea-ice retreat during the Arctic warming.

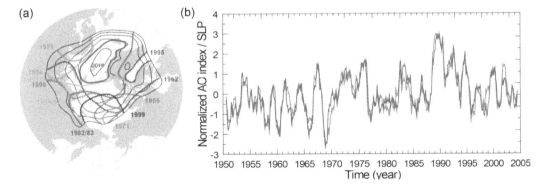

FIGURE 5.7 **(See color insert.)** (a) AOCR and (b) the normalized AO and negative average SLP of AOCR. **Source:** Data adapted from Zhao et al.[33]

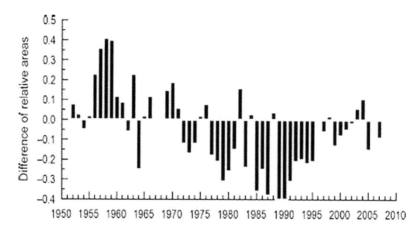

FIGURE 5.8 The AO spatial variation index of January of every year.
Source: Data adapted from Zhao et al.[34]

References

1. Thompson, D.W.J.; Wallace, J.M. The Arctic Oscillation signature in the wintertime geopotential height and temperature fields. Geophys. Res. Lett. **1998**, *25* (9), 1297–1300.
2. Joint Institute for the study of the Atmosphere and Ocean, Arctic Oscillation (AO) time series, 1899 -June 2002, http://jisao.washington.edu/data/aots/.
3. NOAA, Climate Indicators - Arctic Oscillation. http://www.arctic.noaa.gov/detect/climate-ao.shtml.
4. NOAA Climate Prediction Center, Arctic Oscillation (AO).http://www.cpc.ncep.noaa.gov/products/precip/CWlink/ daily_ao_index/ao.shtml.
5. Thompson, D.W.J.; Wallace, J.M. Annular modes in the extratropical circulation, Part I:Month-to-month variability. J. Climat. **2000**, *13* (5), 1000-1016.
6. Overland, J.E.; Wang, M. The Arctic climate paradox: The recent decrease of the Arctic Oscillation. Geophys. Res. Lett. **2005**, *32* (6), L06701. doi: 10.1029/ 2004GL021752.
7. National Snow and Ice Data Center, Arctic Oscillation, http://nsidc.org/arcticmet/glossary/.
8. NEWxSFC, Arctic Oscillation, http://www.cpc.ncep.noaa.gov/research_papers/ncep_cpc_atlas/8/figures/temp/, http://www.cpc.ncep.noaa.gov/research_papers/ncep_cpc_atlas/8/figures/precip/.
9. Department of Environment and Natural Resources, Natural Climate Fluctuations, http://www.enr.gov.nt.ca/_live/pages/wpPages/soe_natural_fluctuations.aspx.
10. Thompson, D.W.J.; Wallace, J.M.; Hegerl, G.C. Annular modes in the extratropical circulation. Part II: Trends J. Climat. **2000**,*13* (5), 1018-1036.
11. Thompson, D.W.; Wallace, J.M. Regional climate impacts of the Northern Hemisphere annular mode. Science **2001**, *293* (5527), 85-89.
12. Barnston, A.G.; Livezey, R.E. Classification, seasonality and persistence of low-frequency atmospheric circulation patterns. Mon. Weather Rev. **1987**, *115* (6), 1083-1126.
13. Hurrell, J.W. Decadal trends in the North Atlantic Oscillation region temperatures and precipitation. Science **1995**, *269* (5224), 676-679.
14. Walker, G.T.; Bliss, E.W. World weather V. Mem. Roy. Meteorol. Soc. **1932**, *4* (36), 53-84.
15. Stephenson, D.B.; Wanner, H.; Bronnimann, S.; Leterbacher, J. The history of scientific research on the North Atlantic oscillation. In *The North Atlantic Oscillation: Climate Significance and Environmental Impact;* Hurrell, J.W., Ed.; American Geophysical Union: Washington, DC, 2003; 37-50.

16. Hurrell, J.W.; Kushnir, Y.; Ottersen, G.; Visbeck, M. An overview of the North Atlantic oscillation. In *The North Atlantic Oscillation: Climate Significance and Environmental Impact;* Hurrell, J.W., Ed.; American Geophysical Union: Washington, DC, 2003; 1-35.

17. Ambaum, M.H.P.; Hoskins, B.J.; Stephenson, D.B. Arctic oscillation or North Atlantic oscillation? J. Climat. **2001**, *14* (16), 3495-3507.

18. Monahan, A.H.; Fyfe, J.C.; Flato, G.M. A regime view of Northern Hemisphere atmospheric variability and change under global warming. Geophys. Res. Lett. **2000**, *27* (8), 1139-1142.

19. Deser, C. On the teleconnectivity of the Arctic oscillation. Geophys. Res. Lett. **2000**, *27* (6), 779-782.

20. Itoh, H. True versus apparent arctic oscillation. Geophys. Res. Lett. **2002**, *29* (8), 109.1-109.4. doi: 10.1029/2001 GL013978.

21. Christiansen, B. Comment on "True versus apparent arctic oscillation." Geophys. Res. Lett. **2002**, *29* (24), 2150. doi: 10.1029/2002GL016051.

22. Hassol, S.J. *Impacts of a Warming Arctic,* Arctic Climate Impact Assessment (ACIA) Overview Report; Cambridge University Press: Cambridge, UK, 2004; p 8.

23. Comiso, J.C. A rapidly declining perennial sea ice cover in the Arctic. Geophys. Res. Lett. **2002**, *29* (20), 1956. doi: 10.1029/2002GL015650.

24. Rothrock, D.A.; Zhang, J.; Yu, Y. The Arctic ice thickness anomaly of the 1990s: a consistent view from observations and models. J. Geophys. Res. **2003**, *108* (C3), 3083. doi: 10.1029/2001JC 001208.

25. Arfeuille, G.; Mysak, L.A.; Tremblay, L.B. Simulation of the inter-annual variability of the wind-driven Arctic sea- ice cover during 1958-1998. Clim. Dyn. **2000**, *16* (2-3), 107-121.

26. Kwok, R.; Cunningham, G.F.; Wensnahan, M.; Rigor, I.; Zwally, H.J.; Yi, D. Thinning and volume loss of the Arctic Ocean sea ice cover: 2003-2008. J. Geophys. Res. **2009**, *114* (C7), C07005. doi: 10.1029/2009JC005312.

27. Tucker, W.B.; Weatherly, J.W.; Eppler, D.T.; Farmer, L.D.; Bentley, D.L. Evidence for rapid thinning of sea ice in the western Arctic Ocean at the end of the 1980s. Geophys. Res. Lett. **2001**, *28* (14), 2851-2854.

28. Rigor, I.G.; Wallace, J.M.; Colony, R.L. On the response of sea ice to the Arctic Oscillation. J. Clim. **2002**, *15* (18), 2648-2668.

29. Rigor, I.G.; Wallace, J.M. Variations in the age of Arctic sea-ice and summer sea-ice extent. Geophys. Res. Lett. **2004**, *31*, L09401. doi: 10.1029/2004GL019492.

30. Rampal, P.; Weiss, J.; Marsan, D.; Bourgoin, M. Arctic sea ice velocity field: general circulation and turbulent like fluctuations. J. Geophys. Res. **2009**, *114* (C10), C10014. doi: 10.1029/2008JC005227.

31. Hannachi, A.; Joliffe, I.; Stephenson, D. Empirical orthogonal functions and related techniques in atmospheric science: a review. Int. J. Climatol. **2007**, *27* (9), 1119-1152. doi: 10.1002/ joc.1499.

32. Dommenget, D.; Latif, M. A cautionary note on the interpretation of EOFs. J. Clim. **2002**, *15* (2), 216-225.

33. Zhao, J.; Cao, Y.; Shi, J. Core region of Arctic oscillation and the main atmospheric events impact on the Arctic. Geophys. Res. Lett. **2006**, *33* (22), L22708. doi: 10.1029/ 2006GL027590.

34. Zhao, J.; Cao, Y.; Shi, J. Spatial variation of the Arctic Oscillation and its long-term change. Tellus **2010**, *62A* (5), 661-672.

6

Asian Monsoon

Congbin Fu
Institute of Atmospheric Physics, Chinese Academy of Sciences

Introduction

Asian monsoon is a component of the global climate system. The name "monsoon" is derived from the Arabic word "*mausim*" meaning "season," large-scale seasonal reversals of dominant wind direction and distinct rainy and dry seasons of climate over the Indian subcontinent, which was named "Indian monsoon." Later on, it became clear that the monsoon is a much larger and complex system that affects the weather and climate over a large domain of tropical and subtropical Asia. A more general term "Asian monsoon" is widely used in climatology.

Classical monsoon theory believes that the seasonal change of land–ocean thermal contrast forced by the seasonal change of solar radiation is the major cause of the formation of a monsoon system. In summer, land is warmer than the ocean and a thermal depression forms over land. The airflow, called summer monsoon, brings the moist air from the ocean into the continent where there are more clouds, higher humidity, and more rainfall; while in winter, land is cooler than the ocean and a high-pressure system forms over land. The airflow, called winter monsoon, brings cold and dry continent air-mass from higher latitudes to the tropical and subtropical continent where there are relatively fewer clouds, low humidity, and less rainfall. Research also points out the role of dynamic and thermal dynamic effects of the Tibetan Plateau in the formation of Asian monsoon. However, the term "monsoon" is very often used to refer to its rainy phase only, or the summer monsoon.

Major Features of Asian Monsoon Climate and Its Two Sub-Systems: South Asian Monsoon and East Asian Monsoon

The Asian monsoon has been classified into two subsystems: the South Asian monsoon and the East Asian monsoon. These two subsystems are linked to each other, but with different features. The South Asian monsoon, or Indian monsoon, is a tropical system affecting the Indian subcontinent and surrounding regions. In summer, the southwest monsoon comes from the Indian Ocean and Arabian Sea. The onset of southwest monsoon usually occurs at the end of May or early June in southern

India and then advances rapidly northward. In mid-July, it spreads over the whole Indian Peninsula. In early September, the southwest monsoon begins to retreat southeastward. The dominant period of southwest monsoon, from June through September, is the rainy season of India when the rainfall is 75% of the annual total. Around September, with the sun retreating south, the northern land mass of the Indian subcontinent begins to cool off rapidly. When a high-pressure system begins to build over northern India, the Indian Ocean and its surrounding region remain warm. This causes the cold wind to sweep down from the Himalayas and the Indo-Gangetic Plain toward the vast expanse of the Indian Ocean and south of the Deccan Peninsula. This is known as the northeast monsoon. From November to April, the India subcontinent enters into a dry season when the northeast monsoon dominates.

The East Asian monsoon is a tropical and subtropical system that affects large parts of Indo-China, the Philippines, China, Korea, and Japan. The East Asian summer monsoon comes from tropical and subtropical oceans. The onset of the East Asian summer monsoon has been generally recognized to start in mid-May over the South China Sea and then moves northward and eastward. It reaches the Yangtze River basin in mid-June and further across the Yellow River basin in late July to early August when the summer monsoon enters into its peak period. The East Asian summer monsoon begins to retreat in September. The beginning and ending of the rainy season in East Asia are closely linked with the advance and retreat of the summer monsoon.

The rain occurs in a concentrated belt that stretches east-west over China and tilts east–northeast over Korea and Japan. Such seasonal rain is known as *Meiyu* in China, *Changma* in Korea, and *Bai-u* in Japan. Different from the South Asia monsoon, the East Asian winter monsoon is stronger than the summer monsoon. It originates from the cold and dry continent air-mass of the Siberian high pressure system with dominant northwest or northeast flows. The East Asian winter monsoon is often accompanied by cold ocean waves, snow, and strong wind. Figure 6.1 presents the long-term mean precipitation (mm/day) and wind (m/s at 850 hPa) in monsoon Asia region in January and July.[1,2]

Variability of Asian Monsoons

Asian monsoon is characterized by high variability at various time scales, from synoptic, intraseasonal, interannual, decadal, and interdecadal, centennial to geological scales. On a synoptic scale, the advance of summer monsoon normally has several phases: onset, burst, break, surge, and retreat, and so forth. Recent research concluded that the onset of the Asian summer monsoon is composed of three sequential stages, that is, the Bay of Bengal, South China Sea, and Indian monsoon onsets. On an interannual scale, there is a close link between Asian monsoon and ENSO, the strongest signal of a tropical atmospheric–ocean system. Some analysis and numerical modeling studies indicate that during the warm phase of ENSO, the South Asian monsoon tends to be weaker than during the cold phase of ENSO, but the East Asian monsoon–ENSO relationship is somewhat different from that of the South Asian monsoon. Recent research also indicates the weakening of such a relationship in the last several decades. The newly discovered IOD, a mode of interannual variability of sea surface temperature of the Indian Ocean, may contribute to the weakening or modifying of the ENSO–Asian monsoon relationship. On an interdecadal scale, the so-called PDO, the out of phase oscillation between the central to western North Pacific and the central to eastern tropical Pacific, a prominent feature of North Pacific sea surface temperature variability is believed to contribute to the Asian monsoon variation and to modify the ENSO–monsoon relationship. On the geological scale, the strengthening of the Asian monsoon has been linked to the uplift of the Tibetan Plateau after the collision of the Indian subcontinent and Asia around 50 million years ago. Based on records from the Arabian Sea and the record of wind-blown dust in the Loess Plateau of China, many geologists believe the Asian monsoon first enhanced around 8 million years ago. More recently, plant fossils in China and new long duration sediment records from the South China Sea led to timing of the Asian monsoon starting 15–20 million years ago and linked to early Tibetan uplift.

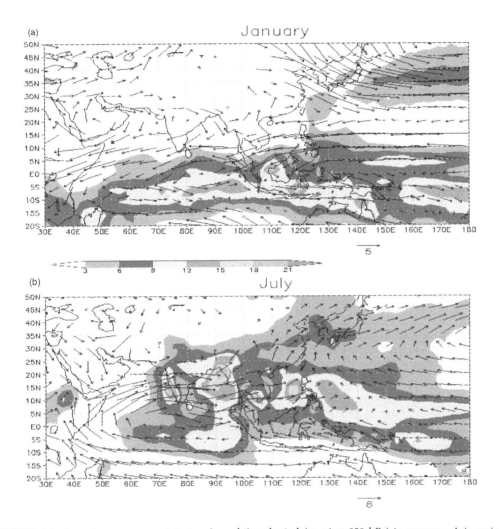

FIGURE 6.1 Long-term mean precipitation (mm d-1) and wind (m s-1 at 850 hPa) in monsoon Asia region, (a) January and (b) July. The scale for precipitation is in shadow. The scale for wind is indicated by the arrow below the figure.
Source: Adapted from CMAP,[1] NCEP.[2]

Human–Monsoon Interaction

The behaviors of Asian monsoons under global warming have been examined by both observation and climate modeling. It is likely that both Asian summer and winter monsoons have become weaker as the Northern Hemisphere warms. Recent studies also suggest that anthropogenic aerosols related to air pollution in Asia may also contribute to the monsoon variation and changing rainfall patterns.

The region affected by the monsoon system, the so-called monsoon Asia, supports the largest population on Earth, being home to 3.6 billion people. Almost all aspects of societal and economic activities in monsoon Asia are critically dependent on the monsoon climate and its variability. It has direct impacts on water resources, air quality, and occurrence of climate-related disasters and indirectly on agriculture, industry, health, urban life, and ecosystem services. Monsoon rainfall provides the major water resources for the region to support human beings and ecosystems, especially in the development of agriculture. Related to high variability of climate, the monsoon Asia region is

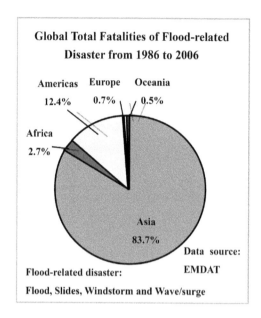

FIGURE 6.2   Total global fatalities from flood-related disasters from 1986 to 2006.
Source: EMDAT, The international Disaster Database, Center for Research on the Epidemiology of Disaster (CRED) supported by USAID.

characterized by high frequency and intensive climate-related disasters, such as floods and drought, which have profound impacts on the food security, water security, and daily life of the people as well as the sustainable development of the region. For example, during the period from 1986 to 2006, the fatalities of flood-related disasters in Asia, including floods, landslides, windstorm, and wave/surge, are 83.7% of global total (Figure 6.2). Prediction of Asian monsoon activities and their impacts on human society has long been an important task for the meteorological agencies of the countries in this region.

Global Monsoon

In the past several decades, it has been more widely accepted that monsoonal climate appears not only in Asia but also in Australia, Africa, South and North America, and the western North Pacific. A new concept of global monsoon was proposed to represent a response of the coupled atmosphere–land–ocean–cryosphere–biosphere system to annual variation of solar radiative forcing.

References

1. Xie, P.; Arkin, P.A. Global precipitation: A 17-year monthly analysis based on gauge observations, satellite estimates, and numerical model outputs. Bull. Am. Meteor. Soc. **1997**, *78* (11), 2539–2558.
2. Kalnay, E.; Kanamitsu, M.; Kistler, R.; Collins, W.; Deaven, D.; Gandin, L.; Iredell, M.; Saha, S.; White, G.; Woollen, J.; Zhu, Y.; Leetmaa, A.; Reynolds, R.; Chelliah, M.; Ebisuzaki, W.; Higgins, W.; Janowiak, J.; Mo, K.C.; Ropelewski, C.; Wang, J.; Jenne, R.; Joseph, D. The NCEP/NCAR 40-year reanalysis project, Bull. Am. Meteor. Soc. **1996**, *77* (3), 437–471.

Bibliography

Chang, C.-P; Ding, Y.; Lau, N.-C.; Johnson, R.H.; Wang, B.; Yasunari, T. *The Global Monsoon System: Research and Forecast*; World Scientific Publishing Co. Pte. Ltd.: Singapore, 2011; 594 pp.

Gao, Y.X. *Several Issues of East Asian Monsoon*; Science Press of China: Beijing, 1962; 278 pp.

Ramage, C.S. *Monsoon Meteorology*; Academic Press, 1971; 296 pp.

Wang, B. Ed; *The Asia Monsoon*; Springer-Praxis Books: Berlin, 2006; 787 pp.

7

Atmospheric Acid Deposition

Gene E. Likens
*Cary Institute of
Ecosystem Studies
University of Connecticut*

Thomas J. Butler
*Cary Institute of
Ecosystem Studies
Cornell University*

Introduction

"Acid rain" is the popular term for acid deposition, which includes both wet and dry deposition from the atmosphere. Wet deposition includes all forms of precipitation: rain, snow, sleet, hail, fog, and cloud water. Dry deposition, which can generally range from 20% to 70% of total acid deposition, includes acidifying particles and gases that adsorb to surfaces or are sometimes taken up by foliage. Some forms of acid deposition are natural, resulting from volcanic emissions (e.g., hydrochloric and sulfuric acids), forest fires (nitric and organic acids), and lightning (nitric acid). Acid deposition is of concern when additional acidity is added to the atmosphere from human-made (also known as anthropogenic) emissions of sulfur dioxide (SO_2) and nitrogen oxides ($NO_x = NO + NO_2$), largely from the burning of fossil fuels for power generation, industrial processes, transportation, heating, and cooking. These sources far exceed natural sources of acidity in many industrialized countries.

Anthropogenic acid deposition was first investigated in 1852 by Robert Angus Smith, an English chemist who studied the chemistry of rain in and around Manchester, and other industrialized cities in England and Scotland. Rain in these industrialized cities was discovered to contain sulfuric acid.[1] However, it was not until the 1960s that acid rain became an environmental issue in northern Europe[2] and North America,[3–5] and was recognized as a widespread regional problem, rather than just a local concern.

Sulfuric acid is formed in the atmosphere by

$$SO_2 + H_2O \rightarrow H_2SO_4 \leftrightarrow H^+ + HSO_4^-$$

$$\leftrightarrow 2H^+ + SO_4 =, \text{or}$$

$$2SO_2 + O_2 + 2H_2O \rightarrow 2H_2SO_4 \text{ etc.}$$

and nitric acid by

$$NO_2 + H_2O \rightarrow HNO_3 \leftrightarrow H^+ + NO_3^- \text{ or}$$

$$NO_2 + OH \rightarrow HNO_3 \text{ etc.}$$

HNO_3 can also form an aerosol that can be dry-deposited. These are the simplest reactions, but other more complex reactions can also produce these acids.

Acidity in precipitation is measured as pH, which is the negative log of the hydrogen ion (H^+) concentration. For every 1-unit change in pH, the acidity, or hydrogen ion concentration, changes 10-fold. For example, a pH of 5.0 = 10 μeq/l (microequivalents per liter of solution) of H^+, a pH of 4.0 = 100 μeq H^+/l, etc. Thus, acidity increases as pH decreases. Natural background levels of precipitation tend to be around pH 5.1–5.2 (8–6 μeq H^+/l)[6], but during the mid-1970s, pH yearly averages were on the order of 4.1 (80 μeq H^+/l) to 4.3 (50 μeq H^+/l) in northwestern Europe and the northeastern United States and southeastern Canada, respectively.[7] Thus, acidity in precipitation was about 8–10 times background levels. Individual events as low as pH 2.4 were recorded during this period, which would be hundreds of times above background values.

Major Emission Sources and Trends

The major anthropogenic sources of SO_2 emissions are coal and oil-fired electric-generating plants, metal smelters, and industrial fuel combustion. Major anthropogenic sources of NO_x are vehicles, electric-generating facilities, and industrial fuel combustion. Trends for the United States are shown in Figure 7.1a and b. European and North American emissions of SO_2 and NO_x have declined sharply over the past 40 years. However, emissions of SO_2 and NO_x have increased in eastern Asia (e.g., China and India), where industrial expansion and electric generation have expanded rapidly in recent decades. Emission controls in this region were limited through 2000 (Figure 7.2a and b). In 2005, the total global emissions of SO_2 were ~115 million metric tons/year, with China accounting for 28% of the total, 5% for India, 15% for North America, and 10% for Central and Western Europe.[9] Total global anthropogenic NO_x emissions for 2010 were estimated at ~77 million metric tons with China, India, North America, and Central and Western Europe accounting for 17%, 8%, 24%, and 12%, respectively.[10] Eastern North America, Europe, and eastern Asia are areas that have been impacted by acid deposition due to several factors such as deposition history and the presence of weathering-resistant bedrock and soils that heighten sensitivity in these regions (see Figure 7.3).

Trends in Acid Deposition

Changes in emissions of SO_2 and NO_x, in most instances, are accompanied by comparable changes in acid deposition in regions downwind of the emission sources. For example, declines in emissions from the eastern and Midwestern United States in recent decades are reflected in significant declines in H^+ concentration. One of the best examples of this change is from Hubbard Brook Experimental Forest (HBEF) located in the White Mountains, New Hampshire, and far from large emission sources. The HBEF has the longest, integrated record of precipitation and stream water chemistry in the world, with records that extend back to 1963.[14] This long-term record shows an over 80% decline in H^+ concentration, from a 5-year average of 77 μeq H^+/l (pH = 4.11) from 1966 to 1970, to a 5-year average of 12 μeq H^+/l (pH 4.92) from 2009 to 2014 (Figure 7.4).

The National Atmospheric Deposition Program (NADP) tracks changes in precipitation chemistry in the United States through the NTN and AIRMoN networks.[15] The trend in acidity of precipitation in the United States is clearly shown by comparing Figure 7.5a with Figure 7.5b.

The U.S. Clean Air Amendments of 1990 included provisions to reduce acid deposition by decreasing SO_2 and NO_x emissions, mainly from power plants. These emission reductions were first phased in after 1994. The effectiveness of these emission reductions is well documented. From 1994 to 2017, pH in

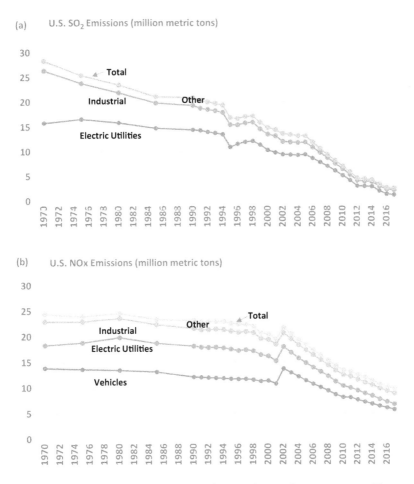

FIGURE 7.1 Annual (a) SO_2 and (b) NO_x emissions in the United States from 1970 to 2017. The significant drop in SO_2 emissions from 1994 to 1995 is the result of the implementation of Title IV (in 1995) of the Clean Air Act amendments of 1990. The large increase in NO_x vehicle emissions from 2001 to 2002 is the result of a change in emission estimate methodology[8], as opposed to a large increase in NO_x emissions. (Adapted from EPA National Emissions Inventory.[8].)

precipitation increased (acidity decreased) from about 4.2 to 5.2 in the northeast, from 4.7 to 5.3 in the southeast, and from 4.4 to 5.4 in the Midwest. These represent declines in H^+ concentrations of 72, 15, and 46 µeq H^+/l, respectively, or declines of 90%, 75%, and 90%, respectively. Reduced emissions also simultaneously reduced dry deposition of acid-generating species.[16,17] More recent air pollution rules, including mandated reductions in NO_x from vehicles, have continued to reduce emissions and further decrease acid deposition in the United States.[18], if these rules are not blocked by the current administration. On a similar time scale, reductions in emissions from clean air legislation and the resulting reduction in acid deposition have also occurred in Canada and Europe (Figure 7.2), and further emissions reductions are anticipated.[19,20]

Environmental Effects of Acid Deposition

Acid deposition became an environmental issue when it was first linked to the loss of fish populations in the 1960s in remote areas of northern Europe where no local sources of pollution were evident.[2] As further research into acid deposition occurred, clear impacts became evident that extended to terrestrial

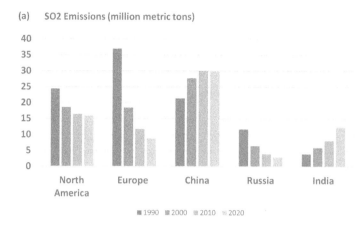

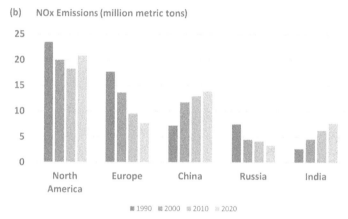

FIGURE 7.2 Trends in **(a)** SO_2 emissions and **(b)** NO_x emissions for selected regions based on model results and current legislation, from 1990 to 2020. (Adapted from Smith et al.,[9] and Cofala et al.[10].)

FIGURE 7.3 A global perspective of the extent of acid deposition (hatched areas) and areas most likely to be impacted because of the sensitivity of the landscape to acid inputs (shaded areas). (Adapted from Likens et al.,[11] Rhode et al.,[12] Kuylenstierna et al.[13].)

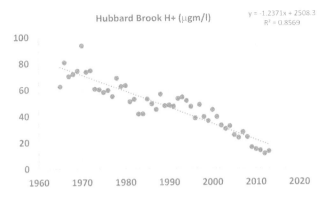

FIGURE 7.4 Long-term record of annual volume-weighted H⁺ concentration in precipitation at HBEF. (Derived from Likens [21].)

ecosystems, human-made structures, and indirectly (largely through the inhalation of very small particles) to human health. Ecosystem impacts occur where the landscape is acid sensitive, which results from a combination of factors that includes precipitation amount, vegetation type and density, the resistance of bedrock to weathering, and the ambient acidity and acid neutralization potential of soils. Geology and soils in these sensitive areas are resistant to weathering and therefore do not have enough readily available base cations, such as calcium (Ca^{2+}) and magnesium (Mg^{2+}), to neutralize soil water acidity. Other sensitive areas include highly weathered tropical soils under the influence of high rainfall rates.[13] The overlap between areas receiving acid deposition and those that are acid sensitive is illustrated in Figure 7.3.

Aquatic Effects

In the 1960s, Svante Odén, a Swedish soil scientist, suggested that the widespread acidification of Swedish lakes and rivers was linked to SO_2 emissions originating in Great Britain and Central Europe. Acid rain became an issue in North America in the 1960s when the phenomenon was discovered in the northeastern United States by Gene Likens et al.[3,5] and was linked to the loss of fish populations in remote acid-sensitive lakes in eastern Canada [22] and the Adirondack Mountains of New York.[23]

Many fish species prefer waters that are above pH 6, although there is a range of acid tolerance among different fish species. There are typically 5–6 fish species in Adirondack Mountain lakes with pH values of 6 and above, but only 1 or 2 fish species when pH values are less than 5.[24] This pattern of lower species diversity with increasing acidity (and lower pH) is observed throughout the food web, including plants, insects, and other invertebrates that are food sources for fish.[25,26] Lakes with a low pH, and low acid-neutralizing capacity (ANC) (e.g., low calcium ion (Ca^{2+}) concentration), often have high dissolved aluminum (Al^{3+}) concentrations, which are toxic to both plants and animals. Aluminum is generally bound and immobilized in nonacidic soils. However, aluminum becomes 1,000 times more soluble as pH declines from 5.6 to 4.6.[27] Aluminum toxicity leads to fish mortality from deterioration of fish gills and impairment of respiration.

Another complication for aquatic ecosystems, especially in cool temperate climates where snowfall is common, is episodic acidification. During winter, snowpack accumulates; then with the first major thaws, much of the acidity stored in the snowpack leaches out preferentially in concentrated form into streams and lakes. This acid shock can result in fish kills and mortality to other aquatic species. Increased release of dissolved aluminum from soils can further increase the toxicity of the waters by this acid pulse entering streams and lakes. Episodic acidification can also occur in non-snowy environments, where large rainstorms lead to rapid runoff and little time for acid neutralization of these runoff waters by watershed soils.

(a)   Hydrogen ion concentration as pH from measurements made at the Central Analytical Laboratory, 1994

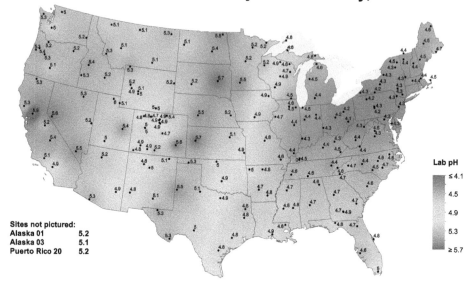

Sites not pictured:
Alaska 01   5.2
Alaska 03   5.1
Puerto Rico 20   5.2

Lab pH
≤ 4.1
4.5
4.9
5.3
≥ 5.7

National Atmospheric Deposition Program/National Trends Network
http://nadp.slh.wisc.edu

(b)   Hydrogen ion concentration as pH from measurements made at the Central Analytical Laboratory, 2017

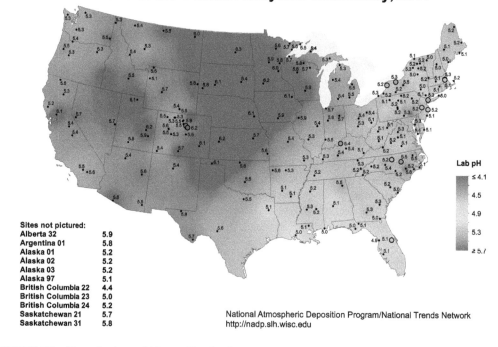

Sites not pictured:
Alberta 32   5.9
Argentina 01   5.8
Alaska 01   5.2
Alaska 02   5.2
Alaska 03   5.2
Alaska 97   5.1
British Columbia 22 4.4
British Columbia 23 5.0
British Columbia 24 5.2
Saskatchewan 21   5.7
Saskatchewan 31   5.8

Lab pH
≤ 4.1
4.5
4.9
5.3
≥ 5.7

National Atmospheric Deposition Program/National Trends Network
http://nadp.slh.wisc.edu

FIGURE 7.5 (**See color insert.**) National levels of pH in precipitation for (**a**) 1994 and (**b**) 2017. (Adapted from NADP[15].)

Terrestrial Effects

In the 1970s and 1980s, forest health was in decline in parts of Europe and eastern North America, especially in acid-sensitive areas. Acid deposition was considered a factor, but linking acid deposition to forest decline was complicated by many other environmental factors. Tree vigor can be compromised by drought, insect infestation and disease, and other forms of air pollution such as high levels of atmospheric ozone, and excess fertilization from nitrogen deposition. The latter stressors (ozone and excess nitrogen) are also related to acid deposition since NO_x is a precursor to ozone formation and nitrogen deposition, as well as acid deposition.

There is substantial evidence supporting the hypothesis that acid deposition is at least one major factor in forest decline, both directly and indirectly. Red spruce, common at high elevations, is directly damaged when calcium is leached from needles by acidic deposition. This impact has been linked to the loss of cold hardiness and freezing damage. High-elevation forests are also exposed to high levels of acidity from clouds and fog, which often have acidity levels many times higher than precipitation in the same area.[28]

Acid deposition on poorly buffered soils removes secondary plant nutrients such as calcium and magnesium, which can already be in short supply. The inability of base-poor soils to neutralize acid deposition can lead to increased levels of Al^{3+} in soil waters, which can be toxic to plant roots and aquatic organisms (as mentioned earlier). High aluminum concentrations and low calcium concentrations in soil waters have been correlated with sugar maple decline in Canada and the northeastern United States.

While ammonia (NH_3) and ammonium $\left(NH_4^+\right)$ are considered basic substances, they can also lead to soil acidification when microbial nitrification occurs. Nitrification is the conversion of NH_3 or NH_4^+ to NO_3^- (nitrate). The simplest form of this conversion is as follows:

$$2NH_4^+ + 3O_2 \rightarrow 4H^+ + 2H_2O + 2NO_2^-$$

(The H^+ increases acidity)

$$2NO_2^- + O_2 \rightarrow 2NO_3^-.$$

NH_3 and NH_4^+ are largely derived from agricultural activity, including fertilizer application, but mainly from the production of livestock waste. Ammonia, unlike SO_2 and NO_x emissions, has not shown large regional declines, and in many agricultural areas, NH_3 emissions are increasing.[29, 30]

Conclusion: Acid Deposition Today and in the Future

Legislation has led to large declines in emissions of SO_2 and NO_x in Europe and North America, which has resulted in a substantial reduction in acid deposition. However, a reduction in acid deposition has not always produced the expected recovery in ecosystems. Many decades of acid deposition have depleted acid-sensitive watersheds and their soils of the acid-buffering capacity necessary to reduce acidifying impacts on water and forest ecosystems.[31] While some recovery has occurred (e.g., rising pH levels in formerly acidified lakes and streams in Europe and North America),[32] these regions over time have become more acid sensitive and therefore continue to be impacted, even though the amount of acid deposition has declined. Recovery in these ecosystems will be a slow process, particularly in the terrestrial components. Decades of acid deposition may require decades of significantly reduced deposition for more complete recovery to occur.[5]

While acid deposition has declined in both Europe and North America, emissions of SO_2 and NO_x (and NH_3) have been increasing in rapidly developing areas such as Southeast Asia. China, India (see Figure 7.2a and b), and other southeastern Asian countries are experiencing rapid growth in fossil fuel use for electric generation, industrial development, and transportation, but have limited controls on the associated SO_2 and NO_x emissions (although since 2007, China has begun reductions in SO_2 from power plants).[33] Acidification from SO_2 and NO_x emissions is partially being neutralized by large amounts of alkaline particulates, which are also being released from some of these same facilities.[13] However, because of the health-related problems associated with these multiple forms of air pollution, it will be important to control particulate,[5] as well as SO_2 and NO_x emissions. If only particulate emissions were reduced, there would be less atmospheric ANC and levels of acid deposition would rise rapidly in this area of the world.[34] In summary, continued reductions in emissions of SO_2 and NO_x will be necessary in areas that have experienced significant acid deposition in the past, and in areas where the potential exists for increasing acid deposition in the future.

References

1. Smith, R.A. *Air and Rain, the Beginning of a Chemical Climatology*; Longmans Green: London, 1872.
2. Odén, S. *The Acidification of Air Precipitation and Its Consequences in the Natural Environment*; Bulletin of the Ecological Research Communications NFR. Translation Consultants Ltd.: Arlington, VA, 1968.
3. Likens, G.E.; Bormann, F.H.; Johnson, N.M. Acid rain. *Environment* **1972**, *14*, 33–40.
4. Likens, G.E.; Bormann, F.H. Acid rain: A serious regional environmental problem. *Science* **1974**, *184* (4142), 1176–1179.
5. Holmes, R.T.; Likens, G.E. *Hubbard Brook the Story of a Forest Ecosystem*; Yale University Press: New Haven, CT & London, 2016. 271 pp.
6. Likens, G.E.; Keene, W.C.; Miller, J.M.; Galloway, J.N. Chemistry of precipitation from a remote, terrestrial site in Australia. *J. Geophys. Res.* **1987**, *92* (D11), 13299–13314.
7. Likens, G.E.; Wright, R.F.; Galloway, J.N.; Butler, T.J. Acid rain. *Sci. Am.* **1979**, *241* (4), 43–51.
8. EPA National Emissions Inventory, 2017. https://www.epa.gov/air-emissions-inventories/2017-national-emissions-inventory-nei-data (accessed March 27, 2019). Data accessed from Average Annual Emissions, within website.
9. Smith, S.J.; van Aardenne, J.; Klimont, Z.; Andres, R.J.; Volke, A.; Delgado, A. Anthropogenic sulfur dioxide emissions: 1850–2005. *Atmos. Chem. Phys.* **2011**, *11*, 1101–1116. doi: 10:5194/acp-11-1101-2011.
10. Cofala, J.; Amann, M.; Mechler, R. *Scenarios of World Anthropogenic Emissions of Air Pollutants and Methane up to 2030*; International Institute for Applied Systems Analysis (IIASA): Laxenburg, Austria, 2005 http://webarchive.iiasa.ac.at/rains/global_emiss/global_emiss.html (accessed March 29, 2019).
11. Likens, G.E.; Butler, T.J.; Rury, M. Acid Rain. In *Encyclopedia of Global Studies*; Anheier, H.; Juergensmeyer, M., Eds.; Sage Publications, Inc.: Los Angeles, London, New Delhi, Singapore, Washington, DC, 2009.
12. Rhode, H.; Dentener, F.; Schulz, M. The global distribution of acidifying wet deposition. *Environ. Sci. Technol.* **2002**, *36*, 4382–4388.
13. Kuylenstierna, J.C.I.; Rhodhe, H.F.; Cinderby, S.; Hicks, K. Acidification in developing countries: Ecosystem sensitivity and the critical load approach on a global scale. *Ambio* **2001**, *30*, 20–28.
14. Likens, G.E.; Buso, D.C. Dilution and the elusive baseline. *Environ. Sci. Tech.* **2012**, *46* (8), 4382–4387. doi: 10.1021/ es3000189.
15. NADP. National Atmospheric Deposition Program (NRSP-3). 2019. NADP Program Office Wisconsin State Laboratory of Hygiene, 465 Henry Mall, University of Wisconsin, Madison, WI 53706. http://nadp.slh.wisc.edu/.
16. Burns, D.A. 2011. The National Acid Precipitation Assessment Program Report to Congress 2011: An Integrated Assessment. http://ny.water.usgs.gov/projects/NAPAP/.
17. Butler, T.J.; Likens, G.E.; Vermeylen, F.M.; Stunder, B.J. The impact of changing nitrogen oxide emissions on wet and dry nitrogen deposition in the northeastern USA. *Atmos. Environ.* **2005**, *39*, 4851–4862.

18. USEPA. 2012. 2010 Progress Report-Environmental and Health Results. www.epa.gov/airmarkets/progress/ARPCAIR10_02.html#litigation (accessed June 2012).

19. Amann, M.; Bertok, I.; Cofala, J.; Gyarfas, F.; Heyes, C.; Klimont, Z.; Schöpp, W.; Winiwarter, W. Baseline Scenarios for the Clean Air for Europe (CAFÉ) Programme, submitted to the European Commission Directorate General for Environment 2005. https://ec.europa.eu/environment/archives/cafe/activities/pdf/cafe_scenario_report_1.pdf.

20. Canada-United States Air Quality Agreement Progress Report. www.ec.gc.ca/Publications/4B98B185-7523-4CFF-90F2-5688EBA89E4A%5CCanadaUnitedStatesAirQualityAgreementProgressReport2010.pdf.

21. Likens, G. 2016. Chemistry of Bulk Precipitation at Hubbard Brook Experimental Forest, Watershed 6, 1963 - present. Environmental Data Initiative. doi: 10.6073/pasta/573d0e1eb5d1ca541005f58146b95d19.

22. Beamish, R.J.; Harvey, H.H. Acidification of the La Cloche Mountain Lakes, Ontario, and resulting fish mortalities. *J. Fish. Res. Bd. Canada* **1972**, *29* (8), 1131–1143.

23. Schofield, C.L. Acid precipitation: Effects on fish. *Ambio* **1976**, *5*, 228–230.

24. Driscoll, C.T.; Lawrence, G.B.; Bulger, A.J.; Butler, T.J.; Cronan, C.S.; Eager, C.; Lambert, K.F.; Likens, G.E.; Stoddard J.L.; Weathers, K.C. Acid deposition in the northeastern United States: Sources and inputs, ecosystem effects, and management strategies. *BioScience* **2001**, *51* (3), 180–198.

25. Schindler, D.W.; Mills, K.H.; Malley, D.F.; Findlay, D.L.; Shearer, J.A.; Davies, I.J.; Turner, M.A.; Linsey, G.A.; Cruikshank, D.R. Long-term ecosystem stress: The effects of years of experimental acidification on a small lake. *Science* **1985**, *228* (4706), 1395–1401.

26. Confer, J.L.; Kaaret, T.; Likens, G.E. Zooplankton diversity and biomass in recently acidified lakes. *Can. J. Fish. Aquat. Sci.* **1983**, *40* (1), 36–42.

27. Johnson, N.M.; Driscoll, C.T.; Eaton, J.S.; Likens, G.E.; McDowell, W.H. Acid rain, dissolved aluminum and chemical weathering at the Hubbard Brook Experimental Forest New Hampshire, USA. *Geochim. Cosmochim. Acta* **1981**, *45*, 1421–1438.

28. Weathers, K.C.; Likens, G.E.; Bormann, F.H.; Bicknell, S.H..; Bormann, B.T.; Daub Jr., B.C.; Eaton, J.S.; Galloway, J.N.; Keene, W.C.; Kimball, K.D.; McDowell, W.H.; Siccama, T.G.; Smiley, D.; Tarrant, R. Cloudwater chemistry from ten sites in North America. *Environ. Sci. Technol.* **1988** *22* (8), 1018–1026.

29. Galloway, J.N.; Townsend, A.R.; Erisman, J.W.; Bekunda, M.; Cai, Z.; Freney, J.R.; Martinelli, L.A.; Seitzinger, S.P.; Sutton, M.A. Transformation of the nitrogen cycle: Recent trends, questions, and potential solutions. *Science* **2008**, *320*, 889–892.

30. Butler, T.J.; Vermeylen, F.; Lehmann, C.M.; Likens, G.E.; Puchalski, M. Increasing ammonia concentration trends in large regions of the USA derived from the NADP/AMoN Network. *Atmos. Environ.* **2016**, *146*, 132–140.

31. Likens, G.E.; Driscoll, C.T.; Buso, D.C. Long-term effects of acid rain: Response and recovery of a forest ecosystem. *Science* **1996**, *272*, 244–246.

32. Stoddard, J.L.; Jeffries, D.S.; Lükewille, A.; Clair, T.A.; Dillon, P.J.; Driscoll, C.T.; Forsius, M.; Johannessen, M.; Kahl, J.S.; Kellogg, J.H.; Kemp, A.; Mannio, J.; Monteith, D.T.; Murdoch, P.S.; Patrick, S.; Rebsdorf, A.; Skjeikvåle, B.L; Stainton, M.P.; Traaen, T.; van Dam, H.; Webster, K.E.; Wieting, J.; Wilander, A. Regional trends in aquatic recovery from acidification in North America and Europe. *Nature* **1999**, *401*, 575–578.

33. Li, C.; Zhang, Q.; Krotkov, N.A.; Streets, D.G.; He, K.; Tsay, S.-C.; Gleason, J.F. Recent large reduction in sulfur dioxide emissions from Chinese power plants observed by the Ozone Monitoring Instrument. *J. Geophys. Res. Lett.* **2010**, *37*, L08807.

34. Liu, M.; Huang, X.; Song, Y.; Tanf, J.; Cao, J.; Zhang, X.; Zhang, Q.; Wang, S.; Xu, T.; Kang, L.; Cai, X.; Zhang, H.; Yang, F.; Wang, H.; Yu, J.Z.; Lau, A.K.H.; He, L.; Huang, X.; Duan, L.; Ding, A.; Xue, L.; Gao, J.; Liu, B.; Zhu, T. Ammonia emission control in China would mitigate haze pollution and nitrogen deposition, but worsen acid rain. *PNAS* **2019**, *116* (16), 7760–7765.

8

Atmospheric Circulation: General

Thomas N. Chase
and David Noone
*University of Colorado
at Boulder*

Introduction

The Earth as a whole is in approximate radiative equilibrium. The amount of energy from the sun is balanced by the emission of infrared radiation from the Earth. However, there is an uneven distribution of energy. Because of the spherical shape of the Earth, sunlight directly illuminates the tropical regions, while close to the poles solar energy strikes the Earth at an oblique angle reducing the amount of energy received per unit area. In the extreme case, the winter pole receives no sunlight at all. The reflectivity of Earth depends on the cloud cover, other atmospheric constituents, and variations in the characteristics of the surface (snow, land, vegetation, and liquid ocean), further influencing the total amount of solar energy input and also further modifying the spatial distribution of energy. On the other hand, the amount of infrared radiation emitted by the Earth is more uniform and depends on only the average temperature of the atmosphere, which varies by about 15% from equator to pole (290 K in the tropics to 250 K at the pole). These two features taken together describe a net surplus of energy in the tropics (where energy gained from the sun dominates) and a net energy deficit near the polar regions (when loss of energy by infrared emission dominates). For energy balance, there is therefore a requirement for energy transport from the tropics to the polar regions. The winds that provide this required energy transport in the atmosphere are the atmospheric general circulation. A brief, historical overview of the concept of the general circulation is given by Lorenz,[1] and a more technical account of atmospheric motions is given by Schneider.[2]

Figure 8.1 shows the long-term mean sea level pressure and surface wind for July and January. For many centuries, it has been known that surface winds are predominantly easterly (wind coming from the east) along the equator and westerly (coming from the west) in the midlatitudes (between about 35° and 55° N and S). The subtropics, centered around 30° N and S, are characterized by low wind speeds (known as the doldrums due to the slow travel for sail-powered ships) and are usually free from poor weather and clouds. The very consistent easterlies in the tropics provide reliable travel for trade and are known as the trade winds. Alongside the wind patterns, atmospheric pressure at sea level is generally low along the equator, high in the subtropics, and low again near the midlatitudes. These features reflect

(a) January

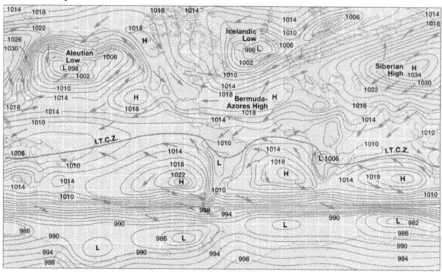

(b) July

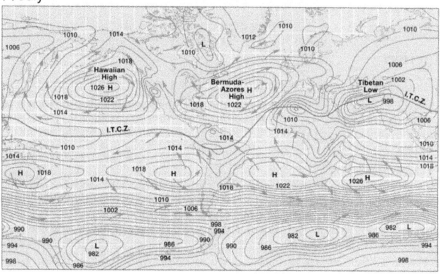

FIGURE 8.1   Long-term average sea-level pressure and wind vectors for: (**a**) January, (**b**) July.
Source: From Aguado & Burt.[3]

and characterize many aspects of the average atmospheric general circulation and are explained by considering the combination of overturning of the atmosphere with conservation of momentum.

Overturning Circulation

Pressure measures the atmospheric mass and, on average, is the force exerted by the mass of the atmosphere per unit area. The global average pressure at the Earth's surface is 986 hPa. Pressure at sea level is typically around 1000 hPa, while the pressure at higher altitudes is lower. Pressure decreases with altitude because the fraction of the atmospheric mass above a point at high altitude is smaller than the fraction of the mass above a point at low altitude. Air can be treated as an ideal gas and so there are relationships among pressure, temperature, and volume. When air is heated at constant pressure, the volume increases (i.e., for a constant mass, density decreases). In the tropics, heating of the atmosphere is associated with the radiative imbalance (and is caused by latent heat realized during precipitation, which is associated with the energy excess in the tropics that had caused previous evaporation at the Earth's surface). The expansion of the column of air in the tropics is synonymous with ascending motion. With expansion, the pressure half way up the column will be higher than a shorter column away from the equator, and the difference in pressure from south to north will force the air at the top of the column poleward. Air moving out of the tropical column is a loss of mass, and so the surface pressure is lower in the tropics than outside the tropics. Air near the surface will be forced to move from high to low pressure and converge toward the equatorial zone from both hemispheres. This region of convergence is also the region of most intense tropical thunderstorm convection and is identified as the Intertropical Convergence Zone (ITCZ). The ITCZ encompasses land masses in the summer months in both hemispheres due to the greater heating of land relative to oceans, which causes an onshore flow moving the ITCZ inland. These continental scale onshore flows are referred to as monsoon circulations and are associated with regions of intense seasonal precipitation and moist land ecosystems such as rainforests.

Poleward flow aloft and equatorward flow at the surface describes a thermally direct overturning circulation cell (the Hadley Cell). Since the poleward moving air has been heated, it transports energy poleward. Radiative cooling (net emission of radiation to space) at higher latitudes ensures that the equatorward moving air has less energy. The sum of these two components provides a net transport of energy out of the tropics, partially meeting the energy transport requirement for radiative balance.

The winds can be described by combining the motion of the overturning with consideration of conservation of momentum on the rotating Earth. Specifically, if one views the Earth rotating from space, a small volume of air that is at rest near the equator is moving at a speed given by the tangent speed of the Earth in solid body rotation. If this air mass is moved poleward at constant altitude, this velocity is conserved. Because the Earth is spherical, the distance between the Earth's surface and the rotating axis is smaller and the tangential speed of the Earth's surface will be smaller than the speed of the air mass. In which case, an observer at the surface will see the air moving eastward (termed westerly). Therefore, air in the upper troposphere (about 10–15 km above the surface) in the subtropics will be westerly because of the upper poleward branch of the overturning cell. This westerly wind region is the subtropical jet stream. The same argument can be applied to an air mass initially in the subtropics that moves to the west as it converges toward the equator (i.e., easterly trade winds). This phenomenon was described succinctly by George Hadley in 1735,[4] and the apparent force that gives rise to the east-west flow was named after Gaspard-Gustave Coriolis in the nineteenth century—the Coriolis force.[5]

While the tropical overturning circulation transports energy from the equator to the subtropics initially, the relationship between surface pressure and wind reveals a conundrum in the need for energy export from the subtropics to further poleward—as is required. For the Earth's atmosphere outside of the tropics, the Coriolis force tends to become equal and opposite to the pressure force and a balance is approximately reached (geostrophic balance). The conservation of momentum requires that the Coriolis force acts perpendicular to the direction of air motion (to the right in the Northern Hemisphere and to the left in the Southern Hemisphere). At the poleward edge of the subtropics, the decrease in pressure toward the pole provides a force that is balanced by a Coriolis force directed to the equator, which in turn requires the wind to be westerly. Indeed, westerly winds outside the tropics are seen in Figure 8.1 to be approximately parallel to the isobars as predicted by geostrophic balance. The component of the wind vectors not parallel to the isobars is due to the slowing of wind speeds affected by surface friction. Upper level winds are more nearly geostrophic.

If this were the entirety of the process, the energy transported poleward from the equator would accumulate in the subtropics and the winds would be purely westerly, preventing further transport of energy toward the poles because the westerly flow would simply transport the energy eastward. This is not the case, however.

As energy continues to accumulate, the continual increase in the pressure difference between the subtropics and higher latitudes eventually becomes sufficiently great that instability in the westerly flow causes the development of continental-size waves embedded in the westerly flow which are associated with frontal weather systems that are composed of low pressure centers near the Earth's surface. Air circulates around the low pressure center, again with the flow almost parallel to the lines of constant pressure when friction is negligible (generally about 1–2 km above the Earth's surface). In the region of the low pressure system, warm moist air is moved poleward on the eastern side and colder and drier polar air travels equatorward of the western side, the sum of the two processes providing a net transport of energy poleward. As such the transport of energy required to meet the demands of radiative balance is met in the midlatitudes by the sequence of high- and low-pressure systems continually being generated and dissipated in the region of large temperature gradients and locally strong jet streams.

Poleward Energy Transport by the General Circulation

Atmospheric motions are fundamentally caused by an excess of energy near the equator and an energy deficit near the poles. This gradient in energy is reflected in temperature and moisture (warm and moist tropics, cold and dry polar regions). The atmospheric circulation exists physically as a mechanism to limit these gradients in spite of continual forcing. Figure 8.2 shows the zonally averaged northward energy transport for the globe. The maximum energy transport poleward in both hemispheres by the atmosphere occurs in midlatitudes at about 40°. This reflects the mixing action of midlatitude cyclonic (frontal) systems discussed in the previous section. Ocean transport of energy is of smaller magnitude with maxima in the sub-tropics.[6] A basic property of the overall atmospheric flow is that it is "chaotic" and therefore unpredictable beyond a week or two.[7]

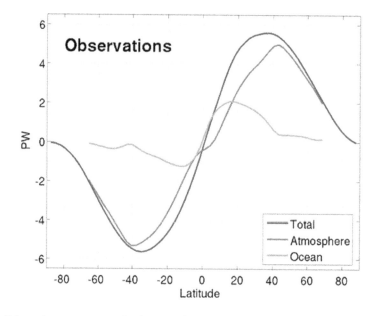

FIGURE 8.2   Poleward energy transport by the general circulation by atmosphere, ocean, and the sum of the two.
Source: Modified from Trenberth & Caron.[6]

Role of the Hydrological Cycle in the General Circulation

The fundamental role of the water molecule in the general circulation cannot be overstated. While H_2O in all its phases is universally recognized as fundamental for the existence of life in general and human activities such as agriculture in particular, from the perspective of the atmospheric circulation water substance plays a central role in major energy transport mechanisms. This is expressed by the latent heating due to phase changes. Energy is needed to evaporate water from a wet land surface or ocean, and this energy is later released in the upper atmosphere if enough cooling occurs to cause condensation, resulting in a net upward energy transport. The radiative properties of the atmosphere dictate that the atmosphere is mostly heated from below with energy transfer from the Earth's surface to the lower atmosphere. Approximately three quarters of the net radiation arriving at the Earth's surface (Figure 8.3) is transferred back to the atmosphere as latent heat (energy transfer due to the phase changes of H_2O that would dominant over ocean areas) as opposed to sensible heat (direct heating of the lower atmosphere) and infrared radiation.

Figure 8.4 shows the partitioning of atmospheric heating by process, altitude, and latitude and indicates that latent heating (third panel from top) is the major process supplying energy to the atmosphere. Compared to sensible heating (fourth panel from top and labeled boundary layer heating), latent heating is of considerably larger magnitude and is realized through a much greater depth of the atmosphere. Net radiative heating indicates mostly a cooling effect due to radiative losses to space. The top panel, total diabatic heating, represents the sum of the other three panels.

The role of the hydrological cycle in the general circulation of the future is of immense interest because three of the largest positive feedback mechanisms (processes that add to the magnitude of the original forcing) to warming due to increased CO_2 involve the H_2O molecule, one feedback for each of its three phases. The water vapor feedback (increased atmospheric water vapor content due to increased temperature and therefore increased saturation point) is responsible for the largest part of the simulated global warming signal.[10,11] Increased cloudiness due to the increased water vapor is also thought to amplify global warming, though the precise magnitude remains poorly known because the net response is a function of the exact composition and altitude of changing clouds. The ice albedo feedback, where melting ice leads to decreased reflection of solar energy and therefore increased energy absorption, is also recognized as a positive feedback to global warming.[12]

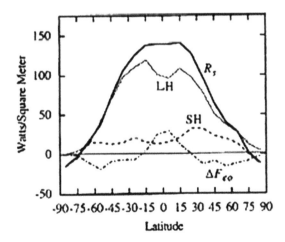

FIGURE 8.3 Annual average surface energy balance over the lobe as a whole indication the proportion of net radiation (R_s) returned to the atmosphere as sensible heating (SH), latent heating (LH) by latitude. ΔF_{eo} represents energy transferred horizontally out of the land ocean below surface column.
Source: From Hartmann.[8]

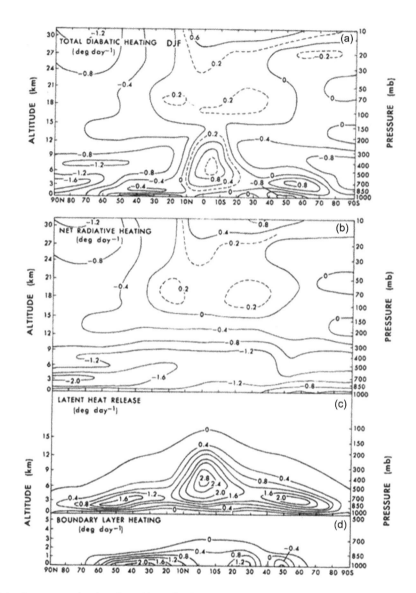

FIGURE 8.4 Comparison by altitude and latitude of: (a) total diabatic heating due to radiation + sensible + latent heating, (b) net radiative heating, (c) latent heat, and (d) sensible heat.
Source: From Peixoto & Oort.[9]

Asymmetric Flow: Teleconnection Patterns, Monsoons

We have previously discussed the symmetric overturning circulation and the asymmetric waves super-imposed upon it that take the form of frontal systems at the Earth's surface. Other asymmetric parts of the general circulation that are important in both medium- to long-range weather forecasting and for understanding changes in the regional climate on time scale from decades to many thousands of years are atmospheric teleconnection patterns. These are defined as quasi-periodic fluctuations in part of the atmospheric-ocean system that then excite waves which propagate large distances away from the initial disturbance. El Nino-Southern Oscillation (ENSO) is one such phenomenon that involves tele-connections and that begins with a disturbance in the equatorial Pacific but subsequently has remote impacts on temperature and precipitation elsewhere on the globe. A number of other teleconnection patterns have been identified[13,14] that change the general atmospheric circulation over periods of years to decades. A discussion of more recently discovered teleconnection patterns is given in Thompson and Wallace.[15]

A second major class of asymmetric motions are monsoon circulations that are defined classically as a reversal of winds from onshore to offshore following the heating and cooling of land surfaces with season (i.e., an annual oscillation). Tropical monsoon systems are linked to the ITCZ in the sense that the heating of land surfaces in the summer draws the ITCZ over the land surface and so shifts the latitude of the rising branch of the tropical overturning circulation.

Another class of atmospheric circulation related to monsoons by the same physical process of differential heating of land and adjacent ocean is the regional sea breeze. Such locally confined circulations are vital, e.g., for the initiation of convection over the numerous islands in the tropical oceans that feed large parts of the ITCZ.

Human Effects: CO_2, Aerosols, Land Cover Change

Human activity alters the general circulation in a variety of ways. The increase in atmospheric CO_2 leads to a general atmospheric warming and consequently changes the global hydrological cycle while also reducing the speed of the tropical atmospheric circulations. These effects will necessarily involve changes in atmospheric circulations globally.[16]

Aerosols, particles resulting from combustion and residing in the atmosphere and that tend to reflect solar energy therefore disrupting the surface energy balance, are also implicated in atmospheric circulation changes.[17]

There is evidence that human-induced land-cover changes alter the general circulation.[18,19] For example, tropical deforestation tends to reduce evapotranspiration and increase albedo creating links to both the distribution of net surface radiation and the hydrological cycle both of which affect pressure patterns and winds. New diagnostic techniques, such as the use of isotope ratios of water as tracers, are proving useful in exploring linkages between the water cycle and both ambient general circulation and perturbations due to humans.[20]

Conclusion

The general motion of the Earth's atmosphere arises from differential energy production with excess energy in the tropics and near the earth's surface. These energy gradients are reflected in pressure relations that drive a circulation which transports energy upward and poleward on average. Deviations from symmetry include baroclinic waves (frontal systems), teleconnection patterns, and monsoon systems. We have emphasized the fundamental processes of energy conversion and transport that define the general circulation and how these relate to the hydrological cycle. Specifically, the atmospheric and oceanic general circulation can be well described by basic laws of physics including Newton's laws of motion applied to air on a rotating spherical Earth, the definition of air as an ideal gas, conservation of energy, and conservation of mass keeping in mind that this is a nonlinear system. We also described some ways in which human activity affects the general circulation through changes that are imposed on the distribution of energy associated with the radiative properties of the atmosphere (including changes to the concentration of gases like carbon dioxide and methane and aerosol particulates), and changes to the properties of the land surface (reflectivity of different types of surfaces and transfer of water from the landscape to the atmosphere).

The features of the general circulation described above are well understood in a basic sense. However, how the atmospheric circulation might change with future perturbations and the precise magnitude of the many feedbacks within the climate system remain without complete description.

References

1. Lorenz, E.N. A history of prevailing ideas about the general circulation of the atmosphere. Bull. Am. Meteorol. Soc. **1983**, *64* (7), 730–733.
2. Schneider, T. The general circulation of the atmosphere. Annu. Rev. Earth Planet. Sci. **2006**, *34*, 655–688, DOI: 10.1146/annrev.earth.34.031405.125144.
3. Aguado, E.; Burt, J.E. *Understanding Weather and Climate*, 6th Ed.; Prentice Hall: New Jersey, 2013; 576 pp. ISBN-13: 9780321769633©2013.
4. Hadley, G. Concerning the cause of the general trade winds. Phil. Trans. Royal Soc. **1735**, *39*, 58–62.
5. Coriolis, C.G. Sur les equations du movement relative des systemes de corps. J. De L'ecole Royale Polytechnique **1835**, *15*, 144–156.
6. Trenberth, K.E.; Caron, J.M. Estimates of meridional atmosphere and ocean heat transports. J. Climate **2001**, *14* (16), 3433–3443. doi: http://dx.doi.org/10.1175/1520-0442(2001)014<3433:E0M AA0>2.0.C0;2
7. Lorenz, E.N. Deterministic non-periodic flow. J. Atmos. Sci. **1963**, *20* (2), 130–141.
8. Hartmann, D.L. *Global Physical Climatology*; Academic Press: San Diego, 1994; 411 pp.
9. Peixoto, J.P.; Oort, A. *Physics of Climate*; American Institute of Physics: New York, USA, 1992; 520 pp.
10. Arrhenius, S. On the influence of carbonic acid in the air upon the temperature of the ground. Lond. Edinb. Dublin Philos. Mag. J. Sci. (Fifth Series) **1896**, *41* (251), 237–275.
11. Held, I.M.; Soden, B.J. Water vapor feedback and global warming. Annu. Rev. Energy Environ. **2000**, *25* (1), 441–475.
12. Solomon, S.; Qin, D.; Manning, M.; Chen, Z.; Marquis, M.; Averyt, K.B.; Tignor, M.; Miller, H.L., Eds. Contribution of working group I to the fourth assessment report of the intergovernmental panel on Climate Change. In *Climate Change 2007: The Physical Science Basis;* Cambridge University Press: Cambridge, UK and New York, USA, 2007; 996 pp.
13. Namias, J. Thirty-day forecasting: A review of a ten-year experiment. Meteorol. Monogr. (American Meteorological Society) **1953**, *2* (6), 24–25.
14. Barnston, A.G.; Livzey, R.E. Classification, seasonality and persistence of low-frequency atmospheric circulation patterns. Mon. Weather Rev. **1987**, *115* (6), 1083–1126.
15. Thompson, D.W.J.; Wallace, J.M. Annular modes in extratropical circulation. Part I: Month to month variability. J. Climate **2000**, *13* (5), 1000–1016.
16. Held, I.M.; Soden, B.J. Robust responses of the hydrological cycle to global warming. J. Climate **2006**, *19* (21), 5686–5699.
17. Ramanathan, V.; Chung, C.; Kim, D.; Bettge, T.; Buja, L.; Kiehl, J.T.; Washington, W.M.; Fu, Q.; Sikka, D.R.; Wild, M. Atmospheric brown clouds: Impacts on South Asian climate and hydrologic cycle. Proc. Natl. Acad. Sci. **2005**, *102* (15), 5326–5333.
18. Eltahir, E.A.B. Role of vegetation in sustaining large-scale atmospheric circulations in the tropics. J. Geophys. Res. **1996**, *101* (D2), 4255–4268.
19. Lee, E.; Chase, T.N.; Rajagopalan, B. Highly improved predictive skill in the forecasting of the North East Asian summer monsoon. Water Resour. Res. **2008**, *44* (10), W10422. doi: 10.1029/2007WR006514.
20. Noone, D.; Galewsky, J.; Sharp, Z.; Worden, J.; Barnes, J.; Baer, D.; Bailey, A.; Brown, D.; Christensen, L.; Crosson, E.; Dong, F.; Hurley, J.; Johnson, L.; Strong, M.; Toohey, D.; Van Pelt, A.; Wright, J. Properties of air mass mixing and humidity in the subtropics from measurements of the D/H isotope ratio of water vapor at the Mauna Loa Observatory. J. Geophys. Res.-Atmos. **2011**, *116* (D22), 113. doi: 10.1029/2011JD015773.

9

Circulation Patterns: Atmospheric and Oceanic

Marcia Glaze Wyatt
University of Colorado
at Boulder

Introduction

Climate: The Big Picture—A Global Perspective

Energy is delivered to Earth by the Sun. In turn, through climate processes, Earth absorbs and partitions available energy among various reservoirs—ocean, ice, land, and atmosphere—and at various levels within each. On different timescales, from seasonal to millennial, stored heat energy is moved via a variety of climate processes to new latitudes, longitudes, depths, and altitudes; even to different reservoirs. In general, heat moves from where there is more to where there is less. Along the journey, some heat escapes, emitted to space. How the incoming absorbed (solar) energy matches up to the outgoing emitted infrared (heat) energy determines Earth's radiative top-of-the-atmosphere (TOA) energy balance, which, in turn, influences the climate system's response. This end result is global; yet regional atmospheric and oceanic circulation patterns, globally intercommunicating to one degree or another, with positive feedbacks countered by negative ones on seasonal to decadal timescales, account for much of this outcome, an outcome that maintains relative stability in the climate system in relation to forcings imposed upon it.

Climate Circulations: Spatiotemporal Traits—Timescale Matters

All circulation patterns fluctuate, most with preferred timescales of variability. Different response times of components to an imposed forcing or disturbance, and nonlinear interactions among those components, leave the subsystem in a state of disequilibrium. Hence, the components never achieve collective stability. Therefore, they are varying constantly. The complexity of the subsystem, with extreme sensitivity to details regarding perturbations and boundary conditions, ensures that the resulting oscillatory behavior is not rigidly periodic. Because the fluctuations are not temporally regular, the oscillatory character is best described as quasi-periodic. Most circulation patterns vary at more than one "quasi-regular" frequency, often including a low-frequency component. Preferred frequencies can be internally generated or internally generated with excitation from, or entrainment by, an external source.

Characteristic processes and traits tend to repeat for a given pattern; yet, again, due to the nonlinear nature of the subsystems, no two snapshots in time will be identical. In addition, teleconnections—climate features or responses highly correlated with, and distantly forced by, an associated regional circulation pattern—may coexist with a given pattern for years and then wane or disappear, or disappear only to return to its former relationship at a later date, a further reminder of climate's complexity.

Debate exists over whether or not climate, with its numerous interacting circulation patterns, is itself chaotic. In the mathematical sense, deterministic chaos refers to behavior that is acutely sensitive to initial conditions. Slight differences in initial conditions result in vastly divergent outcomes in chaotic systems. Were one to know with certainty each element of initial conditions, the outcome could be predicted. Although this knowledge is unattainable, a chaotic system is largely, but not entirely, unpredictable. Nor is the system without order. Although a complex, nonlinear system can be chaotic, and often is, it can also merely reflect behavior that appears chaotic—hence the debate. Weather tends to be chaotic. Climate, on the other hand, with its longer timescale, global energy-balance modulation over time, and negative feedbacks buffering the system from extreme disequilibrium, makes the situation less clear. Timescale makes the difference in nature. On short-term timescales, parts tend to operate individually; yet with time, parts begin to interact. Collective behavior evolves. A network architecture emerges, one with its own set of "rules," underpinning the observation that the whole is not equivalent to the sum of its parts. Nature is replete with examples; climate may well be one.[1,156] Chaotic or not, it is clear that the climate system is complex; its behavior, nonlinear. Significantly, it involves components that behave chaotically, but, as a whole, is not without order.

Taming the Discussion: Chapter Focus and Approach

Over 30 large-scale modes, or leading patterns, of climate variability have been identified across the globe. Most modes are spatially local or regional; yet with many of these, their impacts are hemispheric to global. Some patterns are dominated by atmospheric processes; some by oceanic ones. Some involve sea ice. Coupling between systems (ocean, ice, and atmosphere) and between levels within a system is common, particularly for low-frequency oscillations.

Complex dynamics of and between the numerous circulation modes renders their complete description unfeasible in limited space. Thus, the goal here is to weave together overviews of select individual regional circulation patterns. Discussion touches on their geographically governed dynamics, their individual behaviors, and their interconnectivity. The abridged inventory of patterns described here is divided into two broad categories: circulation patterns of the tropics and those of the extratropics. Subcategories feature individual ocean and atmospheric patterns, including their hemispheric-to-global interconnectivity, with timescale and latitude of interactions playing significant roles in resulting collective behaviors.

Circulation Patterns of the Tropics

Geographically Governed Dynamics

Zonal and meridional dynamics co-occur in all tropical basins. Basin width determines which process dominates. Both processes redistribute heat regionally, and to some degree, globally. The zonal mode draws from energy shuttled into the tropics; the meridional mode (MM) receives its marching orders from the extratropics. Equatorial annual-mean zonal asymmetry is at the root of both zonal mode and MM.

Annual delivery of solar energy to the tropics is symmetric; yet the mean-annual zonal profile is asymmetric—a prerequisite critical for tropical processes. Three tropical-specific traits play large roles in interrupting zonal symmetry imposed along the equator by the annual-mean delivery of solar radiation: (i) At the geographical equator, the Coriolis effect (the apparent deflection of moving objects due to planetary rotation—a function of latitude) is zero. On either side of the equator, directions of deflection are opposite. (ii) Large-scale winds from each hemisphere converge near, but not necessarily aligned with, the geographical equator. Convergence is marked by a region of intense atmospheric convection known as the Intertropical Convergence Zone (ITCZ). And (iii) planetary-scale subsurface ocean waves (eastward-traveling Kelvin waves and westward-propagating Rossby waves), when excited by anomalous winds, transmit information regarding overlying atmospheric disturbances zonally across the equatorial basin. Within months (the exact time involved is dependent upon basin width), the equatorial subsurface adjusts to overlying atmospheric shifts. In turn, modified subsurface dynamics affect surface conditions. Ocean adjustment time is shorter in the tropics than elsewhere on the planet. Two aspects, unique to the tropics, make possible this rapid oceanic adjustment of subsurface waves to overlying atmospheric anomalies: First, latitudinally confined, eastward-directed Kelvin ocean waves travel swiftly and exist only along the equator, and second, although westward-propagating ocean Rossby waves appear at all latitudes, their velocity is fastest in the near-equatorial zone.

Together, this collection of tropical-specific properties generates within all tropical basins a mean state of zonal asymmetry in sea surface temperatures (SSTs), sea surface heights (SSHs), sea-level-pressure (SLP) patterns, and depth of thermocline (a thermocline is the boundary between mixed upper-ocean layer and cold water at depth; the layer in which temperatures decrease rapidly with depth). Prevailing wind directions for a given tropical sector, along with basin width, influence basin-specific details. Zonal and meridional deviations from this asymmetric mean state occur with relative rapidity.

In context of these traits that characterize all tropical basins, differences distinguish regional personality. These differences are scripted by basin size and surrounding landmass configuration, topography, and land cover. Per consequence, each basin is home to a unique set of oceanic–atmospheric circulation patterns. All basins interconnect with coupled ocean–atmospheric dynamics telegraphing basin-mode signatures. Fast positive and lagged negative feedback responses emerge in their communication, tamping extremes in the collective behavior of the unified tropical system.[1]

Zonal (East–West) Modes of Coupled Ocean–Atmosphere Variability in the Tropics

El Niño–Southern Oscillation (ENSO)

El Niño–Southern Oscillation (ENSO) is a dominant tropical Pacific regional mode of interannual coupled ocean–atmosphere climate variability, manifested most markedly as changes in SSTs in the eastern equatorial Pacific (EEP). ENSO interannually deviates from the tropical Pacific mean state, fluctuating between two end-member phases: a cool phase, La Niña, and a warm phase, El Niño.

The mean state of the tropical Pacific is characterized in the west by the Western Warm Pool (WWP), a region of warm SSTs, high SSHs, a deep thermocline, and strong overlying atmospheric convection from which air rises and diverges both meridionally and zonally into Hadley and Walker circulations, respectively. In contrast, typifying the east is a "cold tongue" of cool SSTs, supplied by cold water upwelled from beneath an underlying shallow thermocline, and a lower SSH than in the west. Strong easterly winds co-occur with this zonally nonuniform profile. Land configuration plays a role in shifting the ITCZ north of the equator.[2] This position draws the southeasterly trades across the geographical equator, transporting negative vorticity (spin) into the region, thereby strengthening ITCZ convective activity.

At the helm of the ENSO, coupled ocean–atmosphere system is an equatorially circumnavigating configuration of SLP centers. Atmospheric mass is exchanged zonally among these centers-of-action (COAs) within this globe-encircling configuration—a phenomenon known as the Southern Oscillation (SO). Zonal exchange of atmospheric mass along the equator manifests as fluctuations in the normalized difference in SLP between high- and low-pressure centers of the SO. SLP differences between a low-pressure region over north-central Australia [near Darwin, Australia (12°28′S; 130°51′E)] and a high-pressure zone over the central South Pacific [Tahiti (17°40′S; 149°25′W)] are reflected in the SO index (SOI), one metric used to define ENSO phase and intensity. A negative SOI describes the ENSO warm phase (El Niño), whereas a positive SOI, a cool one (La Niña). Surface waters slosh east–west in concert with the changing SLP gradient associated with the fluctuating SOI, influencing subsurface, and overlying features in their wake. Adjustments of the subsurface ocean and overlying atmosphere to the equatorial zone's SLP redistribution occur regionally. Their impact is global.

ENSO is superimposed upon the seasonal cycle. It is within that cycle that the ENSO system finds itself compromised, susceptible to the onset of El Niño. Seasonally, trade winds naturally ebb and flow, governed by seasonal changes in solar insolation and consequent land-sea temperature contrasts, particularly in the Southern Hemisphere. Southeasterly trades strengthen in austral winter (boreal summer). In the austral spring into summer (boreal fall into winter), the southeasterlies abate. The ITCZ migrates southward. In the east, upwelling of cool water from depth subsides as the thermocline surface lowers, thereby sequestering cold temperatures at depth. SSTs in the east rise accordingly. Zonal asymmetry along the equator diminishes. The coupled ENSO system is destabilized, now vulnerable to an El Niño.

Although El Niño events rarely conform to a typical profile, a standard or canonical El Niño can be described. Hints of one first surface in December [thus its name "El Niño" (Christ child) in reference to the time of year] can be seen in decreased SLP of the high-pressure system over Tahiti and in increased SLP of the low-pressure system over Darwin. Once started, an incipient El Niño may reverse course and disappear. But if it survives seasonally competing processes imposed on its development during the austral fall (boreal spring), and the seasonally related southeasterlies fail to strengthen, a brewing El Niño generally matures.

Whatever their trigger, departures from the tropical Pacific mean state are initially magnified through positive feedback responses in the coupled ocean–atmosphere system via the Bjerknes feedback mechanism,[3,4] where zonal gradients of SSTs and SLPs reinforce one another through shifts in strength and zonal position of the overlying Walker circulation. Simultaneously, the interplay of subsurface downwelling Kelvin (propagation to east) and upwelling Rossby (lagged; propagating to west) ocean waves—excited by atmospheric shifts and resulting subsurface interaction—initially amplifies, then ultimately damps, or even reverses the sign of, the excursions. Together, the Bjerknes feedback mechanism and the lagged Kelvin–Rossby interaction describe a leading model for ENSO, a delayed oscillator system.[5–7]

ENSO can be envisioned as a heat pump,[8–11] driven by a recharge–discharge mechanism.[12–15] The ENSO system gains heat during non-El Niño years and exports it poleward during El Niños. Sequential adjustments in thermocline slope, due to ocean–atmosphere coupling and Kelvin–Rossby wave interactions, execute this outcome. During ENSO-neutral and cool phases, the Walker circulation is shifted far to the west. Convection is fueled by warm SSTs in the WWP. Strong southeasterlies push surface water west; the thermocline shallows in the east. Cold water from depth upwells in the EEP, drawing heat down into the ocean. Below the surface, ocean heat from both hemispheres follows thermocline gradients. Ocean mass and heat head equatorward. The system gains heat.[12] Background conditions favor an El Nino.

During an El Niño, the reverse occurs. Southeasterlies weaken. Warm water from the WWP spreads eastward; a weakened Walker circulation follows. The Hadley circulation contracts and intensifies. Atmospheric heat is funneled poleward. Atmospheric reorganization is communicated to the subsurface Kelvin and Rossby ocean waves. In turn, an interplay between the waves, over time, restructures subsurface gradients such that ocean mass and heat are exported poleward, most to the Northern Hemisphere.[12–17] Heat is lost from the system.

At the surface, El Niño alters global weather patterns. Dry conditions replace moisture displaced from the western tropical Pacific. Upper-level subtropical jets on either side of the equator strengthen and shift equatorward, in particular the jet within the winter hemisphere. Jet flow is extended eastward. Weather systems are redirected accordingly. Regions may experience warmth where it is usually cool, and moisture where it is usually dry. Full onset of an El Niño typically takes hold around March, often persisting 18–24 months thereafter.

Teleconnection patterns associated with ENSO phases can persist for decades and then wane or disappear, at least temporarily. Examples can be found in relationships between ENSO and adjacent basin processes. At times, El Niño events suppress North Atlantic hurricanes; at other times, they do not.[18] Similar inconsistency is seen in the relationship between the El Niños and the Indian summer monsoonal (ISM) rains.[19] Multidecadal changes in background conditions of the various basins contribute to shifting interbasin relationships.[1] At times, equatorial communication is enhanced among tropical basins; at other times, the subtropical relationships are stronger. Even extratropical patterns add their fingerprint, directly and indirectly, to the various interbasin tropical interactions.[20,21] Most interbasin communication is through the atmosphere, although atmospherically induced changes in subsurface ocean dynamics play strong roles, as well.

No longer do scientists see the ENSO system in isolation. Nor do they view El Niño as having one personality. The idea of a canonical El Niño (East Pacific (EP) El Niño) was dethroned when a previously unrecognized "flavor" of El Niño emerged. This "new" El Niño, more frequently occurring since the late 1990s, manifests with warm SSTs in the equatorial central Pacific (CP El Niño).[22] The Walker circulation, a key factor in scripting interbasin interactions, is a slave to the migrating warm waters, and in turn, these atmospheric circulation shifts trigger changes in subsurface Kelvin and Rossby wave dynamics. In this intertwined way, El Niño finds its relationships with other patterns nonstationary.[23–25]

The question of why easterlies occasionally subside or strengthen, and why the interannual oscillation period is irregular, stirs debate. Source water from the subtropics,[26–28] and/or velocity of subtropical water transport,[29,30] could play roles in subsurface water temperature feeding the cold tongue, altering the zonal SST gradient and thereby influencing easterly wind strength. In general, it is suggested that whatever the mechanism, it likely involves the decreased/increased temperature gradient between SSTs of the WWP and the subsurface waters feeding the cold tongue in the east, a condition whose causes could be local or global, external or internal, short term or long term.

Atlantic Niño and Benguela Niño

Mean-state traits of the tropical Atlantic are similar to those in the Pacific: a coupled ocean–atmosphere system with prevailing easterly winds, warm water pooled in the west, and a cold tongue of upwelled deep water in the East Equatorial Atlantic (EEA). Impact of seasonal shifts, augmented by geographical influence from the huge lobe of Africa just north of the EEA, alternately amplify and minimize the mean-state asymmetry.

As with the ENSO system, overlying SLP variations and interplay in the equatorial Atlantic between perturbation-generated subsurface planetary-scale waves—Kelvin and Rossby waves—result in growth, then demise, of a warm event. In the Atlantic, there are two types of warm events: the Atlantic Niño, associated with northeasterly winds, and the Benguela Niño, associated with southeasterly winds. While timing of onset and location of SST and precipitation anomalies differ slightly between the two modes, similarities dominate. With either type, when a warm tropical Atlantic event occurs, easterlies subside; SSTs increase in the east.[31] The zonal atmospheric circulation—the Walker circulation—shifts eastward, weakens, and is replaced by a strengthened meridional atmospheric component—the Hadley circulation. An intensified Hadley cell carries heat poleward and strengthens the Atlantic sector's subtropical jet. Due to the impact of a narrower basin width on interacting sub-surface waves, events in the Atlantic are weaker, shorter, and more frequent than in the Pacific. Despite this, the end results of an El Niño and an Atlantic equatorial warm event are similar: redistribution of heat from the tropics poleward through oceanic and atmospheric circulation changes, with regional and remote impacts.

Unlike the Benguela Niño, the Atlantic Niño may have ties to SST anomalies in the extratropics of the North Atlantic[32]; it may also be remotely influenced by El Niño; yet this relationship with El Nino is inconsistent.[33] There are times when dynamics of the Atlantic shift in apparent response to an El Niño event; within months, an Atlantic Niño follows: 1997/1998, for example. Yet, at other times, no relationship is apparent; the El Niño of 1982/1983 was such a case. Reasons for the contradicting responses along the Atlantic equator to similarly large El Niño events appear linked to opposing influences: Cooling of the equatorial Atlantic, fueled through Bjerknes ocean–atmosphere feedbacks, competes with a few different pathways of El Niño-imposed warming[34] (i) the tropospheric temperature mechanism[35,36]; (ii) a sometimes El Nino-excited feature – the Pacific North American (PNA) teleconnection pattern – warming the North Tropical Atlantic (NTA) region, with impacts on the equator[37]; and (iii) an atmospheric bridge,[38,39] related to changes due to weakening of the Walker circulation over the equatorial Pacific.

Zonal dynamics along the Atlantic equatorial region negatively feedback onto Pacific equatorial dynamics; yet this relationship is inconsistent, yielding to shifting ocean patterns on multidecadal timescales. Multidecadally paced Atlantic-sector ITCZ migrations, associated with low-frequency variability in the Atlantic (Atlantic Multidecadal Oscillation or Variability: AMO or AMV), reroute atmospheric communication pathways between the tropical Atlantic and the equatorial Pacific.[40] Thus, during warm phases of the AMO, tropical forcing from latitudes in the NTA (10°N–20°N) tends to dominate over forcing from the Atlantic equator. In either case—equator or NTA—a warm Atlantic tends to promote cool ENSO events.

Indian Ocean Dipole (IOD) or Indian Nino

Equatorial zonal interannual variability of a coupled ocean–atmosphere system also exists in the Indian Ocean (IO)—the Indian Ocean Dipole (IOD),[41-43] but its character is not fully analogous to its Atlantic and Pacific counterparts. Although the IOD is comparable in fundamental ways to tropical zonal variability in the Atlantic and Pacific Oceans, the geography surrounding the IO imposes significant differences—a discontinuous eastern boundary and the vast landmass of India to the north. The former allows direct oceanic communication with the Pacific Basin to its east. The latter—a southern extension of the Indian subcontinent—blocks east–west oceanic flow within the IO Basin north of 25°N. The imposing influence of this landmass scripts prevailing wind regimes of monsoonal flow: northeasterly in the boreal winter to southwesterly in the boreal summer. These seasonally reversing winds within the northern IO tend to mask the equatorial zonal IOD mode.

Equatorial IO conditions differ from those in the monsoon-dominated regions northward. A weak zonal wind component lingers over the equatorial IO when monsoons are strong. Yet during transitions from one monsoonal flow to the other—that is, during boreal spring and fall—equatorial westerlies intensify. The end result of these diverse wind patterns across the IO is a mean equatorial state of warm water pooled in the eastern basin, below which the thermocline is deep and above which the zonal convection of a Walker cell is strengthened. These mean conditions—a reversed profile of the Pacific's—supply abundant moisture to Indonesia and little to East Africa. Interannual fluctuations occur between amplified and weakened expressions of this equatorial mean state.

The first signs of a developing IOD event emerge characteristically in the late boreal spring or early summer. If the seasonally weak equatorial westerly winds intensify, the equatorial mean state is amplified. A negative (cool) IOD has the potential to develop, delivering more moisture to Indonesia and less to East Africa. On the other hand, if the equatorial westerly winds fail, and are replaced by equatorial anomalous easterlies and southeasterlies, a positive (warm) IOD may take hold. Zonal asymmetry collapses. The SST-anomaly gradient reverses. Atmospheric convection shifts west. Strong rains pelt East Africa. Indonesia and Australia are unusually dry. An "event" typically persists until the following winter.

Teleconnections reach beyond the surrounding region, some decadally modulated or, in some cases, either enhanced or damped by interactions with teleconnections from ENSO[44,45] or with processes in the tropical Atlantic.[46,47] Although the IOD is an independent intrinsic coupled atmospheric–oceanic mode of variability of the IO,[41,42] warm ENSO events often co-occur with positive IOD events and cool ENSO events with negative IOD. ENSO has been posited as both[48,49] being a potential trigger for[50–52] and potentially impacted by the IOD.[53–55]

Meridional (North–South) Modes of Coupled Ocean–Atmosphere Variability in the Tropics

Meridional Modes (MMs)

MMs of the tropics involve latitudinal migration of the ITCZ in relatively swift response to winds and associated cross-equatorial SST anomalies. SST anomalies north and south of the equator vary independently, distinguishing a MM as a gradient mode, not a dipolar one.[56,57] MMs are dominantly interannual. They are multidecadally modulated by oceanic influence.[58] While cross-equatorial warming—the distinguishing feature of MMs—occurs in all three basins, details differ due to geography. Those occurring in the Atlantic [the Atlantic MM (AMM)] and Pacific [Pacific MM (PMM)][56] are similar to one another. That discussion follows.

Atlantic Meridional Mode (AMM) and Pacific Meridional Mode (PMM)

In the Atlantic tropics, the AMM dominates over the Atlantic Niño/Niña mode. In the Pacific, the opposite is true: ENSO dominates over the PMM. Relative basin widths of the two oceans are the reason: The Pacific's wider basin strengthens the zonal component. This is due to basin width's influence on the interplay between subsurface oceanic Rossby and Kelvin waves. But for both oceans, the MMs are variable coupled ocean–atmosphere phenomena in the tropics that are excited by boreal winter extratropical atmospheric variability over their northern basins. In essence, these MMs connect the Northern Hemisphere extratropics to the deep tropics.[56,59] They serve as conduits through which an atmospherically forced signal, generated in the Northern Hemisphere's subtropics, is carried to tropical latitudes.

In the tropics, the original atmospheric signal evolves into a fully coupled ocean–atmospheric dynamic. Wind-induced latent heat flux leads to a summer-centered coupled ocean–atmospheric response in tropical SST anomalies. The tropical Atlantic illustrates this response well: Warm tropical SST anomalies north of the equator generate a meridional SST gradient across the mean latitude of the ITCZ. Strongly sensitive to small meridional SST gradients, the ITCZ migrates into the warmer Northern Hemisphere, where the Coriolis influence imparts a westerly component to the anomalous southerly cross-equatorial flow. Prevailing northeasterlies weaken in response. Surface ocean evaporation slows. In turn, SSTs rise. Warmer SSTs promote more cross-equatorial southerly flow, with this thread of positive feedbacks—the WES (wind–evaporation–SST) mechanism—reinforcing the original signal.[2,59–61]

AMM and PMM impact regional temperature, precipitation, and hurricane[58,62] regimes, with a potential impact extending to zonal tropical modes, for example, ENSO.[63] These phenomena, in turn, reflect a longer-timescale component.[58,156]

Indian Ocean Basin Mode (IOBM)

Unlike the Atlantic and Pacific basins, the IO has no extratropical ocean from which to receive its midlatitude signal. Land dominates its northern extratropical boundary. Yet a MM, of sorts, does exist. That mode is the IO Basin Mode (IOBM).[64] In the IO, it is the dominant mode of interannual variability, characterized by full-basin warming or cooling in its positive and negative phases, respectively. Decay of the zonal IO mode—the IOD—may contribute to a developing IOBM; yet IOBM's evolution is tied largely to ENSO, with an El Niño triggering a warm IOBM event through an atmospheric bridge[65]: El Niño-related weakening and eastward migration of the Walker circulation reduces winds over the IO, thereby decreasing ocean surface evaporation and suppressing cloud cover. A reduction in this cloud cover enhances shortwave radiative heating. SSTs increase. Rossby westward-propagating downwelling ocean waves, triggered by El Niño-related atmospheric shifts, deepen the thermocline in the southwest IO, further amplifying positive SST anomalies and associated atmospheric convection north of the equator. Enhanced atmospheric convection induces cross-equatorial winds ushering water vapor from the Southern Hemisphere into the northern basin, enhancing the ISM and introducing an additional positive feedback.[66,67]

IO warming supports anomalous anticyclonic flow in the northwest Pacific. Through this influence, the IOBM, initially triggered by, and then sustained by, remote forcing of El Niño, in turn, accelerates the demise of El Niño and its transition to La Niña.[49,55,68] Yet, even with El Niño's termination in the boreal winter, warming in the IO Basin continues several months after El Niño ends. This is known as the "capacitor effect."[64,69,70] The IO basin-scale warming, originally a passive response to El Niño, takes on a life of its own, with air–sea interactions internal to the IO Basin extending the IOBM's lifetime, impacting summer monsoonal activity,[64,69,70] with hemispheric implications.[67]

Interconnectivity in the Tropics: Interbasin Communication among Interannually Varying Circulation Patterns

A growing body of research supports the existence of strong connectivity among interannually fluctuating tropical circulation patterns.[1] On interannual timescales, the atmosphere is dominant in transmitting interbasin communication, albeit via different mechanisms.[71–73] The ocean plays a role, as well.[40] Atmospheric shifts trigger subsurface Kelvin and Rossby ocean waves. They modify thermocline gradients through competing downwelling and upwelling actions, in accordance with atmospheric orders. In turn, thermocline depth modifies ocean surface conditions. These feed back onto the atmosphere, which may act regionally, or hemispherically, through planetary-scale Kelvin and Rossby atmospheric waves.

On decadal to multidecadal timescales, the ocean's influence plays a greater role than the atmosphere's influence. For example, between the Pacific and Indian basins, communication is directed through the porous boundary of the Indonesian archipelago, a conduit known as the Indonesian throughflow (ITF). ITF exchange of ocean volume and heat varies multidecadally,[74] with impacts on both oceans. In the Atlantic, pronounced multidecadal variability is rooted in circulation strength of the Atlantic meridional overturning circulation (AMOC), associated with shifts in atmospheric and oceanic COAs, cross-equatorial heat flow, and ITCZ position. In sum, oceanic influence is critical in its impact on background conditions[40,75,76] that modulate interannual patterns and the strength of their remote influence on multidecadal timescales.

In general, it can be argued that El Niño sows seeds of its own demise through its influence on adjacent basins. This can be seen in terms of interdecadal timescales in Figure 9.1: El Niño events mostly warm the tropical and equatorial latitudes within the IO and Atlantic Ocean; however, as previously discussed, its impact on the equatorial Atlantic is inconsistent, due to competing dynamics. In turn, warming in the tropical latitudes of the Atlantic and Indian Oceans indirectly cools the Pacific equatorial zone through negative feedback responses that damp, or terminate, an El Nino event. With ENSO's huge effect on global climate and global heat distribution, these negative feedbacks from adjacent oceans, triggered by the tropical Pacific's influence on them, highlight the climate system's long-term role in enhancing climate stability across the planet.

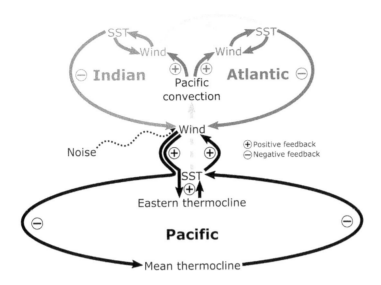

FIGURE 9.1 (See color insert.) Interconnectivity among interannually varying tropical circulation patterns is illustrated through this schematic. The black loop [Bottom set of loops] represents internal feedbacks in the Pacific: Fast positive feedbacks are represented by short arrows; delayed negative feedbacks, by the long arrows. Interbasin feedbacks (positive) from the Pacific onto both the Atlantic and the IOs are shown by the short arrows in center of loop (Pacific convection). In contrast, negative feedbacks onto the Pacific from both adjacent oceans are represented by the upper loop, left side arrows (Indian Ocean) and upper loop, right side arrows (Atlantic Ocean). Note the positive feedback from the Atlantic onto the IO is shown in center-to-center right arrows at very top of upper loop. The overall result of the tropical interconnectivity is a warming tropical Pacific that warms adjacent oceans; in turn, through both direct and indirect means, mostly through the atmosphere, both the warmed Atlantic and the warmed IO negatively feed back onto the Pacific, leading to a termination in a warm ENSO phase, sometimes leading to a La Nina. (Reprinted with permission from AAAS: Cai et al. (2019).[1])

Circulation Patterns of the Extratropics

Geographically Governed Dynamics

The Coriolis effect is nonzero away from the equator and throughout the extratropics. Its parameter, which is same-signed throughout a given hemisphere, increases with latitude. In part, because of Coriolis-related peculiarities, the travel of Kelvin waves in the extratropics is restricted to poleward (equatorward) flow within a narrow zone along the western (eastern) coastlines of continents or cyclonically around a closed boundary. Kelvin waves cannot travel in the open ocean along lines of latitude in the extratropics. Only westward-propagating Rossby waves can, with slower velocities at higher latitudes. A consequence of these dynamics is twofold: (i) Communication of overlying atmospheric changes to the subsurface ocean is propagated only toward the west and (ii) time required for the Rossby waves to travel the cross-basin takes years to decades in the mid-to-high latitudes. This renders adjustment of the subsurface ocean to overlying large-scale wind field slow in the extratropics. Extratropical ocean "dynamic memory" is long; persistence of a climate signal at the mid-to-high latitudes is thereby extended.

In addition to the extratropical traits responsible for modifying climate signals and their transmission, equator-to-pole atmospheric transport of heat is similarly amended. The atmosphere of each hemisphere is broken into three large-scale convection cells, which is due to the Earth's rotation and the consequent Coriolis effect. Related rising, sinking, diverging, and converging basin-scale airflow within these cells establish semipermanent surface high- and low-pressure regions and associated belts of prevailing winds, thus governing fundamental aspects of regional circulation patterns.

Extratropical circulation patterns are more active during winter. Two distinct phenomena emerge: First, seasonally sequestered thermal anomalies re-emerge, enhancing conditions for ocean–atmosphere coupling in the extratropics,[77] and second, at high latitudes, stratospheric dynamics merge with tropospheric processes, amplifying and extending atmospheric impact. Winter behavior is implicit in the discussion of these modes, unless otherwise stated.

Dominantly Atmospheric Circulations

Annular Modes and Related Extratropical Circulation

Shifts of atmospheric mass between the polar region and midlatitudes drive leading modes of extratropical climate variability in both the Northern and Southern Hemispheres. These patterns are called annular modes (AMs), so-named in reference to a set of defining hemispheric features—deep, ring-like (annular), zonally symmetric structures of indices related to anomalous atmospheric flow—for example, zonal means of geopotential heights, SLP, and zonal wind fields. In the Northern Hemisphere, this mode manifests as the Northern AM (NAM); its Southern Hemisphere counterpart is the Southern AM (SAM).[78–80] Geography in each hemisphere endows each with unique traits; yet both modes share fundamental characteristics.

Meridional contrasts of atmospheric pressure and temperature fuel strength of westerly winds that encircle the globe along a subpolar belt centered at about 55°–65° latitude. As polar atmospheric mass shifts equatorward and poleward—on timescales ranging from weekly to seasonally, with a low-frequency component apparent—the westerly wind belt strengthens and weakens, shifting poleward and equatorward with the meridional migrations.

The AM index reflects these meridional migrations. Lower-than-normal SLP over the poles (~60°–90°) and a globally circumnavigating belt of stronger-than-normal westerlies, centered at about 55°–60°, characterize the high-index, or positive, polarity. Cold air is held tightly within the wind belt. Hence, temperatures at the poles are colder-than-normal; those in the mid-latitudes, warmer. Reverse conditions define the low-index, or negative, polarity.

Variability in AM polarity impacts surface temperature, precipitation patterns, SSTs, lower-stratospheric ozone content, and distribution of sea ice. In turn, AMs appear to respond to changes in sea-ice distribution, with other potential influences, including solar variability,[81,82] SST anomalies, Eurasian snow cover,[83–85] and atmospheric chemistry.[86] In short, anything that makes the polar-air mass colder will strengthen the bounding belt of westerly winds, nudging the AM toward its high-index phase. Anything that makes the polar air mass warmer will weaken the winds, pushing the AM into its low-index phase.

AMs exist year-round; yet their behavior differs according to season. AMs have an active season and an inactive one. During the inactive months, there is a tropospheric AM and a stratospheric one. The two act independently. During active seasons, they act as one.

To understand this change, we look at the polar stratosphere. Year-round, it hosts an annulus of strong westerly winds. This belt of winds is the polar-front jet. It circumnavigates the polar region, centered at about 65° latitude. This wind-contained feature is the polar vortex. For much of the year, it is strictly a stratospheric phenomenon. Yet in the active season—in boreal winter for the NAM and late austral spring into austral summer for the SAM—the vortex extends downward into the troposphere, where it merges with the tropospheric westerly wind belt. The stratosphere and the troposphere thereby couple.[87]

When stratospheric–tropospheric coupling engages, stratospheric perturbations influence activity in the troposphere, from the poles to the equator. In turn, tropospheric processes impact the stratosphere, as well. This vertical two-way coupling imparts a low-frequency component[32] to the AM's short-term timescale of variability. Exactly how is a topic of debate, but some studies suggest a tie to dynamics along western boundary currents (WBCs) and their extensions, where on decadal-plus timescales, low-frequency flux of ocean heat out of the ocean along the WBC extension excites convective activity along associated atmospheric fronts, triggering conditions ripe for sudden stratospheric warmings (SSWs).[88–90] SSWs can weaken the polar vortex.

What are these SSWs? While they may increase in number on a multidecadal timescale, with temporal ramifications, they are not an uncommon occurrence, varying in number and impact interannually. We examine them in context of the polar vortex, beginning with timing.

Vortex integrity relies on timing. As the stratospheric vortex seasonally begins to couple with tropospheric winds, the vortex is vulnerable to perturbations. Perturbations delivered at this fragile time can weaken the vortex. Perturbations that generate SSWs often come from Rossby atmospheric waves related to strong convection. A common source is tropical convection.[91] If the Rossby waves hit the vortex, they can distort its symmetry, displace it, or even break through its encircling jet-wind belt. As the waves descend, if within the vortex, compression upon descent can cause temperatures of the contained polar air to increase as much as 50° Celsius or more within a matter of days. These are the SSWs. Thermal gradients between polar and mid-latitude air masses shrink. Per consequence, westerly flow in the stratospheric polar-front jet decelerates, or even temporarily reverses direction. The vortex dramatically weakens. Polar air spills equatorward.

If such an event occurs early enough in the seasonal tropospheric–stratospheric coupling, the vortex may have time to regain strength and recover before the active season ends. Subsequent SSW assaults may then actually strengthen the vortex by imparting westerly angular momentum to the wind belt. In this case, the AM will assume a high-index polarity. In contrast, if the SSW event occurs later in the season, the vortex strength may remain weak throughout the remainder of its active period. In this latter situation, the AM is pushed into a low-index polarity, with polar-air excursions into the mid-latitudes frequent. It is all a matter of timing in how the vortex will react, and thus how the mid-latitude climate will evolve.

While tropical convection directly can spawn the planetary-scale waves that trigger the SSWs that can compromise vortex integrity, it can also impose damage to the vortex indirectly. It does this circuitously through interaction with Rossby atmospheric waves and a wind pattern centered over the tropics.

Background context may be useful here: Rossby waves traveling in the upper atmosphere are vulnerable to being reflected off sharp thermal contrasts between air masses, or topography, especially in the Northern Hemisphere. They are jettisoned vertically upward. Ultimately, these large-scale winds descend, propagating meridionally as they do. Some head equatorward. At the equator, the reflected Rossby waves encounter an equatorial upper-atmospheric pattern—the quasi-biennial oscillation (QBO), a dynamic intricately entangled with those of ENSO. While QBO behaviors can be found in the troposphere and the mesosphere, the QBO behavior relevant to polar-vortex strength hovers near the tropopause, where tropospheric deep tropical processes have set the stage for complex interaction. This mostly lower-stratospheric phenomenon, rooted in vertically propagating Kelvin and Rossby waves from tropical convection,[92] fluctuates between two phases of downward-propagating wind flow. In one phase, the QBO wind flow direction is westerly; in the other, easterly—the stronger, longer-lived phase of the two. Directions reverse approximately every 22–28 months,[93] with consequences for the polar vortex.

For the approaching Rossby waves, QBO direction matters. Due to particulars related to upper-atmosphere tropical waveguides, entry into the tropical system depends upon QBO direction: A westerly QBO stream welcomes the waves; an easterly one typically rejects them. Encountering the QBO easterly phase sends the large-scale upper-atmosphere waves back toward the vortex, where they can impart drag on the westerlies and thereby weaken the vortex.

This QBO–Rossby wave interaction impacts both vortex and tropical convection, giving explanation to the observation that tropical convection is often enhanced (suppressed), and the vortex often weak (strong), during QBO-east (west) phases. The reason lies with wave interaction.[94] If the Rossby waves intrude on the tropical wave guide, they thermodynamically lower the tropopause, thereby suppressing tropical convection. This typifies the QBO-west phase. The vortex is spared further assault. If the QBO-east phase is in play, and Rossby waves are sent back toward the poles, tropical convection thrives; yet the vortex is vulnerable—a vivid example of interconnectivity between poles and equator.

Atlantic Regional Expression of the NAM: The North Atlantic Oscillation (NAO)

While AMs are longitudinally hemispheric in their influences, regional differences do occur, likely due to interference from coexisting patterns. In the Northern Hemisphere, manifestation of the NAM is most pronounced in the North Atlantic sector. This North Atlantic-centered expression of the NAM is the North Atlantic Oscillation (NAO). Both NAM and NAO characterize similar temporal and spatial leading modes of Northern Hemispheric variability. Whereas NAM is hemispheric, NAO is confined to the North Atlantic region. Atmospheric mass redistribution between subpolar and subtropical latitudes manifests as changes in the normalized SLP difference between the North Atlantic atmospheric COAs: the subpolar Icelandic Low (IL) and the subtropical Azores High (AH). A positive SLP anomaly difference represents the high index, or positive polarity, of NAO, and a negative SLP anomaly difference indicates the low index, or negative polarity, of the NAO. Low-frequency modulation of interannual-to-interdecadal variability is apparent.[95,96] A characteristic tripolar pattern of wind-generated SST anomalies within the North Atlantic is generated by higher-frequency fluctuations of the NAO,[97,98] whereas a more uniform, basin-wide SST anomaly pattern, likely involving large-scale ocean dynamics,[98–101] distinguishes the low-frequency component, with significant teleconnection impact.[100,101]

Pacific and North American Regional Expression of the NAM: The PNA Pattern

The NAM also extends its influence into the North Pacific, although not as conspicuously as in the North Atlantic. In fact, SLP fluctuations over the North Pacific and North Atlantic exhibit no significant correlation. In the context of the annular-mode paradigm of meridional seesaw exchange of atmospheric mass,[102] this observation is unexpected. One would expect the Atlantic and Pacific COAs to be correlated positively with one another and both negatively correlated with the polar region. Reconciliation might be found in a coexisting strong mode of variability, with pronounced presence in the North Pacific that masks the NAM signature in the Pacific. That hypothesized role may lie with the PNA pattern.[37]

PNA is a prominent upper-atmospheric mode of low-frequency variability involving anomalous atmospheric pressure. Located at mid-tropospheric heights over the extratropics, influence on its behavior is not confined to the mid-to-high latitudes. Strongly associated with the Aleutian Low (AL) in the North Pacific, the PNA is also modified by tropical processes. Associated with jet-stream tracks, modified by underlying topography and constraints of vorticity conservation, the PNA is linked to the flow of storm systems (wave trains) across the Northern Hemisphere and via this route, may set up a zonal seesaw relationship between the AL in the Pacific and its counterpart in the North Atlantic—the IL, thereby competing with the NAM's signature in the North Pacific.

Intensity and location of weather systems, particularly those affecting North America in the boreal winter, are most impacted by the PNA. Liu et al. (2007)[103] attempted to tease apart behavior of the AL and NAM. Teleconnected influence of each individual COA differed from the teleconnections resulting from the combination of their influences. This observation is consistent with the idea of coexistence of patterns and the consequent possible masking of the NAM signature in the North Pacific. Along these lines, if the PNA signature is removed from data grid points, a statistically significant positive correlation emerges between the AL and the IL,[102] further supporting the competing-mode hypothesis. Similar interfering dynamics are suspected between the SAM and the PNA counterpart, the Pacific South American (PSA) pattern.[104]

Although the PNA is strongly associated with the North Pacific via the AL, its scope is hemispheric.

Pacific Expression of NAM: The North Pacific Oscillation (NPO)

More regional in nature is the North Pacific's analogue to the NAO, namely, the North Pacific Oscillation (NPO).[105] In the North Pacific area, including the most northwestern sector of North America and the eastern coastal areas of Eurasia, the NPO is the most important teleconnection mode. As with the NAO, NPO can be described as a north–south seesaw of atmospheric mass between subpolar and subtropical regions, with the normalized SLP anomaly difference between the regions measuring the polarity and strength of the pattern's interannual-to-interdecadal fluctuations.

Multidecadal Modulation of Regional Patterns Related to NAM

Originally described in 1932, NPO was said to result primarily from geographical shifts of the mean position of the AL.[105] Indeed, it is observed that this low-SLP COA (the AL) migrates to the west and north when weak (increased SLP) and to the east and south when strong (decreased SLP), the most extreme shifts occurring on multidecadal timescales.[106] Spatial scale of NPO influence fluctuates similarly.[107] During decadal-plus intervals, the footprint of the NPO appears confined to the North Pacific Basin. The AL is weak and skewed west of its mean position. During these "regional" time spans, ENSO plays a strong role in the NPO behavior, with pronounced influence on AL. In contrast, during "hemispheric" intervals, the reach of NPO influence spreads far beyond the Pacific Basin. The AL is strong and shifts southeast of its mean position. ENSO forcing on AL is minimal. Instead, extratropical forcing of the low-pressure system is dominant. With an intensified AL, the hemispheric PNA pattern is more pronounced.

NAO, too, shows indications of vacillating on low-frequency timescales between more regionally confined intervals and more hemispherically spanning ones.[108] Not unexpectedly, details of NAO/NPO high-frequency variability and associated teleconnected impacts related to precipitation, surface temperature, ocean-gyre activity, and sea-ice patterns vary accordingly at the low-frequency tempo.

Geographical shifts of the Arctic, subpolar, and subtropical atmospheric COAs on decadal-plus timescales can explain much of the observed low-frequency behavior. Evidence suggests indirect oceanic influence on atmospheric COAs.[109–112] Wintertime stratospheric–tropospheric coupling appears to be a critical component underlying the low-frequency signal, connecting multidecadal variability related to the AMV (aka AMO) to the winter NAO/NAM.[113] Low-frequency solar variability, too, may play a role.[106,114]

Brief Introduction to the SAM and Its Relationship to Antarctica

The SAM, many of its features analogous to the NAM, overlies relatively simple geography. The absence of landmass in the midlatitudes around 60°S conveys unparalleled strength to the SAM-associated annular structure of the mid-to-high-latitude westerly winds. These winds drive the Antarctic Circumpolar Current (ACC) in the Southern Ocean—a globally girdling ocean current that connects all three ocean basins and is fundamental to global intercommunication of climate signals. Through upper-atmospheric communication (via PSA), this southernmost atmospheric pattern, the SAM, whose influence strongly impacts Antarctic temperatures, sea-ice dynamics, and associated deepwater formation, appears to be linked to tropical dynamics, particularly ENSO in the Pacific sector.[104,115,116]

Dominantly Oceanic Circulations

Atmospheric circulations tend to be dominated by stochastic, high-frequency behavior. In contrast, dominantly oceanic circulations operate at low-frequency timescales. Water's high heat capacity, in conjunction with long time spans required for both extratropical subsurface Rossby wave travel and wintertime vertical upper-ocean mixing, contributes to the ponderous pace of oceanic processes. At these low-frequency timescales, the ocean can exert direct and indirect influences on the winter atmosphere, thereby adding persistence to overlying atmospheric circulation. This influence is reflected in a low-frequency modulation of that atmospheric circulation's frequency and intensity at interdecadal-to-multidecadal timescales, especially at specific locations within the extratropics—that is, WBCs and their extensions (WBCE), where at certain timescales, flux of ocean heat is outward, ocean-to-atmosphere flux (OAF). Via this special situation, the ocean can force the atmosphere,[89,90,117] in contrast to the more typical extratropical atmosphere-to-ocean surface forcing.

In addition, as noted in the section on tropical circulation patterns, the ocean modifies hemispheric and global teleconnections associated with the atmospheric patterns.

Atlantic Multidecadal Oscillation (AMO)

Ocean heat transport is northward at all latitudes of the Atlantic Ocean. This cross-equatorial ocean heat transport is unique to the Atlantic, generating a dipole of SST anomalies between the North and South Atlantic basins. This SST distribution is not static. Proxy and instrumental records support the existence of an intrinsic low-frequency mode of variability, characterized by a multidecadally (50–80 years) varying monopolar pattern of SST anomalies averaged across 0°–60°N and 75°–7.5°W. It is known as the[118–122] AMO, sometimes referred to as AMV. Climate impacts of the AMO-related SST anomalies are regional—North America and Eurasia are particularly impacted. AMO's reach can be hemispheric, as well, extending to the North Pacific.[119,120,123,124]

Recent studies introduce the notion that AMO operates at a second timescale of variability, a 20- to 30-year quasi-regular period.[125–128] Both the higher-frequency mode and the more well-known low-frequency mode elude full description, although both appear related to fluctuations in strength of the AMOC,[96,127–132] as is suggested by model studies incorporating an interactive ocean[129,133] and by variations in instrumental,[134] and in proxy[135] records of the interhemispheric SST dipole pattern. Mechanistic details for the AMO–AMOC connection remain under debate.[129,136–139]

Westward propagation of SST anomalies, related to ocean-gyre COA patterns, is associated with the hypothesized higher-frequency AMO mode.[138] Also associated are sea-level variations[126] along the North American and European coast lines. One idea is that the signal is generated from an internal ocean mode within the mid-to-high latitudes of the North Atlantic, perhaps excited by low-frequency atmospheric noise.[138] From the North Atlantic, the thermal anomalies propagate into the Arctic.[125,140]

In contrast, studies support a strong active role for Arctic involvement in the lower-frequency mode.[128,138,139] This more slowly paced pattern might involve a salinity oscillation between the Arctic and high latitudes of the North Atlantic.[128,139,141] According to modeled results, this proposed oscillation is associated with saline Rossby modes in the Arctic, which perhaps are excited by variability in processes that influence freshwater balance in the region (e.g., Atlantic inflow, sea-ice changes, or variability in river runoff).[128] The approximate 60-year rhythm found in 20th-century AMO SST anomalies and Arctic/Atlantic salinity oscillations can be found in numerous patterns: (i) the Atlantic core water temperature,[142,143] (ii) the Arctic sea-ice extent in the Western Eurasian Arctic Shelf Seas (Greenland, Barents, Kara Seas),[144–146] (iii) basin-scale meridional temperature gradient (MTG),[147] (iv) basin-scale westerly wind strength,[144–148] and (v) surface temperature over Eurasia.[147,148]

Potentially related to the salinity oscillations and the abovementioned set of processes is evidence for an approximate 60-year variability in the geographical positioning of the axis between SLP COAs (i.e., IL and AH) overlying the AMO-related SST anomaly pattern. This axis realignment, with its low-frequency component projected onto the NAO, is hypothesized to influence the North Atlantic's high-latitude salinity concentration, and by extension, the multidecadal character of the AMOC strength and AMO polarity,[109,120] with hemispheric and/or global indications.[132,145,146,156]

Pacific Decadal Oscillation

The Pacific Decadal Oscillation (PDO) is an oceanic pattern characterized by a boreal-winter, basin-wide, quasi-periodically varying SST-anomaly pattern north of 20°N in the North Pacific.[149,150] A positive PDO hosts a pattern of cool SST anomalies in the vast central/western sector of the basin and warm El Nino-like anomalies in the tropical sector and along the west coast of North America. A negative PDO pattern is reversed. A related pattern is the Interdecadal Pacific Oscillation (IPO),[151] which includes both the North and South Pacific Oceans and of which the PDO is likely a part. Both the PDO and the IPO have been shown to display two dominant quasi-periodicities, one of 15–30 years and another of 50–70 years. Both are intimately coupled with the overlying atmospheric circulation, which involves the AL and the Hawaiian (or Pacific) High (HH).

Longitudinal/latitudinal migrations of these COAs covary with phases of the PDO, the oceanic partner of the atmospheric mode, NPO. Low-frequency shifts are more pronounced than higher-frequency migrations. With a positive PDO, an intensified AL shifts east and south; the strengthened HH moves west and south.[114] Movements are opposite for a negative PDO. Repositioned COAs can yield unexpected results. Consider the behavior of El Nino-associated phenomena. They can differ according to AL position, and by extension, PDO phase. An example can be found in fluctuating sea-ice growth in the Bering Sea. When PDO is in its negative phase, El Nino events correlate with decreased ice extent. In contrast, when PDO is in its positive phase, Bering Sea-ice extent increases during El Nino events. This is due to the repositioning of AL and the consequent placement of accompanying winds.[152]

Repositioned COAs can also modify the extent and the consequences of PDO-related teleconnections, many of which involve ENSO. The multidecadal component of PDO modulates the frequency and intensity of ENSO.[153–156] During a positive PDO, warm ENSO events increase in both intensity and frequency. During a negative PDO, cool ENSO events are favored.

With PDO modulation of ENSO due to COA migrations, remote teleconnections of ENSO can build in impact due to cumulative effects. Some PDO-modulated ENSO teleconnections impact multidecadal variations in the freshwater balance of the North Atlantic. Impacts occur on differing timescales. They include the (i) alteration of hemispheric-scale precipitation patterns (increased precipitation in the tropical Atlantic with La Nina[157] and decreased with El Nino),[158] (ii) the basin-scale flow of relatively freshwater out of the Pacific (increased fresh Pacific water to the Arctic[152,159] and decreased to the IO during El Nino), and (iii) a possible increase in occurrence of positive IOD events with El Nino. With them comes a decadal-scale-lagged suppression of salt delivery to the Atlantic from the IO (Agulhas leakage), with implications for AMOC strength.[160,161] The opposite occurs with La Nina.

PDO low-frequency control of ENSO teleconnections also influences sea-ice formation regions off the Antarctic coasts [more in the Weddell Sea (Atlantic sector) during El Nino and more in Ross Sea (Pacific sector) during La Nina], with delayed potential impact on strength of the AMOC. Delivery of upper-ocean heat by the North Pacific subtropical gyre to the WBC extension (WBCE) is similarly paced.[162]

From the examples outlined earlier, it can be seen that the Atlantic and Pacific multidecadally varying ocean modes, each through indirectly teleconnected influences, appear to modify the other at decadal-plus timescales, influencing their temporal relationships. That temporal relationship throughout at least the 20th century shows that the 50- to 80-year mode of PDO has occurred in quadrature with that of the AMO.[156,163] In other words, their phases show an approximate offset by a quarter of a period. Their relationship is correlated with drought patterns in the contiguous United States. Observation indicates that drought has been most extreme during positive phases of the AMO, whereas the accompanying phase—positive or negative—of PDO has scripted the distribution of drought-impacted regions, as is shown in Figure 9.2.[164] The record of the Colorado River Basin flow over the last century appears consistent with this observation.[165] In addition, 500-year-long proxy records reflect a similar cadence in the Colorado River Basin stream flow, suggesting a strong and persistent natural component for this observation.[166]

Antarctic Circumpolar Wave

Identified in 1996,[167] the Antarctic circumpolar wave (ACW), a product of polar–subtropical temperature contrasts, consists of SLP and SST anomalies in a wave number-two structure that propagates in the Southern Ocean around Antarctica. Its interannual quasi-periodicity of 4–5 years is modulated on decadal, interdecadal, and multidecadal timescales. The ACW propagates westward in the vicinity

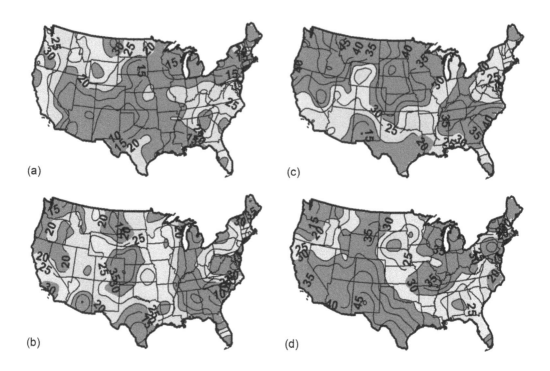

FIGURE 9.2 Drought frequency (in percent of years) in the contiguous United States for warm and cool (positive and negative) low-frequency regimes of the AMO and the PDO. Drought conditions in red. (a) +PDO and −AMO; (b) −PDO and −AMO; (c) +PDO and +AMO; and (d) −PDO and +AMO. Note: Drought is most widespread when AMO is in its warm (positive) state (Figures 9.2c, d). The distribution of drought is modulated with the phase of PDO (more in northwest and north-central and southeast United States with +PDO and +AMO; more southwest and midwest with −PDO and +AMO). The "stadium wave" in Figure 9.3 gives insight into relative phasing between AMO and PDO. (Reproduced with permission: [figure 5 in McCabe et al. (2004)][164] Public domain.)

of the eastward-flowing, SAM-related, hemispherically encircling ACC. Influenced by the fast east-wardly flowing ACC, net motion of the westward ("upstream")-propagating ACW is eastward, yet at a slower rate than that of the ACC. The ACW makes a full trip around the globe at southern latitudes (~65°S–30°S) in 8–9 years. Weather patterns in southern regions of Australia, South America, and Africa are influenced by the ACW.

Influence of and on the ACW extends to the tropics, where it appears to engage in two-way communi-cation with ENSO. The atmosphere transmits tropical information to high southern latitudes, imprint-ing an ENSO footprint upon the ACW. Piggybacked upon the ACW, the original ENSO footprint is slowly modified as it circumnavigates the Antarctic continent. ACW, through various bifurcations equatorward, returns the amended ENSO signature to the tropics, and the return routing of the signal is guided by the phase of PDO.[168,169]

Hemispheric Network of Circulation Patterns

Hemispheric Propagation of Multidecadal Climate Signal through Network of Climate Modes

On multidecadal timescales, individual expression of regional circulation patterns may yield to collective behavior (network behavior), as suggested by relatively recent research.[156] According to this hypothesis, synchronization—the matching of rhythms—and local coupling give rise to a low-frequency signal that propagates across the Northern Hemisphere, and possibly the globe[170] through a lead-lag sequence of atmospheric and oceanic circulations. This propagation signal, nicknamed the "stadium wave,"[145,146,156] is shown in Figure 9.3. Reconstructed components (RCs) represent the low-frequency component of each network index that is collectively shared by all indices in the evaluated climate network. The network is not limited to indices shown in this plot. Sea ice in the Eurasian Arctic Shelf Sea region[140,145,146,156] is thought to play a pivotal role in propagating this signal and synchronizing circulation patterns. In turn, this collective, slowly varying signature furnishes the modulating low-frequency background upon which higher-frequency behavior of regional circulations is superimposed.

Numerous studies find that computer model-simulated data differ from observational (instrumental) data; their spatiotemporal patterns differ, with ENSO and the AMO showing significant mismatches. Failure of models to capture the interconnectivity found in climate systems—the interconnected tropics on interannual timescales and hemispheric-signal propagation[170,171] tying numerous climate modes together on multidecadal timescales—suggests computer climate-model representations, as sophisticated and impressive as they are, may be missing critical dynamics in their designs.[e.g., 1,113,171–173]

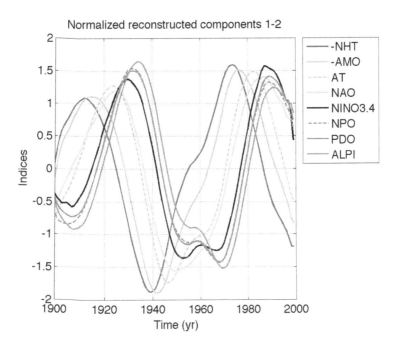

FIGURE 9.3 A low-frequency signal propagates through a network of synchronized winter climate indices. The propagation pattern is termed the "stadium-wave" signal by its original authors[156]—a reference to its communication through a sequence of indices. The plot shows normalized RCs of an eight-index climate network. Indices .... anomalies of the averaged Northern Hemisphere surface temperature (NHT), the AMO, the Atmospheric .... (AT: refers to the zonal component of basin-scale winds), the NAO, (NINO 3.4)—an index .... the PDO, and the AL index. The RC time series shown have been normalized to have a unit .... synchronized at nonzero lags (except for NPO and PDO, whose rescaled RCs are virtually .... ted reconstructions of NHT and AMO are of negative polarity. All indices are winter: DJF; .... own) have been found to participate in this signal propagation—from fish populations .... he Earth's rotational rate, and sea-ice extent.[145,146] The phasing of each understandable in .... climate indices.[145,146] (Reproduced with permission: [figure 2 in Wyatt et al. (2012)][156].)

Conclusion

Earth's regional climate patterns work on a variety of timescales to globally redistribute heat from where there is an excess to where there is a deficit. Earth system components—ocean, atmosphere, land, and sea ice—participate in tailoring characteristics of individual circulation patterns. Impact of individual circulation patterns on weather and climate, on local, regional, hemispheric, and global spatial scales, is modified according to interaction among patterns and timescales.[1,156] Complexity of local and regional details tends to yield to hemispheric order in the long term. This may be attributable, at least in part, to synchronized network behavior of climate indices at low-frequency timescales.[156] Interaction of regional circulation patterns, especially at multidecadal tempos, is often accompanied by characteristic phenomena. Interconnectedness of the Earth's climate network is profound, the foundation of the planet's heat engine.

Acknowledgments

I extend my gratitude to Roger Pielke, Sr., professor emeritus, University of Colorado, Boulder. His encouragement has been instrumental in motivating work related to this project.

References

1. Cai, W.; Wu, L.; Lengaigne, M.; Li, T.; McGregor, S.; Kug, J-S.; Yu, J-Y.; Stuecker, M. F.; Santoso, A.; Li, X.; Ham, Y-G.; Chikamoto, Y.; Ng, B.; McPhaden, M. J.; Du, Y.; Dommenget, D.; Jia, F.; Kajtar, J. B.; Keenlyside, N.; Lin, X.; Luo, J-J.; Martin-Rey, M.; Ruprich-Robert, Y.; Wang, G.; Xie, S.-P.; Yang, Y.; Kang, S. M.; Choi, J-Y.; Gan, B.; Kim, G-I.; Kim, C-E.; Kim, S.; Kim, J-H.; Chang, P. Pantropical climate interactions. *Science* 2019, 363(6430), 1–11. doi: 10.1126/science.aav4236.
2. Xie, S.-P.; Philander, G.H. A coupled ocean-atmosphere model of relevance to the ITCZ in the eastern Pacific. *Tellus A* 1994, 46(4), 340–350. doi: 10.3402/tellusa.v46i4.15484.
3. Bjerknes, J. A possible response of the atmospheric Hadley circulation to equatorial anomalies of ocean temperature. *Tellus* 1966, 18(4), 820–829. doi: 10.1111/j.2153-3490.1966.tb00303.x.
4. Bjerknes, J. Atmospheric teleconnections from the equatorial Pacific. *Mon. Weather Rev.* 1969, 97(3), 163–172. doi: 10.1175/1520-0493(1969)097<0163:ATFTEP>2.3.CO;2.
5. Schopf, P.S.; Suarez, M.J. Vacillations in a coupled ocean-atmosphere model. *J. Atmos. Sci.* 1988, 45, 549–566. doi: 10.1175/1520-0469(1988)045<0549:VIACOM>2.0.CO;2.
6. Suarez, M.J.; Schopf, P.S. A delayed action oscillator for ENSO. *J. Atmos. Sci.* 1988, 45(21), 3283–3287. doi: 10.1175/1520-0469(1988)045<3283:ADAOFE>2.0.CO;2.
7. Battisti, D.S.; Hirst, A.C. Interannual variability in a tropical atmosphere-ocean model: Influence of the basic state, ocean geometry, and nonlinearity. *J. Atmos. Sci.* 1989, 46(12), 1687–1712. doi: 10.1175/1520-0469(1989)046<1687:IVIATA>2.0.CO;2.
8. Sun, D-Z. The heat sources and sinks of the 1986–87 El Niño. *J. Clim.* 2000, 13, 3533–3550, doi: 10.1175/1520-0442(2000)013<3533:THSASO>2.0.CO;2.
9. Sun, D-Z.; Trenberth, K.E. Coordinated heat removal from the equatorial Pacific during the 1986–87 El Niño. *Geophys. Res. Lett.* 1998, 25(14), 2659–2662. doi: 10.1029/98GL01813.
10. Sun, D-Z. A possible effect of an increase in the warm-pool SST on the magnitude of El Niño warming. *J. Clim.* 2003, 16(2), 185–205. doi: 10.1175/1520-0442(2003)016<0185:APEOAI>2.0.CO;2.
11. Sun, D.-Z. The control of meridional differential heating over the level of ENSO activity: A Heat-Pump Hypothesis. In Earth's Climate: The Ocean-Atmosphere Interaction; Geophysical Monograph Series; Wang, C., Xie, S.-P., Carton, J.A., Eds.; American ⬛⬛⬛ ⬛⬛⬛ a numl175/1 doi: 10.1175/1 Washington, DC, 2004, 147, 71–83. doi: 10.1029/147GM04.
12. Brady, E.C. Interannual variability of meridional heat transport in ⬛⬛⬛ of the upper equatorial Pacific Ocean. *J. Phys. Oceanogr.* 1994, 24, 2675–2694. ⬛⬛⬛ (1994)024<2675:IVOMHT>2.0.CO;2.

13. Jin, F-F. An equatorial recharge paradigm for ENSO. Part I: Conceptual model. *J. Atmos. Sci.* 1997, 54, 811–829. doi: 10.1175/1520-0469(1997)054<0811:AEORPF>2.0.CO;2.

14. Wyrtki, K. Water displacements in the Pacific and the genesis of El Niño cycles. *J. Geophys. Res.* 1985, 90(C4), 7129–7132. doi: 10.1029/JC090iC04p07129.

15. Meinen, C.S.; McPhaden, M.J. Observations of warm water volume changes in the equatorial Pacific and their relationship to El Niño and La Niña. *J. Clim.* 2000, 13, 3551–3559. doi: 10.1175/1520-0442(2000)013<3551:OOWWVC>2.0.CO;2.

16. An, S.-I.; Kang, I.-S. A further investigation of the recharge oscillator paradigm for ENSO using a simple coupled model with the zonal mean and eddy separated. *J. Clim.* 2000, 13, 1987–1993. doi: 10.1175/1520-0442(2000)013<1987:AFIOTR>2.0.CO;2.

17. Wyrtki, K.; Wenzel, J. Possible gyre-gyre interaction in the Pacific Ocean. *Nature* 1984, 309(5968), 538–540. doi: 10.1038/309538a0.

18. Kim, H.-M.; Webster, P.J.; Curry, J.A. Impact of shifting patterns of Pacific Ocean warming on north Atlantic tropical cyclones. *Science* 2009, 325(5936), 77–80. doi: 10.1126/science.1174062.

19. Kumar, K.K.; Rajagopalan, B.; Hoerling, M.; Bates, G.; Cane, M. Unraveling the mystery of Indian monsoon failure during El Niño. *Science* 2006, 314, 115–119. doi: 10.1126/science.1131152.

20. Vimont, D.J.; Battisti, D.S.; Hirst, A.C. Footprinting: A seasonal connection between the tropics and mid-latitudes. *Geophys. Res. Lett.* 2001, 28, 3923–3926. doi: 10.1029/2001GL013435.

21. Vimont, D.J.; Wallace, D.J.; Battisti, D.S. The seasonal footprinting mechanism in the Pacific: Implications for ENSO. *J. Clim.* 2003, 16, 2668–2675. doi: 10.1175/1520-0442(2003)016<2668:TSFMIT>2.0.CO;2.

22. Amaya, D.J.; Foltz, G.R. Impacts of canonical and Modoki El Niño on tropical Atlantic SST. *J. Geophys. Res. Oceans* 2014, 119, 777–789. doi: 10.1002/2013JC009476.

23. Federov, A.V.; Philander, S.G. Is El Niño changing? *Science* 2000, 288(5473), 1997–2002. doi: 10.1126/science.288.5473.1997.

24. Izumo, T.; Vialard, J.; Dayan, H.; Lengaigne, M.; Suresh, I. A simple estimation of equatorial Pacific response from windstress to untangle Indian Ocean Dipole and Basin influences on El Niño. *Clim. Dyn.* 2016, 46, 2247–2268. doi: 10.1007/s00382-015-2700-4.

25. McPhaden, M.J. A 21st century shift in the relationship between ENSO SST and warm water volume anomalies. *Geophy. Res. Lett.* 2012, 39(L09706). doi: 10.1029/2012GL051826.

26. Sun, D.-Z.; Zhang, T. A regulatory effect of ENSO on the time-mean thermal stratification of the equatorial upper ocean. *Geophy. Res. Lett.* 2006, 33, L07710. doi: 10.1029/2005GL025296.

27. Guilderson, T.P.; Schrag, D.P. Abrupt shift in subsurface temperatures in the tropical Pacific associated with changes in El Niño. *Science* 1998, 281, 240–243. doi: 10.1126/science.281.5374.240.

28. Kerr, R.A. As the oceans switch, climate shifts. *Science* 1998, 281(5374), 157. doi: 10.1126/science.281.5374.157.

29. McPhaden, M.J.; Zhang, D. Slowdown of the meridional overturning circulation in the upper Pacific Ocean. *Nature* 2002, 415, 603–608. doi: 10.1038/415603a.

30. McPhaden, M.J.; Zhang, D. Pacific Ocean circulation rebounds. *Geophys. Res. Lett.* 2004, 31, L18301. doi: 10.1029/2004GL020727.

31. Zebiak, S.E. Air-sea interaction in the equatorial Atlantic region. *J. Clim.* 1993, 6, 1567–1586. doi: 10.1175/1520-0442(1993)006<1567:AIITEA>2.0.CO;2.

32. Nnamchi, H.C.; Li, J.; Kucharski, F.; Kang, I.-S; Keenlyside, N.S.; Chang, P.; Farneti, R. An equatorial-extratropical dipole structure of the Atlantic Niño. *J. Clim.* 2016, 29, 7295–7311. doi: 10.1175/JCLI-D-15-0894.1.

33. Lubbeck, J.F.; McPhaden, M.J. On the inconsistent relationship between Pacific and Atlantic Niños. *J. Clim.* 2012, 25, 4294–4303. doi: 10.1175/JCLI-D-11-00553.1.

34. Chang, P.; Fang, Y.; Saravanan, R.; Link, J.; Seide, H. The cause of the fragile relationship between the Pacific El Niño and the Atlantic Niño. *Nature* 2006, 443, 324–328. doi: 10.1038/nature05053.

35. Chiang, J.C.H.; Sobel, A.H. Tropical tropospheric temperature variations caused by ENSO and their influence on the remote tropical climate. *J. Climate* 2002, 15, 2616–2631. doi: 10.1175/1520-0442(2002)015<2616:TTTVCB>2.0.CO;2.

36. Chiang, J.C.H.; Lintner, B.R. Mechanisms of remote tropical surface warming during El Niño. *J. Climate* 2005, 18, 4130–4149. doi: 10.1175/JCLI3529.1.

37. Wallace, J.M.; Gutzler, D.S. Teleconnections in the geopotential height field during the Northern Hemisphere winter. *Mon. Weather Rev.* 1981, 109, 784–812. doi: 10.1175/1520-0493(1981)109<0784:TITGHF>2.0.CO;2.

38. Wang, C. An overlooked feature of tropical climate: Inter-Pacific–Atlantic variability. *Geophys. Res. Lett.* 2006, 33, L12702. doi: 10.1029/2006GL026324.

39. Klein, Stephen A.; Soden, Brian J.; Lau, N.-C. Remote sea surface temperature variations during ENSO: Evidence for a tropical atmospheric bridge. *J. Climate* 1999, 12, 917–932. doi: 10.1175/1520-0442(1999)012<0917:RSSTVD>2.0.CO;2.

40. Martin-Rey, M.; Polo, I.; Rodriguez-Fonseca, B.; Losada, T.; Lazar, A. Is there evidence of changes in tropical Atlantic variability modes under AMO phases in the observational record? *J. Clim.* 2018, 31, 515–536. doi: 10.1175/JCLI-D-16-0459.s1.

41. Saji N.H.; Goswami, B.N.; Vinayachandran, P.N.; Yamagata, T.A. Dipole mode in the tropical Indian Ocean. *Nature* 1999, 401, 360–363. doi: 10.1038/43854.

42. Webster, P.J.; Moore, A.M.; Loschnigg, J.P.; Leben, R.R. Coupled ocean-atmosphere dynamics in the Indian Ocean during 1997–1998. *Nature* 1999, 401, 356–360. doi: 10.1038/43848.

43. Yamagata, T.; Behera, S.K.; Luo, J.-J.; Masson, S.; Jury, M.R.; Rao, S.A. Coupled ocean-atmosphere variability in the tropical Indian Ocean, earth's climate. In The Ocean-Atmosphere Interaction; Geophysical Monograph Series; Wang, C., Xie, S.-P, Carton, J.A., Eds.; American Geophysical Union: Washington, DC, 2004, 147, 189–211. doi: 10.1029/147GM12.

44. Annamalai, H.; Murtugudde, R.; Potemra, J.T.; Xie, S-P.; Liu, P.; Want, B. Coupled dynamics over the Indian Ocean: Spring initiation of the zonal mode. *Deep-Sea Res.* 2003, 50, 2305–2330. doi: 10.1016/S0967-0645(03)00058-4.

45. Saji, N.H.; Yamagata, T. Possible impacts of Indian Ocean dipole mode events on global climate. *Clim. Res.* 2003, 25, 151–169. doi: 10.3354/cr025151.

46. Kucharski, F.; Bracco, A.; Molteni, F.; Yoo, J.H. Low-frequency variability of the Indian monsoon-ENSO relationship and the tropical Atlantic: The "weakening" of the 1980s and 1990s. *J. Clim.* 2007, 20, 4255–4266. doi: 10.1175/JCLI4254.1.

47. Rong, X.Y.; Zhang, R.H.; Li, T. Impacts of Atlantic sea surface temperature anomalies on the Indo-East Asian summer monsoon-ENSO relationship. *Chinese Sci. Bull.* 2010, 55, 1397–1408. doi: 10.1007/s11434-010-3098-3.

48. Fischer, A.S.; Terray, P.; Guilyardi, E.; Gualdi, S.; Delecluse, P. Two independent triggers for the Indian Ocean dipole/zonal mode in a coupled GCM. *J. Clim.* 2005, 18, 3428–3449. doi: 10.1175/JCLI3478.1.

49. Kug, J-S.; Kang, I-S. Interactive feedback between ENSO and the Indian Ocean. *J. Clim.* 2006, 16, 1784–1800. doi: 10.1175/JCLI3660.1.

50. Masumoto, Y.; Meyers, G. Forced Rossby waves in the southern tropical Indian Ocean. *J. Geophys. Res.* 1998, 103, 27589–27602. doi: 10.1029/98JC02546.

51. Yu, J.-Y.; Lau, K.M. Contrasting Indian Ocean SST variability with and without ENSO influence: A coupled atmosphere-ocean GCM study. *Meteorol. Atmos. Phys.* 2005, 90(3–4), 179–191. doi: 10.1007/s00703-004-0094-7.

52. Zhang, W.; Wang, Y.; Jin, F-F.; Stuecker, M. F., Turner, A.G. Impact of different El Niño types on the El Niño/IOD relationship. *Geophys. Res. Lett.* 2015, 42(20), 8570–8576. doi: 10.1002/2015GL065703.

53. Izumo, T.; Vialard, J.; Lengaigne, M.; de Boyer Montegut, C.; Behera, S.K.; Luo, J-J.; Cravatte, S.; Masson, S.; Yamagata, T. Influence of the state of the Indian Ocean dipole on the following year's El Niño. *Nat. Geosci.* 2010, 3, 168–172. doi: 10.1038/ngeo760.

54. Annamalai, H.; Xie, S.-P.; McCreary, J.-P.; Murtugudde, R. Impact of Indian Ocean sea surface temperature on developing El Niño. *J. Clim.* 2005, 18, 302–319. doi: 10.1175/JCLI-3268.1.

55. Jourdain, N.C.; Lengaigne, M.; Vialard, J.; Izumo, T.; Gupta, S.S. Further insights on the influence of the Indian Ocean dipole on the following year's ENSO from observations and CMIP5 models. *J. Clim.* 2016, 29, 637–658. doi: 10.1175/JCLI-D-15-0481.1.

56. Chiang, J.C.H.; Vimont, D.J. Analogous Pacific and Atlantic meridional modes of tropical atmosphere-ocean variability. *J. Clim.* 2004, 17, 4143–4158. doi: 10.1175/JCLI4953.1.

57. Enfield, D.B.; Mayer, D.A. Tropical Atlantic sea surface temperature variability and its relation to El Niño-Southern oscillation. *J. Geophys. Res.* 1997, 102(C1), 929–945. doi: 10.1029/96JC03296.

58. Vimont, D.J.; Kossin, J.P. The Atlantic meridional mode and hurricane activity. *Geophys. Res. Lett.* 2007, 34, L07709. doi: 10.1029/2007GL029683.

59. Amaya, D.J.; DeFlorio, M.J.; Miller, A.J.; Xie, S.-P. WES feedback and the Atlantic Meridional Mode: Observations and CMIP5 comparisons. *Clim. Dyn.* 2017, 49, 1665–1679. doi: 10.1007/s00382-016-3411-1.

60. Chang, P.; Ji, L.; Li, H. A decadal climate variation in the tropical Atlantic Ocean from thermodynamic air-sea interactions. *Nature* 1997, 385, 516–518. doi: 10.1038/385516a0.

61. Xie, S.-P. A dynamic ocean-atmosphere model of the tropical Atlantic decadal variability. *J. Clim.* 1999, 12, 64–70. doi: 10.1175/1520-0442-12.1.64.

62. Smirnov, D.; Vimont, D.J. Variability of the Atlantic meridional mode during the Atlantic hurricane season. *J. Clim.* 2011, 24, 1409–1424. doi: 10.1175/2010JCLI3549.1.

63. Chang, P.; Zhang, L.; Saravanan, R.; Vimont, D.J.; Chiang, J.C.H.; Ji, L.; Seidel, H.; Tippett, M.K. Pacific meridional mode and El Niño-Southern oscillation. *Geophys. Res. Lett.* 2007, 34, L16608. doi: 10.1029/2007GL030302.

64. Yang, J.; Liu, Q.; Xie, S-P.; Liu, Z.; Wu, L. Impact of the Indian Ocean SST basin mode on the Asian summer monsoon. *Geophys. Res. Lett.* 2007, 34, L02708. doi: 10.1029/2006GL028571.

65. Klein, S.A.; Soden, B.J.; Lau, N.-C. Remote sea surface temperature variations during ENSO: Evidence for a tropical atmospheric bridge. *J. Clim.* 1999, 12, 917–932. doi: 10.1175/1520-0442(1999)012<0917:RSSTVD>2.0.CO;2.

66. Xie, S-P.; Annamalai, H.; Schott, F.A.; McCreary, J.P. Jr, Structure and mechanisms of South Indian Ocean climate variability. *J. Clim.* 2002, 15, 864–878. doi: 10.1175/1520-0442(2002)015<0864:SAMOSI>2.0.CO;2.

67. Yang, J.; Liu, Q.; Liu, Z.; Wu, L.; Huang, F. Basin mode of Indian Ocean sea surface temperature and Northern Hemisphere circumglobal teleconnection. *Geophys. Res. Lett.* 2009, 36, L19705. doi: 10.1029/2009GL039559.

68. Ohba, M.; Ueda, H. An impact of SST anomalies in the Indian Ocean in acceleration of the El Niño to La Niña transition. *J. Meteorol. Soc. Jpn.* 2007, 85(3), 335–338. doi: 10.2151/jmsj.85.335.

69. Xie, S.-P.; Hu, K.; Hafner, J.; Tokinaga, H.; Du, Y.; Huang, G.; Sampe, T. Indian Ocean capacitor effect on Indo-western Pacific climate during the summer following El Niño. *J. Clim.* 2009, 22, 730–747. doi: 10.1175/2008JCLI2544.1.

70. Du, Y.; Xie, S-P.; Huang, G.; Hu, K. Role of Air-Sea Interaction in the Long Persistence of El Niño-Induced North Indian Ocean arming. *J. Clim.* 2009, 22, 2023–2037. doi: 10.1175/2008JCLI2590.1.

71. McGregor, S.; Timmermann, A.; Stuecker, M.F.; England, M.H.; Merrifield, M.; Jin, F.-F.; Chikamoto, Y. Recent walker circulation strengthening and Pacific cooling amplified by Atlantic warming. *Nat. Clim. Change* 2014, 4, 888–892. doi: 10.1038/NCLIMATE2330.

72. Latif, M.; Barnett, T.P. Interactions of the tropical Oceans. *J. Clim.* 1995, 8, 952–964. doi: 10.1175/1520-0442(1995)008<0952:IOTTO>2.0.CO;2.

73. Li, X.; Xie, S-P.; Gille S.T.; Yoo, C. Atlantic-induced pan-tropical climate change over the past three decades. *Nat. Clim. Change* 2016, 6, 275–279. doi: 10.1038/NCLIMATE2840.

74. Sprintall, J.; Révelard, A. The Indonesian throughflow response to Indo-Pacific climate variability. *J. Geophys. Res.-Oceans* 2013, 119, 1161–1175. doi: 10.1002/2013JC009533.

75. Annamalai, H.; Potemra, J.; Murtugudde, R.; McCreary, J.P. Effect of preconditioning on the extreme climate events in the tropical Indian Ocean. *J. Clim.* 2005, 18, 3450–3469. doi: 10.1175/JCLI3494.1.

76. Lee, S.-K.; Park, W.; Baringer, M.O.; Gordon, A.L.; Huber, B.; Liu, Y. Pacific origin of the abrupt increase in Indian Ocean heat content during the warming hiatus. *Nat. Geosci.* 2015, 8, 445–449. doi: 10.1038/ngeo2438.

77. Alexander, M.A., Deser, C.A. Mechanism for the recurrence of wintertime Midlatitude SST anomalies. *J. Phys. Oceanogr.* 1995, 25, 122–137. doi: 10.1175/1520-0485(1995)025<0122:AMFTRO>2.0.CO;2.

78. Thompson, D.W.J.; Wallace, J.M. The arctic oscillation signature in the wintertime geopotential height and temperature fields. *Geophys. Res. Lett.* 1998, 25(9), 1297–1300. doi: 10.1029/98GL00950.

79. Thompson, D.W.J.; Wallace, J.M. Annular modes in the extratropical circulation, Part I: Month-to-month variability. *J. Clim.* 2000, 13, 1000–1016. doi: 10.1175/1520-0442(2000)013<1000:AMITEC>2.0.CO;2.

80. Thompson, D.W.J.; Wallace, J.M.; Hegerl, G.C. Annular modes in the extratropical circulation. Part II: Trends. *J. Clim.* 2000, 13, 1018–1036. doi: 10.1175/1520-0442(2000)013<1018:AMITEC>2.0.CO;2.

81. Shindell, D.T.; Schmidt, G.A.; Miller, R.L.; Mann, M.E. Volcanic and solar forcing of climate change during the preindustrial era. *J. Clim.* 2003, 16, 4094–4107. doi: 10.1175/1520-0442(2003)016<4094:VASFOC>2.0.CO;2.

82. Kodera, K.; Kuroda, Y. Dynamical response to the solar cycle. *J. Geophys. Res. Atmos.* 2002, 107(D24), 4749. doi: 10.1029/2002JD002224.

83. Cohen, J.; Barlow, M.; Kushner, P.J.; Kazuyuki, S. Stratosphere-troposphere coupling and links with Eurasian land surface variability. *J. Clim.* 2007, 20, 5335–5343. doi: 10.1175/2007JCLI1725.1.

84. Cohen, J. Eurasian Snow Cover Variability and Links with Stratosphere-Troposphere Coupling and Their Potential use in Seasonal to Decadal Climate Predictions. *Science and Technology Infusion Climate Bulletin* 2011, 1–6. www.nws.noaa.gov/ost/climate/STIP/FY11CTBSeminars/jcohen_062211.pdf.

85. Cohen, J.; Furtado, J.C.; Jones, J.; Barlow, M.; Whittleston, D.; Entekhabi, D. Linking Siberian snow cover to precursors of stratospheric variability. *J. Clim.* 2014, 27, 5422–5432. doi: 10.1175/JCLI-D-13-00779.1.

86. Shindell, D.T.; Miller, R.L.; Schmidt, G.A.; Pandolfo, L. Simulation of recent northern winter climate tends by greenhouse-gas forcing. *Nature* 1999, 399, 452–455. doi: 10.1038/20905.

87. Baldwin, M.P.; Dunkerton, T.J. Propagation of the Arctic oscillation from the stratosphere to the troposphere. *J. Geophys. Res.* 1999, 104(D24), 30937–30946. doi: 10.1029/1999JD900445.

88. Schimanke, S.; Korper, J.; Spangehl, T.; Cubasch, U. Multi-decadal variability of sudden stratospheric warmings in an AOGCM. *Geophys. Res. Lett.* 2011, 38, L01801. doi: 10.1029/2010GL045756.

89. Kelly, K.A.; Dong, S. The relationship of western boundary current heat transport and storage to midlatitude ocean-atmosphere interaction. In *Earth's Climate: The Ocean-Atmosphere Interaction*; Geophysical Monograph Series; Wang, C., Xie, S.-P., Carton, J.A., Eds.; American Geophysical Union: Washington, DC, 2004, 147, 347–363. doi: 10.1029/147GM19.

90. Minobe, S.; Kuwano-Joshida, A.; Komori, N.; Xie, S.-P.; Small, R.J. Influence of the Gulf Stream on the troposphere. *Nature* 2008, 452, 206–210. doi: 10.1038/nature06690.

91. Kessler, W.S.; Kleeman, R. Rectification of the Madden-Julian oscillation into the ENSO cycle. *J. Clim.* 2000, 13, 3560–3575. doi: 10.1175/1520-0442(2000)013<3560:ROTMJO>2.0.CO;2.

92. Collimore, C.C.; Martin, D.W.; Hitchman, M.H.; Huesmann, A., Waliser, D.E. On the relationship between the QBO and tropical deep convection. *J. Clim.* 2003, 16, 2552–2568. doi: 10.1175/1520-0442(2003)016<2552:OTRBTQ>2.0.CO;2.

93. Jiang, N.; Neelin, J. D.; Ghil, M. Quasi-quadrennial and quasi-biennial variability in the equatorial Pacific. *Clim. Dyn.* 1995, 12(2), 101–112. doi: 10.1007/BF00223723.

94. Holton, J.R.; Tan, H.-C. The influence of the equatorial quasi-biennial oscillation on the global circulation at 50mb. *J. Atmos. Sci.* 1980, 37, 2200–2208. doi: 10.1175/1520-0469(1980)037<2200:TIO TEQ>2.0.CO;2.

95. Hurrell, J.W. Decadal trends in the North Atlantic oscillation: Regional temperatures and precipitation. *Science* 1995, 269(5224), 676–679. doi: 10.1126/science.269.5224.676.

96. Delworth, T.L.; Zeng, F.; Vecchi, G.A.; Yang, X.; Zhang, L.; Zhang, R. The North Atlantic oscillation as a driver of rapid climate change in the Northern Hemisphere. *Nat. Geosci.* 2016, 9, 509–512. doi: 10.1038/NGEO2738.

97. Deser, C.; Blackmon, M.L. Surface climate variations over the North Atlantic Ocean during winter: 1900–1993. *J. Clim.* 1993, 6, 1743–1753. doi: 10.1175/1520-0442(1993)006<1743:SCVOTN>2.0.CO;2.

98. Eden, C.; Jung, T. North Atlantic interdecadal variability: Oceanic response to the North Atlantic oscillation. *J. Clim.* 2001, 14, 676–691. doi: 10.1175/1520-0442(2001)014<0676:NAIVOR>2.0.CO;2.

99. Bjerknes, J. Atlantic air-sea interaction. *Adv Geophys* 1964, 10, 1–82. doi: 10.1016/S0065-2687(08)60005-9.

100. Kushnir, Y. Interdecadal variations in North Atlantic sea-surface-temperature and associated atmospheric conditions. *J. Clim.* 1994, 7, 142–157. doi: 10.1175/1520-0442(1994)007<0141:IVINAS>2.0.CO;2.

101. Visbeck, M.; Chassignet, E.P.; Curry, R.G.; Delworth, T.L.; Dickson, R.R.; Krahmann, G. The North Atlantic oscillation: Climatic significance and environmental impact. In *The Ocean's Response to North Atlantic Oscillation Variability*; Geophysical Monograph Series; American Geophysical Union, 2003, 134, 113–145. doi: 10.1029/GM134.

102. Wallace J.M.; Thompson, D.W.J. The Pacific center of action of the Northern Hemisphere annular mode: Real or artifact? *J Clim.* 2002, 15, 1987–1991. doi: 10.1175/1520-0442(2002)015<1987:TPCOAO>2.0.CO;2.

103. Liu, J.; Curry, J.A.; Dai, Y.; Horton, R. Causes of the northern high-latitude land surface winter climate change. *Geophys. Res. Lett.* 2007, 34, L14702. doi: 10.1029/2007GL030196.

104. Ding, Q.; Steig, E.J.; Battisti, D.S.; Wallace, J.M. Influence of the tropics on the Southern annular mode. *J. Clim.* 2012, 25(18), 6330–6348. doi: 10.1175/JCLI-D-11-00523.1.

105. Walker, G.T.; Bliss, E.W. World weather V. *Memoirs R. Meteorol. Soc.* 1932, 4(36), 53–84. www.rmets.org/sites/default/files/ww5.pdf.

106. Kirov, B.; Georgieva, K. Long-term variations and interrelations of ENSO, NAO, and solar activity. *Phys. Chem. Earth* 2002, 27(6–8), 441–448. doi: 10.1016/S1474-7065(02)00024-4.

107. Wang, L.; Chen, W.; Huang, R. Changes in the variability of North Pacific oscillation around 1975/1976 and its relationship with East Asian winter climate. *J. Geophys. Res.* 2007, 112, D11110. doi: 10.1029/2006jD008054.

108. Walter, K.; Graf, H.F. On the changing nature of the regional connection between the North Atlantic oscillation and sea surface temperature. *J. Geophys. Res.* 2002, 107(D17), 4388. doi: 10.1029/2001JD000850.

109. Polonski, A.B.; Basharin, D.V.; Voskresenskaya, E.N.; Worley, S.J.; Yurovsky, A.V. Relationship between the North Atlantic oscillation, Euro-Asian climate anomalies and Pacific variability. *Pac. Oceanogr.* 2004, 2(1–2), 52–66. www.researchgate.net/publication/268300577.

110. Grosfeld, K.; Lohmann, G.; Rimbu, N.; Fraedrich, K.; Lunkeit, F. Atmospheric multidecadal variations in the North Atlantic realm: Proxy data, observations, and atmospheric circulation model studies. *Clim. Past* 2007, 3, 39–50. doi: 10.5194/cp-3-39-2007.

111. Sugimoto, S.; Hanawa, K. Decadal and interdecadal variations of the Aleutian low activity and their relation to upper oceanic variations over the North Pacific. *J. Meteorol. Soc. Jpn.* 2009, 87(4), 601–614. doi: 10.2151/jmsj.87.601.

112. Frankignoul, C.; Sennechael, N.; Kwon, Y-Oh.; Alexander, M.A. Influence of the meridional shifts of the Kuroshio and the Oyashio extensions on the atmospheric circulation. *J. Clim.* 2011, 24, 762–777. doi: 10.1175/2010JCLI3731.1.

113. Omrani, N.E.; Keenlyside, N.S.; Bader, J.; Manzini, E. Stratosphere key for wintertime atmospheric response to warm Atlantic decadal conditions. *Clim. Dyn.* 2014, 42(3–4), 649–663. doi: 10.1007/s00382-013-1860-3.

114. Georgieva, K.; Kirov, B.; Tonev, P.; Guineva, V.; Atanasov, D. Long-term variations in the correlation between NAO and solar activity: the importance of north-south solar activity asymmetry for atmospheric circulation. *Adv. Space Res.* 2007, 40, 1152–1166. doi: 10.1016/j.asr.2007.02.091.

115. Rind, D.; Chandler, M.; Lerner, J.; Martinson, D.G.; Yuan, X. Climate response to basin-specific changes in latitudinal temperature gradient and implications for sea-ice variability. *J. Geophys. Res.* 2001, 106, 20161–20173. doi: 10.1029/2000JD900643.

116. Okumura, Y.M.; Schneider, D.; Deser, C.; Wilson, R. Decadal-interdecadal climate variability over Antarctica and linkages to the tropics: Analysis of ice core, instrumental, and tropical proxy data. *J. Clim.* 2012, 25, 7421–7441. doi: 10.1175/JCLI-D-12-00050.1.

117. Xie, S-P. Satellite observations of cool ocean-atmosphere interaction. *Bull. Am. Meteorol. Soc.* 2004, 85, 195–208. doi: 10.1175/BAMS-85-2-195.

118. Kerr, R.A. A North Atlantic climate pacemaker for the centuries. *Science* 2000, 288(5473), 1984–1985. doi: 10.1126/science.288.5473.1984.

119. Enfield, D.B.; Mestas-Nunez, A.M.; Trimble, P.J. The Atlantic multidecadal oscillation and its relation to rainfall and river flows in the continental U. S. *Geophys. Res. Lett.* 2001, 28, 277–280. doi: 10.1029/2000GL012745.

120. Dima, M.; Lohmann, G. A hemispheric mechanism for the Atlantic multidecadal oscillation. *J. Clim.* 2007, 20, 2706–2719. doi: 10.1175/JCLI4174.1.

121. Sutton, R.T.; Hodson, D.L.R. Influence of the ocean on North Atlantic climate variability 1871–1999. *J. Clim.* 2003, 16, 3296–3313. doi: 10.1175/1520-0442(2003)016<3296:IOTOON>2.0.CO;2.

122. Delworth, T.L.; Mann, M.E. Observed and simulated multidecadal variability in the Northern Hemisphere. *Clim. Dyn.* 2000, 16, 661–676. doi: 10.1007/s003820000075.

123. Sutton, R.T.; Hodson, D.L.R. Atlantic Ocean forcing of North American and European summer climate. *Science* 2005, 309, 115–118. doi: 10.1126/science.1109496.

124. Knight, J.R.; Folland, C.K.; Scaife, A.A. Climate impacts of the Atlantic multidecadal oscillation. *Geophys. Res. Lett.* 2006, 33, L17706. doi: 10.1029/2006GL026242.

125. Frankcombe, L.M.; Dijkstra, H.A.; von der Heydt, A. Subsurface signatures of the Atlantic multidecadal oscillation. *Geophys. Res. Lett.* 2008, 35, L19602. doi: 10.1029/2008GL034989.

126. Frankcombe, L.M.; Dijkstra, H.A. Coherent multidecadal variability in North Atlantic sea level. *Geophys. Res. Lett.* 2009, 36, L15604. doi: 10.1029/2009GL039455.

127. Chylek, P.; Folland, C.K.; Dijkstra, H.A.; Lesins, G.; Dubey, M. Ice-core data evidence for a prominent near 20 year time-scale of the Atlantic multidecadal oscillation. *Geophys. Res. Lett.* 2011, 38, L13704. doi: 10.1029/2011GL047501.

128. Frankcombe, L.M.; Dijkstra, H.A. The role of Atlantic-Arctic exchange in North Atlantic multidecadal climate variability. *Geophys. Res. Lett.* 2011, 38, L16603. doi: 10.1029/2011GL048158.

129. Knight, J.R.; Allan, R.J.; Folland, C.K.; Vellinga, M.; Mann, M.E. A signature of persistent natural thermohaline circulation cycles in observed climate. *Geophys. Res. Lett.* 2005, 32, L20708. doi: 10.1029/2005GL024233.

130. Latif, M.; Böning, C.; Willebrand, J.; Biastoch, A.; Dengg, J.; Keenlyside, N.; Schweckendiek, U.; Madec, G. Is the thermohaline circulation changing? *J. Clim.* 2006, 19(18), 4631–4637. doi: 10.1175/JCLI3876.1.

131. Ottera, O.H.; Bentsen, M.; Drange, H.; Suo, L. External forcing as a metronome for Atlantic multidecadal variability. *Nat. Geosci.* 2010, 3, 688–694. doi: 10.1038/ngeo955.

132. Zhang, R.; Sutton, R.; Danabasoglu, G.; Kwon, Y.-O.; Marsh, R.; Yeager, S.G.; Amrhein, D.E.; Little, C.M. A review of the role of the Atlantic meridional overturning circulation in Atlantic multidecadal variability and associated climate impacts. *Rev. Geophys.* 2019, 57(2), 316–375. doi: 10.1029/2019RG000644.

133. Msadek, R.; Frankignoul, C.; Li, L.Z.X. Mechanisms of the atmospheric response to North Atlantic multidecadal variability: A model study. *Clim. Dyn.* 2010, 36(7–8), 1255–1276. doi: 10.1007/s00382-010-0958-0.

134. Keenlyside, N.S.; Latif, M.; Jungclaus, J.; Kornblueh, L.; Roeckner, E. Advancing decadal-scale climate prediction in the North Atlantic sector. *Nature* 2008, 453(7191), 84–88. doi: 10.1038/nature06921.

135. Black, D.; Peterson, L.C.; Overpeck, J.T.; Kaplan, A.; Evans, M.N.; Kashgarian, M. Eight centuries of North Atlantic ocean atmosphere variability. *Science* 1999, 286, 1709–1713. doi: 10.1126/science.286.5445.1709.

136. Vellinga, M.; Wu, P. Low-latitude freshwater influence on centennial variability of the Atlantic thermohaline circulation. *J. Clim.* 2004, 17, 4498–4511. doi: 10.1175/3219.1.

137. Timmermann, A.; Latif, M.; Voss, R.; Grotzner, A. Northern Hemisphere interdecadal variability: A coupled air-sea mode. *J. Clim.* 1998, 11, 1906–1931. doi: 10.1175/1520-0442-11.8.1906.

138. Frankcombe, L.M.; von der Heydt, A.; Dijkstra, H.A. North Atlantic multidecadal climate variability: An investigation of dominant time scales and processes. *J. Clim.* 2010, 23, 3626–3638. doi: 10.1175/2010JCLI3471.1.

139. Jungclaus, J.H.; Haak, H.; Latif, M.; Mikolajewicz, U. Arctic-North Atlantic interactions and multidecadal variability of the meridional overturning circulation. *J. Clim.* 2005, 18, 4013–4031. doi: 10.1175/JCLI3462.1.

140. Polyakov, I.V.; Alexeev, V.A.; Ashik, I.M.; Bacon, S; Beszczynska-Moller, A.; Carmack, E.C.; Dmitrenko, I.A.; Fortier, L.; Gascard, J-C.; Hansen, E.; Holemann, J.; Ivanov, V.V.; Kikuchi, T.; Kirillov, S.; Lenn, Y-D.; McLaughlin, F.A.; Piechura, J.; Repina, I.; Timokhov, L.A.; Walczowski, W.; Woodgate, R. Fate of early 2000s Arctic warm water pulse. *Bull. Am. Met. Soc. (BAMS)* May 2011, 92, 561–566. doi: 10.1175/2010BAMS2921.1.

141. Delworth, T.L.; Manabe, S.; Stouffer, R.J. Multidecadal climate variability in the Greenland sea and surrounding regions: a coupled model simulation. *Geophys. Res. Lett.* 1997, 24(3), 257–260. doi: 10.1029/96GL03927.

142. Polyakov, I.V.; Alekseev, G.V.; Timokhov, L.A.; Bhatt, U.S.; Colony, R.L.; Simmons, H.L.; Walsh, D.; Walsh, J.E.; Zakharov, V.F. Variability of the intermediate Atlantic water of the Arctic ocean over the last 100 years. *J. Clim.* 2004, 17(23), 4485–4497. doi: 10.1175/JCLI-3224.1.

143. Polyakov, I.V.; Bhatt, U.S.; Simmons, H.L.; Walsh, D.; Walsh, J.E.; Zhang, X. Multidecadal variability of North Atlantic temperature and salinity during the twentieth century. *J. Clim.* 2005, 18, 4562–4581. doi: 10.1175/JCLI3548.1.

144. Frolov, I.E.; Gudkovich, A.M.; Karklin, B.P.; Kvalev, E.G.; Smolyanitsky, V.M., Eds.; Arctic and Antarctic Research Institute (AARI). In Climate Change in Euraslan Arctic Shelf Seas: Centennial Ice Cover Observations; St. Petersburg and Springer-Praxis Books in Geophysical Sciences, Praxis Publishing: Russia and Chichester, UK, 2009, 1–165, ISBN 978-3-540-85875-1. doi: 10.1007/978-3-540-85875-1.

145. Wyatt, M.G. A Multidecadal Climate Signal Propagating Across the Northern Hemisphere through Indices of a Synchronized Network. Ph.D. Dissertation. University of Colorado: Boulder, CO, 2012, 201. https://scholar.colorado.edu/geol_gradetds/36.

146. Wyatt, M.G.; Curry, J.A. Role for Eurasian Arctic shelf sea ice in a secularly varying hemispheric climate signal during the 20th century. *Clim. Dyn.* 2014, 42(9–10), 2763–2782. doi: 10.1007/s00382-013-1950-2.

147. Outten, S.D.; Esau, I. A link between Arctic sea ice and recent cooling trends over Eurasia. *Clim. Change* 2012, 110(3–4), 1069–1075. doi: 10.1007/s10584-011-0334-z.

148. Ugryumov, A.I.; Khar'kova, N.V. Modern changes in St. Petersburg climate and atmospheric circulation variations. *Russ. Meteorol. Hydro.* 2008, 33(1), 15–19. doi: 10.3103/S1068373908010032.

149. Mantua, N.J.; Hare, S.R.; Zhang, Y.; Wallace, J.M.; Francis, R.C. A Pacific interdecadal climate oscillation with impacts on salmon production. *Bull. Am. Meteorol. Soc.* 1997, 78, 1069–1079. doi: 10.1175/1520-0477(1997)078<1069:APICOW>2.0.CO;2.

150. Minobe, S. A 50–70-year climatic oscillation over the North Pacific and North America. *Geophys. Res. Lett.* 1997, 24, 683–686. doi: 10.1029/97GL00504.

151. Power, S.; Casey, T.; Folland, C.K.; Colman, A.; Mehta, V. Inter-decadal modulation of the impact of ENSO on Australia. *Clim. Dyn.* 1999, 15, 319–323. doi: 10.1007/s003820050284.

152. Niebauer, H. Variability in Bering sea ice cover as affected by a regime shift in the North Pacific in the period 1947–1996. *J. Geophys. Res.* 1998, 103(C12), 27717–27737. doi: 10.1029/98JC02499.

153. Lin, R.; Zheng, F.; Dong, X. ENSO frequency asymmetry and the Pacific decadal oscillation in observations and 19 CMIP5 models. *Adv. Atmos. Sci.* 2018, 35, 495–506. doi: 10.1007/s00376-017-7133-z.

154. An, S.-I.; Wang, B. Interdecadal change of the structure of the ENSO mode and its impact on the ENSO frequency. *J. Clim.* 2000, 13, 2044–2055. doi: 10.1175/1520-0442(2000)013<2044:ICOTSO>2.0.CO;2.

155. Kravtsov, S. An empirical model of decadal ENSO variability. *Clim. Dyn.* 2012, 39, 2377–2391. doi: 10.1007/s00382-012-1424-y.

156. Wyatt, M.G.; Kravtsov, S.; Tsonis, A.A. Atlantic multidecadal oscillation and Northern Hemisphere's climate variability. *Clim. Dyn.* 2012, 38(5/6), 929–949. doi: 10.1007/s00382-011-1071-8.

157. Schmittner, A.; Appenzeller, C.; Stocker, T.F. Enhanced Atlantic freshwater export during El Nino. *Geophys. Res. Lett.* 2000, 27(8), 1163–1166. doi: 10.1029/1999GL011048.

158. Latif, M.; Roeckner, E.; Mikolajewicz, U.; Voss, R. Tropical stabilization of the thermohaline circulation in a greenhouse warming simulation. *J. Clim.* 2000, 13, 1809–1813. doi: 10.1175/1520-0442(2000)013<1809:L>2.0.CO;2.

159. Gordon, A.L. Interocean Exchange. In *Ocean Circulation and Climate: Observing and Modelling the Global Ocean*; International Geophysics Series; Siedler, G., Church, J., Gould, J., Eds.; Academic Press: San Diego, CA, 2001, 77, 303–314. www.researchgate.net/publication/286229888.

160. Gordon, A.L.; Weiss, R.F.; Smethie, W.M.; Warner, M.J. Thermocline and intermediate water communication between the South Atlantic and Indian Oceans. *J. Geophys. Res.* 1992, 97, 7223–7240. doi: 10.1029/92JC00485.

161. Schouten, M.W.; de Ruijter, W.P,M.; van Leeuwen, P.J.; Dijkstra, H.A. An oceanic teleconnection between the equatorial and southern Indian Ocean. *Geophys. Res. Lett.* 2002, 29(16), 1812. doi: 10.1029/2001GL014542.

162. Hasegawa, T.; Yasuda, T.; Hanawa, K. Multidecadal variability of the upper ocean heat content anomaly field in the North Pacific and its relationship to the Aleutian low and the Kuroshio transport. *Papers Meteorol. Geophys.* 2007, 58, 155–166. doi: 10.2467/mripapers.58.155.

163. Zhang, R.; Delworth, T.L. Impact of the Atlantic multidecadal oscillation on North Pacific climate variability. *Geophys. Res. Lett.* 2007, 34, L23708. doi: 10.1029/2007GL031601.

164. McCabe, G.J.; Palecki, M.A.; Betancourt, J.L. Pacific and Atlantic ocean influences on multidecadal drought frequency in the United States. *Proc. Natl. Acad. Sci.* 2004, 101(12), 4136–4141. doi: 10.1073/pnas.0306738101.

165. Nowak, K.; Hoerling, M.; Rajagopalan B.; Zagona, E. Colorado river basin hydroclimatic variability. *J. Clim.* 2012, 25(2), 4389–4403. doi: 10.1175/JCLI-D-11-00406.1.

166. Woodhouse, C.A; Gray, S.T; Meko, D.M. Updated stream-flow reconstruction for the Upper Colorado river basin. *Water Resour. Res.* 2006, 42, W05415. doi: 10.1029/2005WR004455.

167. White, W.B.; Peterson, R.G. An Antarctic circumpolar wave in surface pressure, wind, temperature, and sea-ice extent. *Nature* 1996, 380(6576), 699–702. doi: 10.1038/380699a0.

168. White, W.B.; Annis, J. Influence of the Antarctic circumpolar wave on El Niño and its multidecadal changes from 1950 to 2001. *J. Geophys. Res.* 2004, 109, C06019. doi: 10.1029/2002JC001666.

169. White, W.B.; Chen, S.-C.; Allan, R.J.; Stone, R.C. Positive feedbacks between the Antarctic circumpolar wave and the global El Niño-Southern oscillation wave. *J. Geophys. Res.* 2002, 107(C10), 3165–3162. doi: 10.1029/2000JC000581.

170. Kravtsov, S.; Grimm, C.; Gu, S. Global-scale multidecadal variability missing in state-of-the-art climate models. *npj Clim. Atmos. Sci.* 2018, 1(34), 1–10. doi: 10.1038/s41612-018-0044-6.

171. Wyatt, M.G.; Peters, J.M. A secularly varying hemispheric climate-signal propagation previously detected in instrumental and proxy data not detected in CMIP3 data base. *SpringerPlus* 2012, 1, 68. http://springerplus.com/content/1/1/68.

172. Kravtsov, S. Pronounced differences between observed and CMIP5-simulated multi-decadal climate variability in the twentieth century. *Geophys. Res. Lett.* 2017, 44, 5749–5757. doi: 10.1002/2017GL074016.

173. Kravtsov, S.; Wyatt, M.G.; Curry, J.A.; Tsonis, A.A. Two contrasting views of multidecadal climate variability in the twentieth century. *Geophys. Res. Lett.* 2015, 41(19), 6881–6888. doi: 10.1002/2014GL061416.

Bibliography

Burroughs, W.J. *Weather Cycles: Real or Imaginary*, 2nd Ed.; Cambridge University Press: Cambridge, UK, 2003, 1–317, ISBN 0 521 52822 4.

Frolov, I.E.; Gudkovich, A.M.; Karklin, B.P.; Kvalev, E.G.; Smolyanitsky, V.M. Arctic and Antarctic Research Institute (AARI). In *Climate Change in Eurasian Arctic Shelf Seas: Centennial Ice Cover Observations*; Philppe Blondel, C., Geology, F.G.S., Eds.; SpringerPraxis Books in Geophysical Sciences, Praxis Publishing: St. Petersburg and Russia Chichester, UK, 2009, 1–165, ISBN 978-3-540-85875-1. doi: 10.1007/978-3-540-85875-1.

Klyashtorin, L.B.; Lyubushin, A.A. *Government of the Russian Federation; State Committee for Fisheries of the Russian Federation; Federal State Unitary Enterprise; Russian Federal Research Institute of Fisheries and Oceanography; Moscow; Cyclic Climate Changes and Fish Productivity*; Dr. Gary D. Sharp Editor of English version of book; Center for Climate/Ocean Resources Study; VNIRO Publishing: Salinas, CA: Moscow, 2007, 1–223, ISBN 978-5-85382-339-6.

Mann, K.H.; Lazier, J.R.N. *Department of Fisheries and Oceans; Bedford Institute of Oceanography; Dartmouth, Nova Scotia; Canada. Dynamics of Marine Ecosystems; Biological-Physical Interactions in the Oceans*, 3rd Ed.; Blackwell Publishing; Malden, MA, Oxford, UK, and Carlton, VIC, Australia, 2006, 1–495. doi: 10.1002/9781118687901.

Pikovsky, A.; Rosenblum, M.; Kurths, J. *Synchronization: A Universal Concept in Nonlinear Sciences*; Cambridge University Press: New York, 2001, ISBN 978-0-521-59285-7, online (2010) ISBN 978 0511 755 743, doi: 10.1017/CBO9780511755743.

10

Dew Point Temperature

John S. Roberts
Rich Products Corporation

Introduction

Dew point temperature is an aspect of moist air below which condensation of water vapor present in the air begins to occur. At the dew point temperature, the relative humidity of moist air is 100%, and further cooling of the air initiates water vapor condensing due to losing the energy it gained during evaporation. Some common examples of dew formation as a result of moist air being cooled below its dew point temperature are observed on windowpanes, on grass in the evening, on a glass containing a cold beverage, on cooling pipes, and morning fog. When the dew point is below freezing temperature, 0°C, sublimation occurs where the water vapor converts directly to frost. When this condition applies, the term frost point temperature is used. The objectives of this entry are to further explain dew point temperature and related concepts and to provide and compare several methods in which dew point temperature is determined.

Definition and Concepts

Dew point temperature is the temperature at which moist air must be cooled at constant pressure and constant humidity ratio for the partial pressure of water vapor in the air to equal the equilibrium water vapor pressure at that same temperature and pressure.[1,2] Some definitions state that it is the temperature where moist air must be cooled under constant pressure and humidity ratio to reach saturation.[3,4] The term saturation can be misleading because it indicates that the air cannot accept more water vapor molecules physically, when in fact it can if temperature increases. Several published texts have stated concerns in using saturation in terms of moist air.[1,5,6] This misunderstanding of saturation of air has led to the misconception of how much water vapor air can "hold." The limitation of water vapor in air is not due to a lack of space in the air or due to any binding of water vapor to oxygen or nitrogen gas molecules, which does not occur. Rather, the limitation of water vapor in air is solely due to the amount of available thermal energy to maintain water in the vapor state in the air.[1] This is shown in the relationship of the equilibrium (saturation) water vapor pressure with temperature. As temperature increases, the equilibrium water vapor pressure increases. In maintaining the use of saturation with respect to air, several texts have defined saturated air as an equilibrium between water vapor in the air and the condensed water phase at the existing temperature and pressure.[3,5] This concept of saturation of air is with respect to equilibrium and should be kept in mind when studying water vapor in air. In addition, moist air is a term typically used when describing the amount of water vapor in air between complete dryness and complete humidification.[3]

The composition of moist air can be described by the humidity ratio, or specific humidity, and the relative humidity. The specific humidity, W, is the ratio of the mass of water vapor, m_v, to the mass of the dry air, m_a:

$$W = \frac{m_v}{m_a} \tag{10.1}$$

The humidity ratio can also be expressed in terms of partial pressures:

$$W = \frac{m_v}{m_a} = \frac{M_v p_v V/RT}{M_a p_a V/RT} = \frac{M_v p_v}{M_a p_a} \tag{10.2}$$

where M is molecular weight, V is volume, p is partial pressure, R is gas constant, T is temperature, subscript v represents water vapor, and subscript a represents dry air. The ratio of the molecular weight of water vapor to the molecular weight of dry air is approximately 0.62198. The vapor pressure of air, p_a, can be expressed as $p - p_v$ from the relationship $p = p_a + p_v$, where p is the total barometric pressure of the moist air. Thus, the equation for humidity ratio can be reduced to

$$W = 0.62198 \frac{p_v}{p - p_v} \tag{10.3}$$

Dew point temperature, T_d, is defined mathematically as the solution T_d, (p, W) of the equation:

$$W_s = (p, T_d) = W \tag{10.4}$$

W_s is the humidity ratio of air at saturation (equilibrium) and can be expressed similar to Equation 10.1:

$$W_s = 0.62198 \frac{p_{vs}}{p - p_{vs}} \tag{10.5}$$

where p_{vs} is the saturation vapor pressure at the dew point temperature. Equation 10.5 reduces to

$$p_{vs}(T_d) = p_v = \frac{pW}{0.62198 + W} \tag{10.6}$$

An important aspect either in determining the dew point or as moist air approaches the dew point during cooling is that both the pressure and the composition of moist air (humidity ratio or absolute humidity) remain constant while the relative humidity increases. The relative humidity is defined as the mole fraction of water vapor in moist air, n_v, to the mole fraction of equilibrium water vapor in air, n_{vs}, at the same temperature and pressure. As partial vapor pressure, p_v, equals $n_v p$ and saturation vapor pressure, p, equals $n_{vs} p$, the relative humidity can be expressed in terms of pressure:

$$\varphi = \frac{p_v}{p_s} \bigg)_{T,p} \tag{10.7}$$

Expressed in words, the relative humidity is the ratio of the amount of moisture in the air to the equilibrium moisture in air at the same temperature and pressure. When the pressure and humidity ratio remain constant, the relative humidity increases as temperature decreases. So as the air cools, the partial pressure of the water vapor remains constant (constant moisture content in the air) and the equilibrium vapor pressure of water decreases as the temperature decreases. Eventually, the air temperature would have decreased such that the equilibrium vapor pressure equals the partial vapor pressure. Thus, as the air cools, the numerator remains constant while the denominator decreases until unity is reached in Equation 10.7. In the case where pressure remains constant, the relative humidity would be a function of both temperature and humidity ratio. As the dew point temperature is solely a function of the humidity ratio (at constant pressure), the dew point temperature is a better indicator of the amount of moisture in the air than relative humidity.[1,5,6]

Determination of Dew Point Temperature

Dew point temperature can be determined by the following means when other properties of moist air are known: from a psychrometric chart, a table on thermodynamic properties of water at saturation, equations, or physical measurements using hygrometers. This section describes how each method is used to determine dew point temperature.

Psychrometric Chart

The easiest means to determine dew point temperature is from a psychrometric chart. Figure 10.1 is a reduced version of the psychrometric chart showing only the dry bulb temperature, humidity ratio, and the saturation (equilibrium) curve. The condition of the moist air can be located on the chart by knowing any two of the following properties: dry bulb temperature, humidity ratio or relative humidity, and wet bulb temperature. Once the state of the moist air has been established on the chart, the dew point temperature of this moist air can be determined by following along constant humidity ratio to the saturation curve. The temperature at this intersection is the dew point temperature.

Tables and Equations

When a psychrometric chart is not available, the dew point temperature can be determined by tables or by equations when the dry bulb temperature is given along with either the relative humidity or the humidity ratio. The state of moist air at the dew point temperature is saturated, 100% relative humidity; thus the partial vapor pressure equals the saturation vapor pressure. The corresponding temperature is the dew point temperature from the saturation vapor pressure–temperature relationship, which can be determined either from tables or from equations. If humidity ratio is given, as expressed in Equation 10.3, the partial pressure of water vapor is equal to the saturation vapor pressure of the moist air and can be directly calculated from Equation 10.6. This direct determination is valid because the humidity ratio remains constant when the dew point temperature is determined, as illustrated in Figure 10.1. When relative humidity is given and is below 100%, the process of determining the partial pressure of the water is more involved because the relative humidity of the moist air is less than the relative humidity of the moist air at the dew point temperature. Once the partial vapor pressure is determined, it can be set equal to the saturation vapor pressure because that is the condition of the moist air at the dew point temperature.

As shown in the expression for relative humidity in Equation 10.7, the saturation vapor pressure–temperature relationship must be known to determine the partial pressure of water vapor. This relationship is also needed to calculate the dew point temperature, as shown in Equation 10.4. The saturation vapor pressure can be found in a table on thermodynamic properties of water at saturation or can be calculated from equations. A table on thermodynamic properties of water at saturation consists of temperature and corresponding saturation water vapor pressure, specific volume, enthalpy, and entropy.[3] Several equations are widely used to express the saturation vapor pressure of water with respect to temperature. One set of equations widely used is given below.[3]

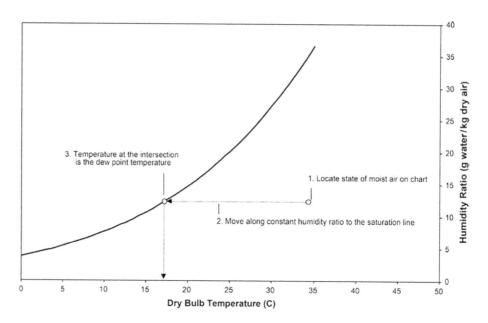

FIGURE 10.1   Dew point temperature determination using a psychrometric chart.

For the temperature range –100°C–0°C:

$$\ln(p_{vs}) = (-5674.5329/T) + (6.3925247)$$

$$+ (-0.009677843)T + (6.22115701 \times 10^{-7})T^2$$

$$+ (2.0747825 \times 10^{-9})T^3 + (-9.484204 \times 10^{-13})T^4$$

$$+ (4.1635019)\ln(T) \tag{10.8}$$

For the temperature range 0–200°C:

$$\ln(p_{vs}) = (-5.8002206/T) + (1.3914993)$$

$$+ (-0.048640239)T + (4.1764768 \times 10^{-5})T^2$$

$$+ (1.4452093 \times 10^{-8})T^3 + (6.5459673)\ln(T) \tag{10.9}$$

where p_{vs} is the saturation vapor pressure, Pa; and T is absolute temperature, K. Another widely used equation to calculate saturation vapor pressure as a function of temperature is the Magnus–Tetens equation:[7]

$$p_{vs} = 0.6105 exp\left[\frac{aT}{b+T}\right] \tag{10.10}$$

where $a = 17.27$, $b = 237.7$, and T is in °C. This equation is an approximation of the variation of saturation vapor pressure with temperature, but its advantage is in its simplicity. The saturation vapor pressure of water can then be used to determine the corresponding dew point temperature either by tables or through equations.

The following equations are expressed to solve for dew point temperature. Equations 10.11 and 10.12 have been commonly used to calculate dew point temperature from saturated vapor pressure.[3]

For the frost point temperatures below 0°C:

$$T_d = 6.09 + (12.608)\ln(p_v) + 0.4959(\ln(p_v))^2 \tag{10.11}$$

and for dew point temperature in the range 0°C to 93°C:

$$T_d = 6.54 + (14.526)\ln(p_v) + 0.7389(\ln(p_v))^2$$
$$+ 0.09486(\ln(p_v))^3 + 0.4569(p_v)^{0.1984} \tag{10.12}$$

where T_d is in °C and p_v is in kPa.

The dew point temperature is often calculated from the Magnus–Tetens equation, Equation 10.10, which can be expressed to solve for dew point temperature as shown below:

$$T_d = \frac{\ln\left(\dfrac{p_v}{0.6105}\right) \cdot b}{a - \ln\left(\dfrac{p_v}{0.6105}\right)} \tag{10.13}$$

A good overview of dew point temperature as it relates to relative humidity is presented along with various equations to illustrate the relationship, such as one based on the Clausius–Clapeyron equation[8]

$$\frac{d\rho_{vs}}{dT} = \frac{L\rho_{vs}}{R_v T^2} \tag{10.14}$$

which can be simplified to the following expression to calculate dew point temperature:

$$T_d = T_{db}\left(1 - \frac{T_{db}\ln(\phi)}{L/R_v}\right) \tag{10.15}$$

where temperature is in Kelvin, L is the enthalpy of vaporization (2.501×10^6 J/kg at $T_{db} = 273.15$ K and 2.257×10^6 J/kg at $T_{db} = 373.15$ K) and R_v is the gas constant for water vapor (461.5 J/K kg).

A new set of empirical equations were introduced to calculate dew point.[9]

For −20 < T < 10°C:

$$T_d = 193.03 + 28.633 p_v^{0.1609}$$

For 10 < T < 180°C $\tag{10.16}$

$$T_d = \frac{3816.44}{23.197 - \ln(p_v)} + 46.13$$

The dew point temperature calculated in Equations 10.15 and 10.16 are in Kelvin.

Table 10.1 shows the comparison of dew point temperatures determined from the different methods described in the preceding text with dry bulb temperature and relative humidity known. Equations 10.12–10.14 produced very similar dew point temperatures and were close to the dew point temperature determined from a table on thermodynamic properties of water at saturation.[3] The difficulty in determining dew point temperature from the table is that the saturation vapor pressures are given in

TABLE 10.1 Comparison of Dew Point Temperatures, T_d, Given Dry Bulb Temperature, T_{db}, and Relative Humidity, φ, Using a Table, a Psychrometric Chart, and Various Equations

T_{db} (°C)	φ(×100) (%)	$P_{vs}\,\varphi$ (Table) (kPa)	p_v ($\varphi = p_{vs}$) (kPa)	T_d (Table) T (°C)	T_d (Psych. Chart) (°C)	T_d (Equation 10.12) (°C)	T_d (>Equation 10.13) (°C)	T_d (Equation 10.14) (°C)	T_d (Equation 10.16) (°C)
9	70	1.1481	0.80367	3.82	4	3.84	3.85	3.86	4.17
29	20	4.0083	0.80166	3.79	4	3.80	3.81	3.46	4.13
19	40	2.1978	0.87912	5.11	5	5.14	5.13	5.00	5.43
19.5	70	2.2683	1.5878	14.43	14	13.93	13.93	13.87	14.12
34	30	5.3239	1.5972	14.48	14	14.02	14.02	13.75	14.21
29.5	50	4.1272	2.0636	17.99	18	18.02	18.04	17.90	18.12
46	29	10.098	2.9283	23.68	24	23.68	23.74	23.37	23.82
35	80	5.6278	4.5022	31.02	31	31.01	31.10	31.00	31.12

whole-number dry bulb temperatures, and therefore interpolation is needed when the saturation vapor pressure lies between two dry bulb temperature entries. The psychrometric chart is useful in assessing approximate dew point temperatures, usually in whole numbers, with reasonable accuracy. But as illustrated in Figure 10.1, the chart's advantage is in the ease of determining the dew point temperature. The results of using Equation 10.16, which was recently offered as an alternative empirical equation in calculating dew point temperature,[9] is also shown in Table 10.1. The dew point temperatures were predicted slightly higher in all calculations than the other methods, more at lower dry bulb temperature conditions.

Methods of Measurement

The gravimetric hygrometer is regarded as a primary standard of measurement of moisture in air because it determines the humidity ratio directly.[3,4] As shown by the psychrometric chart in Figure 10.1, the humidity ratio directly correlates with the dew point temperature. This standard method of measurement consists of passing moist air through a series of three U-tubes filled with desiccant. The amount of water vapor absorbed and the volume of dry gas can be measured precisely under controlled temperature and pressure conditions. This instrument is not used in industrial applications because the equipment is cumbersome, requires special care, and measurement times are long (up to 30 hours for low dew point temperature measurements).[3,4] There are several more practical devices to measure dew point temperature, and the three most commonly used in industry and research are capacitive (aluminum oxide) sensors, saturated salt (lithium chloride) sensors, and optical surface condensation (chilled mirror [CM]) hygrometers.[10,11] These instruments vary in the method of measurement, range of measurement, level of accuracy, and cost. The choice of measurement depends on the environment it will be used and the level of accuracy required. The following sections describe the operation of these devices along with their advantages and disadvantages.

Capacitive (Aluminum Oxide) Sensor

Aluminum oxide sensors are the most widely used, especially in heating, ventilating, and air conditioning (HVAC) applications, where cost is more critical than accuracy. A capacitor consists of two electrodes with a dielectric in between. These capacitive sensors consist of an aluminum substrate, followed by a layer of a hygroscopic material, aluminum oxide, and then followed by a coating of gold film.[10] The aluminum substrate and gold film are the electrodes of the capacitor, and the aluminum oxide acts as the dielectric. Some recent capacitive sensors have a polymer as their dielectric.

Water vapor from moist air absorbs into aluminum oxide, and the amount of absorbed water vapor directly correlates with the capacitance of the sensor. The instrument determines the dew point based on a series of correlations, first on the capacitance-water vapor relationship and then on the temperature-vapor pressure relationship.

Capacitive sensors have the advantage of measuring dew points in extreme temperature conditions. These sensors can measure low frost points, down to $-100°C$ and are able to withstand very high temperatures as found in pressurized heating applications. Capacitive sensors have good sensitivity at low humidity levels; so these sensors are frequently used in petrochemical and power industries.[11] Capacitive sensors are small and portable; so they are often mounted in ducts or walls in industrial processing stream applications. These sensors also have a fast response time, thus making them suitable in applications where a quick read is critical in saving time and product, such as spot testing of gas cylinders.[12] However, capacitive sensors can become saturated in environments where the relative humidity is above 85%.[11] Also, these sensors are based on a secondary measurement, capacitance; so periodic recalibration is necessary.

Saturated Salt (Lithium Chloride) Sensor

Lithium chloride-saturated salt sensors consist of a thin wall metal tube covered with a glass sleeve impregnated with saturated lithium chloride solution. Lithium chloride salt solution is the most desired because at saturation it has a low equilibrium relative humidity of 11% and maintains this humidity over a wide range of temperatures, 11.2% at 0°C and 10.8% at 70°C.[4] This sensor bobbin is wrapped spirally with two wire electrodes, or bifilars, which are connected to an alternating current voltage source.[3] The ionic solution completes the electrical circuit between the wires. There is a direct relationship between the amount of condensation and the amount of current drawn between the two electrodes.

The operation of this sensor is best described by explaining how the dew point is determined from the following two conditions: initial moist air with a partial vapor pressure greater than the saturated salt, and initial moist air with a partial vapor pressure less than the saturated salt. In the first condition, when the initial moist air has a partial vapor pressure greater than that of the saturated salt solution, water vapor from the air will condense onto the sensor due to the vapor pressure gradient. As more water vapor condenses, the salt solution becomes more conductive. The current creates resistive heating, which increases the vapor pressure of the salt solution. The rate of condensation and rate of heating decrease until equilibrium are reached between the partial vapor pressure of the solution and the surrounding ambient moist air. A temperature probe is embedded into the sensor to measure the solution temperature needed to achieve this equilibrium vapor pressure. At this increased sensor temperature, the increased partial vapor pressure of the solution is proportional to the saturated vapor pressure of water at that same temperature to maintain the 11% equilibrium relative humidity of the lithium chloride-saturated solution. So, with the temperature reading, the saturated water vapor pressure can be determined. The partial vapor pressure can then be determined from this saturated water vapor pressure using the correlation shown in Equation 10.7 and based on 11% equilibrium relative humidity of the saturated salt solution. This partial pressure of the solution is the same partial vapor pressure of the ambient moist air. As dew point temperature directly correlates with vapor pressure, the dew point can be determined based on the temperature–vapor pressure relationship. The hygrometer internally calculates all of these relationships to determine dew point. If the initial moist air has a partial vapor pressure less than the saturated salt solution, evaporation would occur until the sensor is at equilibrium with the air. As the solution loses some water to its environment, less current is drawn, and the temperature sensor would record the evaporative cooling.

The limits of this sensor are that moist air conditions below 11% relative humidity cannot be measured because this is the equilibrium vapor pressure of saturated lithium chloride, and dew point above 70°C cannot be measured. As equilibrium has to be reached, the measurement times are longer than with most sensors. Also, exposure to liquid water can wash away the salt crystals in the sensor. These low-cost sensors are typically used in applications where an inexpensive, slow, and moderate accuracy is needed, such as in refrigeration controls, dryers, dehumidifiers, airline monitoring, and pill coaters.[11]

Optical Surface Condensation (CM) Hygrometer

The most direct method to determine dew point temperature is the CM hygrometer and is the most widely used for precise dew points.[4,10] This method of measurement involves a metallic mirrored surface, a mechanism to cool this mirrored surface, a light-emitting device (LED), and a photodetector. The principle of measurement is based on the reflectance of light from the LED by the mirror. The gas (i.e., moist air) is passed over the mirror, and when the temperature of the mirror is above the dew point of the moist air, the photodetector reads the direct reflectance of the light. However, when the mirror is cooled to the dew point, water vapor begins to condense on the mirror surface. The light is then scattered, and the decrease of reflected light intensity is picked up by the detector. The key factors in accurate measurement of the CM hygrometers are controlling the rate of cooling of the mirror, thus controlling the amount of condensing water vapor on the mirror surface, preventing contaminants on the mirror surface, and precise reading of the temperature of the mirror.

The mirrored surface is made of a good conductor, normally of copper or silver, and coated with an inert metal to prevent tarnishing and oxidation. As water vapor can condense at or below the dew point temperature, precise control of the mirror temperature is important not to allow significant dew formation. If significant dew has formed, the humidity ratio will reduce slightly, and the reading will not be the precise dew point temperature of the initial air condition. Modern CM hygrometers cool the mirror using a solid-state heating element, such as a Peltier device, to control cooling of the mirror. High-quality CM hygrometers have a feedback control mechanism where the photo-detector communicates with the Peltier device to maintain the mirror temperature such that the dew drop level is at a minimum and remains constant; thus, the rates of evaporation and condensation are equal.

Contaminants, either insoluble or water soluble, on the mirror surface can scatter light and give false readings of dew formation. When insoluble contamination occurs, the mirror surface has to be cleaned manually. Water-soluble contamination is normally in the form of dissolved salts, and there are several methods to correct this type of contamination. One method developed by General Eastern is known as the Programmable Automatic Contaminant Error Reduction (PACER) system.[10–12] This method chills the mirror for an extended time, so excessive amounts of dew form to dissolve the salts on the surface. Then the system rapidly heats the mirror to evaporate the water. This results in localized regions of salt crystals on the mirror surface while the majority of the surface is clean. The other method that optical CM hygrometers use to correct for contamination is the cycling chilled mirror (CCM) hygrometer. In this method of measurement, the mirror is chilled to the dew point for a short time (5% of the measurement time) and then heated above the dew point.[10] This process is repeated to give a reproducible reading of dew point. Having the mirror at dew point for a limited time reduces the chance of contamination. In both methods, eventually the surface will have to be manually cleaned.

Moisture on the mirror is at equilibrium with the water vapor in the air at one unique temperature; so the measurement of the mirror temperature is a fundamental and primary measurement of dew point temperature.[1,11] A precise temperature sensor, such as a platinum resistance thermometer, measures the mirror temperature at the dew point. CM hygrometers have a dew point measurement accuracy of ±0.2°C.

The advantages of CM hygrometers are in their accuracy, reliability, fast response time, ability for longer continuous and unattended operation due to its self-correcting capabilities, and ability to measure dew points ranging from −70°C to 95°C. However, these instruments can be expensive. Typical applications of the CM hygrometers are for humidity calibration standards, for critical environmental monitoring (i.e., clean rooms, pharmaceutical laboratories, areas where sensitive computers and electronics are located, and environmental chambers), and for monitoring extreme environmental conditions (i.e., heat-treating furnaces, engine test beds).[11]

References

1. Bohren, C.F.; Albrecht, B.A. *Atmospheric Thermodynamics*; Oxford University Press, Inc.: New York, 1998.
2. Mohen, M.J.; Shapiro, H.N. *Fundamentals of Engineering Thermodynamics*, 4th Ed.; John Wiley and Sons, Inc.: New York, 2000.
3. *ASHRAE Handbook Fundamentals*, SI Ed.; American Society of Heating, Refrigeration, and Air Conditioning Engineers, Inc.: Atlanta, GA, 1989.
4. Kuehn, T.H.; Ramsey, J.W.; Threlkeld, J.L. *Thermal Environmental Engineering*, 3rd Ed.; Prentice Hall: Saddle River, NJ, 1998.
5. Williams, J. *The Weather Book,* 2nd Ed.; Vintage Books: New York, 1997.
6. http://www.shorstmeyer.com/wxfaqs/humidity/humidity.html (accessed October 2009).
7. http://www.paroscientific.com/dewpoint.htm (accessed October 2009).
8. Lawrence, M.G. The relationship between relative humidity and the dew point temperature in moist air. Bull. Am. Meteorol. Soc. **2005**, *86* (2), 225–233.
9. Li, S.Q.; Gong, Z.X. Calculations of the state variables of moist air. In *Drying'92*; Mujumdar, A.J., Ed.; Elsevier Science Publishers B.V.: Amsterdam, the Netherlands, 1992.
10. Wiederhold, P.R. *Water Vapor Measurement. Methods and Instrumentation*; Marcel Dekker, Inc.: New York, 1997.
11. http://iceweb.com.au/Analyzer/humidity_sensors.html (accessed October 2009).
12. Larson, K. Humidity measurement update. Control **1990**, June, 49–59.

11

Estimating Arctic Sea-Ice Shortwave Albedo

Ying Qu
*Northeast Normal
University*

Introduction

Arctic sea-ice shortwave albedo is one of the critical parameters for characterizing the surface radiation and energy budget over the Northern Hemispheric cryosphere [1,2], which determines the fraction of surface-reflected shortwave solar radiations, as well as the amounts of heat allocations and exchanges among the atmosphere-ocean-snow/sea-ice system in the polar region [3]. Moreover, the regional and global energy exchange, material transport, and hydrological cycle can also be severely affected by the seasonal and annual variations of Arctic sea-ice albedo [4,5].

The Arctic sea-ice zone was profoundly affected by the global climate change in the past decades [6]. There is evidence that the extent, thickness, and albedo of Arctic sea ice were dramatically changed due to the global warming effect. A variety of observations have demonstrated that the extent of Arctic sea ice in summer decreases significantly [7–9], and the Arctic sea ice becomes thinner [10,11] and darker [12,13]. A decrease in Arctic shortwave albedo led to an increase in solar heat input to the melt ocean water, and further enhanced the heating effects of the Arctic zone by the sea-ice albedo feedback mechanism [14,15]. Thus, it is very important to monitor the spatial and temporal dynamics of Arctic sea-ice albedo under the context of global climate change.

Arctic Sea-Ice Albedo Datasets

The Arctic sea-ice albedo can be measured by various ground instruments, such as pyranometer onboard of ships, autonomous vehicles, and drifting buoys [16]. However, it is not sufficient to represent the spatiotemporal variations of Arctic sea-ice albedo due to their limited footprints [17]. Thus, it has great advantages to estimate the Arctic sea-ice shortwave albedo from satellite observations. In recent decades, a variety of surface albedo datasets were generated from different polar-orbit and geostationary satellite platforms [18], with temporal resolutions from daily to monthly and spatial resolutions from 10 m to 100 km. However, because the Arctic sea ice is rapidly changing with strong reflectance anisotropic properties (strong forward scattering effect), the traditional land surface albedo estimation methods cannot be directly applied for generating sea-ice albedo datasets [17,19]. For simplicity, several studies were derived from the Arctic sea-ice albedo products based on the Lambertian assumption.

TABLE 11.1   Current Available Broadband Albedo Datasets over the Arctic Sea-Ice Region

Name	Spatial Resolution	Temporal Resolution	Temporal Span	Spatial Coverage	Sensors
APP	5 km	Twice-daily	1981–2005	Arctic/Antarctic	AVHRR
APP-x	25 km	Monthly	1982–2004	Arctic/Antarctic	AVHRR
CM-SAF SAL	15 km	Weekly/monthly	2004–present 2009–present	Baseline area/Arctic	AVHRR/SEVIRI
CLARA SAL	0.25°	Pentad/monthly	1982–2015	Global/Arctic/Antarctic	AVHRR
GLASS	1 km/5 km	8 days	1982–2015	Global	AVHRR/MODIS

Source: Data adapted from Qu et al. (2016).

Key et al. [20] developed the Advanced Very High Resolution Radiometer (AVHRR) Polar Pathfinder (APP) albedo product and the APP-extended albedo product [21] from the AVHRR pathfinder data. Riihelä et al. [22] proposed an algorithm for generating the Satellite Application Facility on Climate Monitoring (CM-SAF) Surface Albedo (SAL) product from AVHRR and the Spinning Enhanced Visible and Infrared Imager (SEVIRI) data, and then applied it to generate a long-term global-covered CLoud, Albedo, and RAdiation (CLARA) [23,24] SAL dataset from AVHRR data. For considering the reflectance anisotropic effect of Arctic sea ice, Qu et al. [19] developed a direct-estimation algorithm for generating the Global LAnd Surface Satellite (GLASS) [25] phase-2 broadband surface albedo products from AVHRR and the Moderate Resolution Imaging Spectroradiometer (MODIS) data. Table 11.1 lists the currently available satellite remote sensing datasets. In addition, several algorithms were also developed for estimating Arctic sea-ice albedo, and planned to generate long-term sea-ice albedo products in the future. For example, Peng et al. [26] proposed a method for estimating Arctic sea-ice albedo from the Visible Infrared Imaging Radiometer Suite (VIIRS) data, and Pohl et al. [27] developed a narrow-to-broadband conversion method for estimating Arctic sea-ice albedo from the MEdium Resolution Imaging Spectrometer (MERIS) data.

Direct-Estimation Algorithm

In most of the currently available Arctic sea-ice albedo datasets, the reflectance anisotropic effect of snow and sea ice was ignored, and the surface albedo was estimated based on Lambertian assumption. For addressing this issue, Qu et al. [19] proposed a direct-estimation algorithm for estimating Arctic sea-ice shortwave albedo, and have applied it to generate a long-term, global land/ocean fully covered surface albedo dataset, which is called the GLASS phase-2 broadband albedo product (http://glass-product.bnu.edu.cn/introduction/abd.html).

As seen in Figure 11.1, the shortwave surface albedo of Arctic sea ice was estimated by the following steps: (i) A bidirectional reflectance distribution function (BRDF) database of Arctic sea ice was simulated with the asymptotic radiative transfer (ART) [28,29] and the three-component ocean water albedo (TCOWA) [30] models. In this procedure, a variety of snow/ice and ocean water parameters (snow grain size/density, soot concentrations, volume fraction and radius of brine pocket/air bubble, wind speed/direction, and chlorophyll concentrations) were used for representing different scenarios of Arctic sea ice. (ii) The top of atmosphere (TOA) reflectance of Arctic sea ice was simulated with the atmospheric radiative transfer model, and the corresponding broadband albedo was estimated by the angular integrating [31,32] and narrow-to-broadband conversions [33,34] procedures. (iii) The relationships between TOA reflectance and broadband surface albedo for different solar/view angles were built based on a multivariate linear regression method, which is also called the angular bin regression method (Equations 11.1 and 11.2). The regression coefficients for different solar/view angles were stored in the direct-estimation coefficient look-up table (LUT). (iv) When there are available satellite observations, the shortwave white- and black-sky albedo can be directly estimated based on the LUT

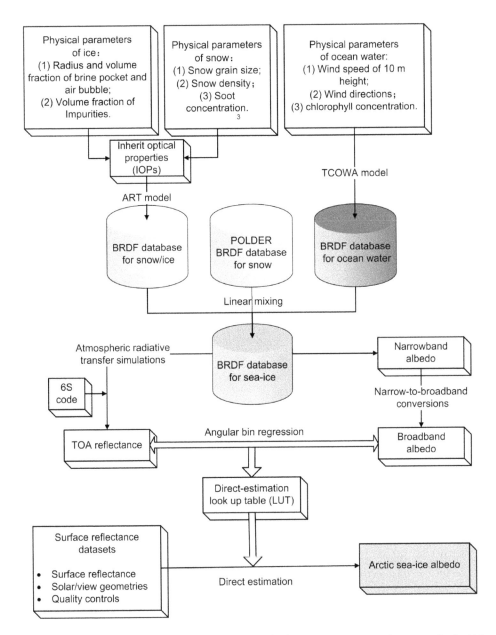

FIGURE 11.1 Flowchart of the direct-estimation algorithm for deriving Arctic sea-ice shortwave albedo. (Adapted from Qu et al., 2016.)

and corresponding solar/view geometries. Thus, the shortwave Arctic sea-ice albedo with high spatial and temporal resolution can be derived from satellite observations, such as MODIS, AVHRR, and VIIRS data.

In the direct-estimation algorithm, the white- and black-sky albedo can be estimated by the following equations [19,35]:

$$\alpha_{\text{wsa}} = m_0\left(\theta_s,\theta_v,\varphi\right) + \sum_{i=1}^{n} m_i\left(\theta_s,\theta_v,\varphi\right)\rho_i^{\text{TOA}}\left(\theta_s,\theta_v,\varphi\right) \tag{11.1}$$

$$\alpha_{bsa}\left(\theta_{bsa}\right)=n_0\left(\theta_{bsa},\theta_s,\theta_v,\varphi\right)+\sum_{i=1}^{n}n_i\left(\theta_{bsa},\theta_s,\theta_v,\varphi\right)\rho_i^{TOA}\left(\theta_s,\theta_v,\varphi\right),\qquad(11.2)$$

where α_{wsa} denotes the shortwave white-sky albedo; $\alpha_{bsa}\left(\theta_{bsa}\right)$ denotes the shortwave black-sky albedo at solar zenith angle of θ_{bsa}; $\rho_i^{TOA}\left(\theta_s,\theta_v,\varphi\right)$ is the TOA reflectance with a solar zenith angle of θ_s, view zenith angle of θ_v, and relative azimuth angle of φ; $m_i\left(\theta_s,\theta_v,\varphi\right)$ and $n_i\left(\theta_{bsa},\theta_s,\theta_v,\varphi\right)$ are the coefficients of the multivariate linear regressions for different solar/view geometries.

The blue-sky albedo (the in situ measured albedo) can be expressed as the linear weighted combination of the black- and white-sky albedo [36], as follows:

$$\alpha_{blue\text{-}sky}=\alpha_{bsa}\cdot\left(1-D\right)+\alpha_{wsa}\cdot D,\qquad(11.3)$$

where $\alpha_{blue\text{-}sky}$ is the blue-sky albedo; α_{bsa} and α_{wsa} are the black-sky albedo and white-sky albedo, respectively; D stands for the fraction of diffuse skylight.

An example of the Arctic sea-ice shortwave albedo derived from the direct-estimation algorithm is shown in Figure 11.2. In this study, the shortwave white-sky albedo was first derived from MODIS data and then was averaged over 5 days from day 166 to day 170 in 2007. From this figure, we can see that the spatial distribution of Arctic sea-ice shortwave albedo can be well presented by the direct-estimation algorithm. The albedo of the sea ice in the edge region is much lower than that in the polar region, and the albedo of snow-covered multiyear sea-ice region near the Greenland ice sheet is relatively higher than that of the first-year sea-ice region near the East Siberian, Beaufort, and Laptev seas.

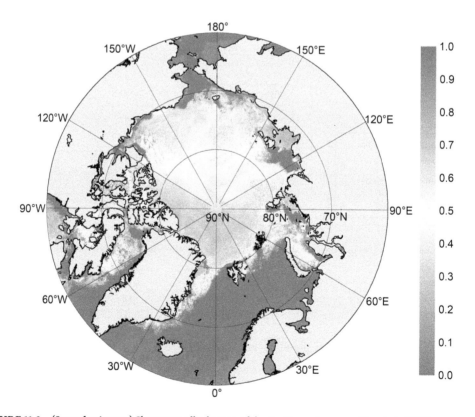

FIGURE 11.2 **(See color insert.)** Shortwave albedo map of the Arctic sea-ice zone. (Qu et al., 2016.)

Figure 11.3 shows the Arctic sea-ice albedo datasets derived by the direct-estimation algorithm and the CLARA SAL product around the Banks, Prince Patrick, and Melville islands, and Kara Sea. The comparison results show the spatial resolution of the dataset derived by the direct-estimation algorithm (1 km) is much higher than that of CLARA SAL product (25 km), which has potential advantages for providing much more accurate information for global climate change studies.

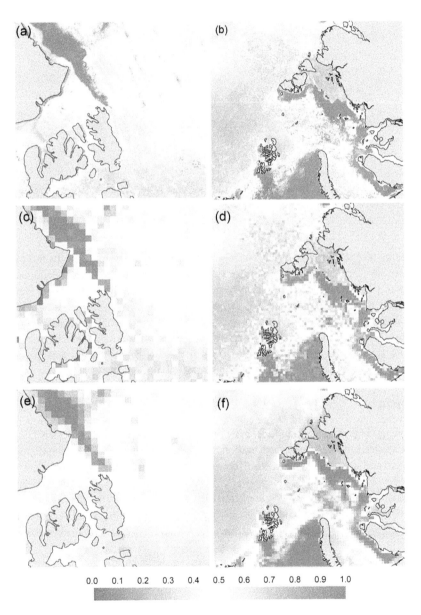

FIGURE 11.3 Comparison between the Arctic sea-ice albedo maps derived by the direct-estimation algorithm and CLARA SAL product. Shortwave surface albedo (white-sky albedo) averaged over a 5-day period from day 166 to day 170 in 2007, where (a) and (b) are the albedo maps derived by the direct-estimation algorithm (1 km); (c) and (d) are the direct-estimated albedo maps resampled to 25 km; (e) and (f) are the albedo maps derived by the CLARA SAL product (25 km); (a), (c), and (e) are the albedo maps around the Banks, Prince Patrick, and Melville islands; (b), (d), and (f) are the albedo maps around the Kara Sea. (Qu et al., 2016.)

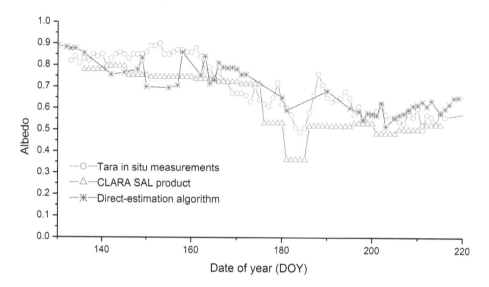

FIGURE 11.4 Comparison of the direct-estimated albedo, CLARA SAL product, and pyranometer measurements of the Tara expedition. (Qu et al., 2016.)

The estimations of Arctic sea-ice albedo were also validated and compared with the pyranometer measurements of the Tara Oceans Polar Circle expedition [16]. The comparison results of the direct-estimated albedo, CLARA SAL product, and pyranometer measurements of the Tara expedition are shown in Figure 11.4. If the measurements of pyranometer were taken as reference values, the albedo derived by the direct-estimation algorithm has an R^2 value of 0.67 and a root mean square error (RMSE) of 0.068, whereas the CLARA SAL product has an R^2 value of 0.81 and an RMSE of 0.097. Compared with the CLARA SAL product, the albedo derived by the direct-estimated algorithm has a relatively lower estimation of RMSE and much higher spatial and temporal resolution.

Conclusions

As a key parameter of surface energy budget in the polar region, a variety of surface albedo datasets over the Arctic sea-ice zone have been derived from satellite observations. In this chapter, the currently available datasets and a recently proposed direct-estimation algorithm for generating Arctic sea-ice shortwave albedo were discussed. The comparison and validation results show that the Arctic sea-ice shortwave albedos derived by the direct-estimation algorithm are more accurate than those estimated based on Lambertian assumption, and can be applied for generating long-term Arctic sea-ice albedo datasets with high spatial and temporal resolution.

Acknowledgments

This research was financially supported by the National Key Research and Development Project (2016YFA0602301), the National Natural Science Foundation of China (41971287, 41601349), the Jilin Province's Science and Technology Development Program (20180520220JH), and the Fundamental Research Funds for the Central Universities (2412019FZ003).

References

1. Perovich D; Grenfell T; Light B, Hobbs P. Seasonal Evolution of the Albedo of Multiyear Arctic Sea Ice. *Journal of Geophysical Research* 2002, 107, 8044.
2. Perovich DK; Polashenski C. Albedo Evolution of Seasonal Arctic Sea Ice. *Geophysical Research Letters* 2012, 39, L08501.
3. Hudson SR; Granskog MA; Sundfjord A; Randelhoff A; Renner AHH; Divine DV. Energy Budget of First-Year Arctic Sea Ice in Advanced Stages of Melt. *Geophysical Research Letters* 2013, 40, 2679–2683.
4. Cohen J; Screen JA; Furtado JC; Barlow M; Whittleston D; Coumou D; Francis J; Dethloff K; Entekhabi D; Overland J.; Jones, J. Recent Arctic Amplification and Extreme Mid-Latitude Weather. *Nature Geoscience* 2014, 7, 627–637.
5. Sévellec F; Fedorov AV; Liu W. Arctic Sea-Ice Decline Weakens the Atlantic Meridional Overturning Circulation. *Nature Climate Change* 2017, 7, 604–610.
6. Johannessen OM; Bengtsson L; Miles MW; Kuzmina SI; Semenov VA; Alekseev GV; Nagurnyi AP; Zakharov VF; Bobylev LP; Pettersson LH. Arctic Climate Change: Observed and Modelled Temperature and Sea-Ice Variability. *Tellus A* 2004, 56, 328–341.
7. Serreze MC; Holland MM; Stroeve J. Perspectives on the Arctic's Shrinking Sea-Ice Cover. *Science* 2007, 315, 1533–1536.
8. Stroeve J; Holland MM; Meier W; Scambos T, Serreze M. Arctic Sea Ice Decline: Faster Than Forecast. *Geophysical Research Letters* 2007, 34, L09501.
9. Stroeve J; Serreze MC; Holland MM; Kay JE; Malanik J; Barrett AP. The Arctic's Rapidly Shrinking Sea Ice Cover: A Research Synthesis. *Climatic Change* 2012, 110, 1005–1027.
10. Lindsay R; Schweiger A. Arctic Sea Ice Thickness Loss Determined Using Subsurface, Aircraft, and Satellite Observations *The Cryosphere* 2015, 9, 269–283.
11. Renner AHH; Gerland S; Haas C; Spreen G; Beckers JF; Hansen E; Nicolaus M; Goodwin H. Evidence of Arctic Sea Ice Thinning from Direct Observations. *Geophysical Research Letters* 2014, 41, 5029–5036.
12. Pistone K; Eisenman I; Ramanathan V. Observational Determination of Albedo Decrease Caused by Vanishing Arctic Sea Ice. *Proceedings of the National Academy of Sciences of the United States of America* 2014, 111, 3322–3326.
13. Riihelä A; Manninen T; Laine V. Observed Changes in the Albedo of the Arctic Sea-Ice Zone for the Period 1982–2009. *Nature Climate Change* 2013, 3, 895–898.
14. Curry JA; Schramm JL; Ebert EE. Sea Ice-Albedo Climate Feedback Mechanism. *Journal of Climate* 1995, 8, 240–247.
15. Moon W; Wettlaufer JS. On the Nature of the Sea Ice Albedo Feedback in Simple Models. *Journal of Geophysical Research: Oceans* 2014, 119, 5555–5562.
16. Vihma T; Jaagus J; Jakobson E; Palo T. Meteorological Conditions in the Arctic Ocean in Spring and Summer 2007 as Recorded on the Drifting Ice Station Tara. *Geophysical Research Letters* 2008, 35, L18706.
17. Qu Y. Sea Surface Albedo. In: Liang S editor. *Sea Surface Albedo. Comprehensive Remote Sensing*; Cambridge, MA: Elsevier; 2017.
18. Qu Y; Liang S; Liu Q; He T; Liu S, Li X. Mapping Surface Broadband Albedo from Satellite Observations: A Review of Literatures on Algorithms and Products. *Remote Sensing* 2015, 7, 990–1020.
19. Qu Y; Liang S; Liu Q; Li X; Feng Y; Liu S. Estimating Shortwave Arctic Sea-Ice Albedo from Modis Data. *Remote Sensing of Environment* 2016, 186, 32–46.
20. Key JR; Wang X; Stoeve JC; Fowler C. Estimating the Cloudy-Sky Albedo of Sea Ice and Snow from Space. *Journal of Geophysical Research: Atmospheres* 2001, 106, 12489–12497.

21. Wang X, Key JR. Arctic Surface, Cloud, and Radiation Properties Based on the AVHRR Polar Pathfinder Dataset. Part I: Spatial and Temporal Characteristics. *Journal of Climate* 2005, 18, 2558–2574.

22. Riihelä A; Laine V; Manninen T; Palo T; Vihma T. Validation of the Climate-SAF Surface Broadband Albedo Product: Comparisons with in Situ Observations over Greenland and the Ice-Covered Arctic Ocean. *Remote Sensing of Environment* 2010, 114, 2779–2790.

23. Riihelä A; Manninen T; Laine V; Andersson K; Kaspar F. Clara-Sal: A Global 28 Yr Timeseries of Earth's Black-Sky Surface Albedo. *Atmospheric Chemistry and Physics* 2013, 13, 3743–3762.

24. Karlsson KG; Anttila K; Trentmann J; Stengel M; Meirink JF; Devasthale A; Hanschmann T; Kothe S; Jääskeläinen E; Sedlar J; Benas N. Clara-A2: The Second Edition of the Cm SAF Cloud and Radiation Data Record from 34 Years of Global AVHRR Data. *Atmospheric Chemistry and Physics* 2017, 17, 5809–5828.

25. Liang S; Zhao X; Yuan W; Liu S; Cheng X; Xiao Z; Zhang X; Liu Q; Cheng J; Tang H; Qu Y. A Long-Term Global Land Surface Satellite (Glass) Dataset for Environmental Studies. *International Journal of Digital Earth* 2013, 6, suppl, 69–95.

26. Peng J; Yu Y; Yu P; Liang S. The VIIRS Sea-Ice Albedo Product Generation and Preliminary Validation. *Remote Sensing* 2018, 10, 1826–1849.

27. Pohl C; Istomina L; Tietsche S; Jäkel E; Stapf J; Spreen G; Heygster G. Broad Band Albedo of Arctic Sea Ice from MERIS Optical Data. *The Cryosphere Discussion* 2019, doi: 10.5194/tc-2019-62.

28. Kokhanovsky AA; Aoki T; Hachikubo A; Hori M; Zege EP. Reflective Properties of Natural Snow: Approximate Asymptotic Theory Versus in Situ Measurements. *IEEE Transactions on Geoscience and Remote Sensing* 2005, 43, 1529–1535.

29. Kokhanovsky AA, Zege EP. Scattering Optics of Snow. *Applied Optics* 2004, 43, 1589–1602.

30. Feng Y; Liu Q; Qu Y; Liang S. Estimation of the Ocean Water Albedo from Remote Sensing and Meteorological Reanalysis Data. *IEEE Transactions on Geoscience and Remote Sensing* 2016, 54, 850–868.

31. Lucht W; Schaaf C; Strahler A. An Algorithm for the Retrieval of Albedo from Space Using Semiempirical BRDF Models. *IEEE Transactions on Geoscience and Remote Sensing* 2000, 38, 977–998.

32. Schaaf C; Gao F; Strahler A; Lucht W; Li X; Tsang T; Strugnell N; Zhang X; Jin Y; Muller J. First Operational BRDF, Albedo Nadir Reflectance Products from Modis. *Remote Sensing of Environment* 2002, 83, 135–148.

33. Liang S. Narrowband to Broadband Conversions of Land Surface Albedo I: Algorithms. *Remote Sensing of Environment* 2001, 76, 213–238.

34. Stroeve J; Box J; Gao F; Liang S; Nolin A; Schaaf C. Accuracy Assessment of the Modis 16-Day Albedo Product for Snow: Comparisons with Greenland in Situ Measurements. *Remote Sensing of Environment* 2005, 94, 46–60.

35. Qu Y; Liu Q; Liang S; Wang L; Liu N; Liu S. Direct-Estimation Algorithm for Mapping Daily Land-Surface Broadband Albedo from Modis Data. *IEEE Transactions on Geoscience and Remote Sensing* 2014, 52, 907–919.

36. Pinty B; Verstraete M; Gobron N; Taberner M; Widlowski J; Lattanzio A; Martonchik J; Dickinson R; Govaerts Y. Coupling Diffuse Sky Radiation and Surface Albedo. *Journal of the Atmospheric Sciences* 2005, 62, 2580–2591.

12

Fronts

Jesse Norris and
David M. Schultz
University of Manchester

Introduction

A front is a boundary that separates different air masses in the Earth's atmosphere. Frontal passages are usually characterized by a change in temperature, moisture content, and wind direction. Changes in cloud cover and type and precipitation occurrence may also be associated with fronts. The process by which a front forms is called frontogenesis and results in an increase in the horizontal temperature gradient across the front. Associated with frontogenesis, cross-frontal circulations result to maintain thermal wind balance, where the horizontal temperature gradient is proportional to the vertical shear of the horizontal wind parallel to the isotherms (contours of constant temperature). A variety of fronts are possible, including surface-based fronts in association with low-pressure systems (cold, warm, and occluded fronts), upper-level fronts associated with the tropopause, and smaller-scale fronts associated with land breezes, sea breezes, and convective storms.

Definition

A front is a boundary separating different air masses in the Earth's atmosphere. An air mass is a volume of air near the surface of the Earth with relatively homogeneous air temperature, moisture content (e.g., relative humidity, mixing ratio, dewpoint temperature), and static stability (a measure of the atmosphere's resistance to vertical air movements). Air masses usually have a horizontal extent of thousands of kilometers and are defined primarily by their source regions, which are functions of latitude (tropical or polar) and underlying surface (continental or maritime). Thus, fronts are usually characterized by gradients in temperature, humidity, or both. Frontal passages are sometimes also indicated by changes in wind direction, minima in pressure, and changes in the sensible weather (cloud cover and type, precipitation occurrence and rate).

Formation

Frontogenesis is the process by which fronts form. It is defined mathematically as an increase in the horizontal temperature gradient over time.[1] The opposite is frontolysis, a decrease in the horizontal temperature gradient over time. A front can form from a broad region of initially weak temperature gradient that is increased as a result of the large-scale wind pattern. Two such wind patterns promote frontogenesis: confluence and horizontal shear (Figure 12.1). Confluence is where two or more airstreams originating far from one another meet and merge into a single airstream (for example, airstreams from the southwest and southeast meeting along a north–south-oriented axis and flowing north; Figure 12.1a), similar to the manner in which rivers merge at a confluence. If the two airstreams have different temperatures, then a front can form roughly along the axis of confluence.[2] In contrast, horizontal wind shear is where wind speed changes in a direction perpendicular to the flow (for example, wind speed changing from southerly to northerly from west to east). If the isotherms are oriented in a way to be rotated by the shear, then a front may form (Figure 12.1b).[3].

Cross-Frontal Circulation

For weather phenomena the size of fronts and larger, the horizontal temperature gradient is proportional to the vertical shear of the horizontal wind parallel to the isotherms, a principle called thermal wind balance. Thus, fronts are accompanied by increasing wind speed with height, which is why surface fronts usually occur in conjunction with the polar-front jet stream, a region of strong winds 5–12 km above the earth's surface.

When frontogenesis occurs, the atmosphere responds in two ways to attempt to maintain thermal wind balance: by increasing the vertical wind shear and by decreasing the horizontal temperature gradient through ascent (and hence adiabatic cooling) of the warm air and descent (and hence adiabatic warming) of the cold air. The resulting horizontal and vertical motions are called the secondary circulation and can be expressed mathematically through the Sawyer–Eliassen equation.[4,5] At the surface, the weak vertical velocities inhibit the frontolytic effects of the adiabatic temperature changes that would weaken the horizontal temperature gradient.

Although most fronts occur near the surface of the Earth, sometimes fronts form aloft near the tropopause.[6] Aloft, the descent of the tropopause due to subsidence along the jet stream can lead to an intensifying horizontal temperature gradient and an upper-level front.

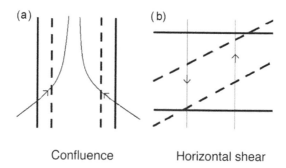

Confluence   Horizontal shear

FIGURE 12.1   Wind patterns favorable for frontogenesis: (**a**) confluence and (**b**) horizontal shear. Thin solid lines represent the streamlines of the flow, thick solid lines represent the temperature contours at an initial time, and thick dashed lines represent the temperature contours at a later time.

Cyclones, Jet Streams, and Fronts

Low-pressure systems known as extratropical cyclones develop along the polar-front jet stream in conjunction with meanders in the jet stream called Rossby waves. Around a low-pressure center, the air flows counterclockwise in the Northern Hemisphere and clockwise in the Southern Hemisphere. Confluence and horizontal shear induced by the wind flow around the cyclone cause the isotherms to rotate, contract, and form two surface fronts (stage I in Figure 12.2), which are often linked to upper-level fronts. The cold front generally extends equator ward from the cyclone center and marks the boundary of the cold air that is transported around the cyclone from the polar region (stage II in Figure 12.2). The warm front generally extends east from the cyclone center and marks the boundary of the warm air that is transported around the low center from the tropics. Fronts typically slope over the colder and more statically stable air mass.

As a low-pressure system deepens, the strengthening winds around the system rotate the cold and warm fronts around the low (Figure 12.2). If the low becomes strong enough, these fronts are wrapped up, and the intervening warm air is lifted aloft, forming a boundary between two cold air masses called an occluded front (stages III and IV in Figure 12.2). Because warm fronts tend to consist of relatively stable air compared to cold fronts, the cold front often advances over the warm-front, forming a warm-type occluded front. Recent research has shown that the warm-type occluded front is the most common structure for occluded fronts.[7] Cyclones with occluded fronts are often intense, with strong surface winds and bands of heavy precipitation embedded within regions of light and moderate precipitation.

Weather Associated with Fronts

Each of these three principal types of fronts (cold, warm, occluded) is associated with a different sequence of weather. Traditional conceptual models of fronts, however, often oversimplify the variety of observed fronts. Consider cold fronts. By definition, cold fronts are associated with near-surface temperatures

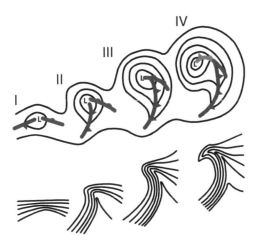

FIGURE 12.2 Four stages in the evolution of an extratropical cyclone in the Northern Hemisphere. The stages are separated by approximately 6–24 hours. Top: thin solid lines represent isobars (contours of constant pressure) at mean sea level and thick solid lines represent fronts using conventional notation (line with triangles represents the cold front, line with semicircles represents the warm front, and line with alternating triangles and semicircles represents the occluded front). Bottom: solid lines represent isotherms (contours of constant temperature) in the lower troposphere near the Earth's surface with warm air to the south and cold air to the north.
Source: Copyright © 2011, American Meteorological Society.

decreasing as warm prefrontal air is replaced by advancing cold air behind the front. Yet, pools of cold, stable air in advance of the front may mask the temperature changes occurring above the surface or cause abrupt warming when replaced by the postfrontal air.[8,9] Conceptual models of cold fronts in the Northern Hemisphere often depict the wind shift from southerly to westerly or northerly coincident with the temperature decrease; yet, many cold fronts occur where the cold air arrives after the wind shift has occurred.[10] Finally, cold fronts are often depicted as having steeply sloping leading edges, forcing strong ascent and thunderstorms. In contrast, many cold fronts are not associated with such convective storms or even a change in cloud type or amount.[11,12] Thus, caution should be exercised in generalizing the specific characteristics of the fronts depicted in idealized conceptual models to real fronts.

A cross-section through a warm front, by comparison, is much more gently sloping than that of a cold front. Warm southerly air is generally depicted as gently ascending over cold easterly flow. Thus, a warm front in the Northern Hemisphere is usually characterized by increasing temperatures and a wind shift from easterly to southerly. This gentle ascent is consistent with the traditional conceptual model depicting low clouds and steady precipitation near the surface rising up to high cirrus clouds well in advance of the warm front. In reality, deep convective storms and regions of heavier precipitation organized in bands may be present in the ascending air.

Occluded fronts are usually depicted as having weak, if any, temperature differences across them. Winds may shift at the frontal passage, but if the occluded front is part of an intense surface cyclone, the shift in wind direction may be minimal. Clouds and precipitation are often not collocated with the surface occluded front. Instead, the heaviest precipitation is associated with warm air aloft that is rising due to the secondary circulations aloft.[13]

Other Fronts and Front-Like Phenomena

In addition to the fronts associated with extratropical cyclones, other types of fronts occur on smaller scales. These fronts often occur near the boundaries between different land-surface characteristics. A sea-breeze front is one example. The sea-breeze front forms during the daytime in the warm season when cool moist air from over a water body moves onshore, replacing warm dry air inland. The sea breeze can be a focal point for convective storms. At night, the reverse land breeze occurs when cool air over the land moves offshore over the warmer water. The land-breeze front tends to be weaker because the temperature gradient between land and water is usually weaker at night than in the daytime. Other similar front-like behavior can be seen across gradients in vegetation types or soil moisture.

Convective storms also produce front-like phenomena. For example, the leading edge of cool outflow from convective storms can produce a gust front as it moves into the warmer humid air mass ahead of the storm.[14] The lifting along gust fronts can produce new convective cells that merge with the original storm, thus creating a longer-lasting storm.

The dryline of the southern and central United States is not a front, but sometimes has front-like properties. The dryline is a climatological boundary between the moist air originating over the Gulf of Mexico and the dry air originating over the southwestern United States.[15] As such, the dryline is primarily identifiable by the strong gradient in moisture across it. Front-like circulations have been observed across the dryline because the dry air tends to be warmer than the moist air late in the day and the moist air tends to be warmer than the dry air at night. The dryline also tends to be associated with stronger gradients when a synoptic-scale low-pressure system is present.[16] The dryline is often the focus for deep convective storms, which are often associated with hail, strong winds, and tornadoes.

Impacts of Fronts

The ascent along fronts can produce heavy and sustained precipitation. For example, the ascending warm air of an extratropical cyclone can have its origins deep from within the subtropics, bringing abundant moisture poleward. Long stretches of cloud can extend over thousands of kilometers ahead of the surface cold front. This flow is referred to as the warm conveyor belt and, when this moist flow rises over the warm front or over mountains, heavy precipitation can result. The mountains of the western United States receive much of their cool-season precipitation from such warm conveyor belts (also called the Pineapple Express or atmospheric rivers in some contexts).[17,18] As such, the precipitation associated with fronts can be integral to water management in some locations.

Fronts can also affect the air quality. Pollutants may be ventilated out of the boundary layer, and fresh clean air may replace the polluted air when fronts pass through a region. However, where a layer of cool stable air may persist near the surface, some fronts may be unable to penetrate this layer and freshen the air. Why some fronts can clear the boundary layer in such situations and others cannot is an active research topic.

Conclusion

Fronts commonly form in conjunction with the development of extratropical low-pressure systems. They are boundaries between air masses and are associated with changes in temperature, humidity, pressure, and clouds and precipitation. Despite their conceptual simplicity, many fronts do not fit this relatively simple framework. For example, wind shifts and temperature gradients may not be coincident, and mixing across the front means that these boundaries are not rigid surfaces.[10,19] Research to improve our understanding of frontal structure and evolution and their relationship to low-pressure systems is ongoing.

References

1. Petterssen, S. Contribution to the theory of frontogenesis. Geofys. Publ. **1936**, *11* (6), 1–27.
2. Bergeron, T. Über die dreidimensional verknüpfende Wetteranalyse I. Geofys. Publ. **1928**, *5* (6), 1–111.
3. Williams, R.T. Atmospheric frontogenesis: A numerical experiment. J. Atmos. Sci. **1967**, *24*, 627–641.
4. Sawyer, J.S. The vertical circulation at meteorological fronts and its relation to frontogenesis. Proc. Roy. Soc. Lond. **1956**, *A234*, 346–362.
5. Eliassen, A. On the vertical circulation in frontal zones. Geofys. Publ. **1962**, *24* (4), 147–160.
6. Keyser, D.; Shapiro, M.A. A review of the structure and dynamics of upper-level frontal zones. Mon. Wea. Rev. **1986**, *114*, 452–499.
7. Schultz, D.M.; Vaughan, G. Occluded fronts and the occlusion process: A fresh look at conventional wisdom. Bull. Amer. Meteor. Soc. **2011**, *92*, 443–466, ES19–ES20.
8. Sanders, F.; Kessler, E. Frontal analysis in the light of abrupt temperature changes in a shallow valley. Mon. Wea. Rev. **1999**, *127*, 1125–1133.
9. Doswell, C.A., III.; Haugland, M.J. A comparison of two cold fronts: Effects of the planetary boundary layer on the mesoscale. Electronic J. Severe Storms Meteor. **2007**, 2, 1–12.
10. Schultz, D.M. A review of cold fronts with prefrontal troughs and wind shifts. Mon. Wea. Rev. **2005**, *133*, 2449–2472.
11. Mass, C.F.; Schultz, D.M. The structure and evolution of a simulated midlatitude cyclone over land. Mon. Wea. Rev. **1993**, *121*, 889–917.
12. Schultz, D.M.; Roebber, PJ. The fiftieth anniversary of Sanders (1955): A mesoscale-model simulation of the cold front of 17–18 April 1953. *Synoptic–Dynamic Meteorology and Weather Analysis and Forecasting: A Tribute to Fred Sanders*, Meteor. Monogr; No. 55, Amer. Meteor. Soc. 2008, 126–143.
13. Novak, D.R.; Colle, B.A.; Aiyyer, A.R. Evolution of mesoscale precipitation band environments within the comma head of northeast U.S. cyclones. Mon. Wea. Rev. **2010**, *138*, 2354–2374.
14. Charba, J. Application of gravity current model to analysis of squall-line gust front. Mon. Wea. Rev. **1974**, *102*, 140–156.

15. Schaefer, J.T. The life cycle of the dryline. J. Appl. Meteor. **1974**, *13*, 444–449.
16. Schultz, D.M.; Weiss, C.C.; Hoffman, P.M. The synoptic regulation of dryline intensity. Mon. Wea. Rev. **2007**, *135*, 1699–1709.
17. Lackmann, G.M.; Gyakum, J.R. Heavy cold-season precipitation in the northwestern United States: Synoptic climatology and an analysis of the flood of 17–18 January 1986. Wea. Forecast. **1999**, *14*, 687–700.
18. Neiman, P.J.; Ralph, F.M.; Wick, G.A.; Lundquist, J.D.; Dettinger, M.D. Meteorological characteristics and overland precipitation impacts of atmospheric rivers affecting the west coast of North America based on eight years of SSM/I satellite observations. J. Hydrometeor **2008**, *9*, 22–47.
19. Schultz, D.M. Perspectives on Fred Sanders' research on cold fronts. *Synoptic-Dynamic Meteorology and Weather Analysis and Forecasting: A Tribute to Fred Sanders,* Meteor. Monogr.; No. 55, Amer. Meteor. Soc. 2008, 109–126.

Bibliography

Browning, K.A. Conceptual models of precipitation systems. Wea. Forecasting **1986**, *1*, 23–41.
Hoskins, B.J. The mathematical theory of frontogenesis. Annu. Rev. Fluid Mech. **1982**, *14*, 131–151.
Hoskins, B.J.; Bretherton, F.P. Atmospheric frontogenesis models: Mathematical formulation and solution. J. Atmos. Sci. **1972**, *29*, 11–37.
Shapiro, M.A., Grønås, S. Eds.: *The Life Cycles of Extratropical Cyclones;* American Meteorological Society: Boston, 1999; 359.
Shapiro, M.A.; Keyser, D. Fronts, jet streams and the tropopause. In *Extratropical Cyclones, The Erik Palmén Memorial Volume;* Newton, C.W.; Holopainen, E.O., Eds.; Amer. Meteor. Soc. 1990, 167–191.

13

Land–Atmosphere Interactions

Somnath Baidya
Roy and Deeksha
Rastogi
University of Illinois

Introduction

Land and atmosphere are integral components of our natural environment. Interactions between land and atmosphere occur through continuous exchange of heat, moisture, momentum, and various gases (Figure 13.1). These exchanges result in various natural phenomena, such as sea breezes and monsoons, at wide ranges of spatial and temporal scales. Human activity such as deforestation, agriculture, and urbanization can modify these exchanges and thus affect weather, climate, water cycle, and other aspects of the natural environment. Hence, land–atmosphere interaction is a topic of interest for scientists and policymakers worldwide.

Processes

The land surface is a major source of energy for the atmosphere. Even though all energy originally comes from the sun, this energy is in the form of shortwaves that cannot be readily absorbed by the atmospheric gases. Some of these solar shortwaves are reflected back but a major part is absorbed by the surface. Some of this energy is used by plants for photosynthesis but the rest heats up the surface. The surface then tries to cool down by transferring energy to the atmosphere by three processes: longwave radiation, conduction, and latent heating. Most of the energy transfer occurs through longwave radiation emitted by the surface. Unlike solar shortwaves, these longwaves are easily absorbed by the atmosphere, especially by clouds and water vapor. The second process is conduction where heat exchange occurs between the surface and the lowest layer of the atmosphere in direct contact with the surface. During the daytime heat flows from the warm surface into the lower atmosphere and is then carried upward by convection. At night, the surface is cooler than the air above and consequently heat flows downward from the atmosphere into the ground. Finally, energy is also transferred in the form of latent heating by evaporation and transpiration from the ground and vegetation. During evaporation and transpiration, liquid water in the soil and vegetation absorbs heat and is released into the atmosphere in the form of water vapor. The heat carried by the vapors is released into the atmosphere when the vapors condense to form cloud droplets.

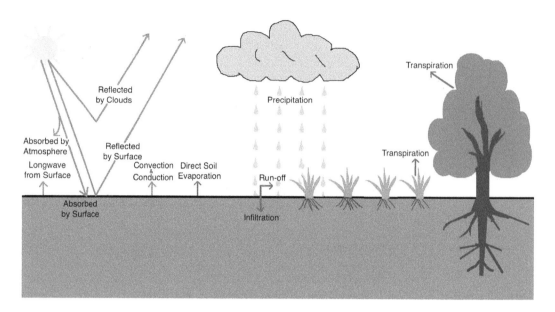

FIGURE 13.1 Schematic diagram representing various processes of energy and water exchange between the land surface and the atmosphere.

The Earth's surface acts as both a source and a sink for moisture in the atmosphere. Moisture is transported as water vapor from the land surface to the atmosphere by evaporation and transpiration. Most of this moisture is returned back to the surface in the form of rain, snow, hail, and freezing rain. A tiny amount of moisture is also transferred back to the surface when some of the water vapor in the lower atmosphere condenses directly onto the surface as dew and frost on cold nights. Some of this moisture seeps into the ground and becomes part of the groundwater system, whereas the rest flow downslope as surface runoff into nearby waterbodies. Even though land covers less than 30% of the Earth's surface, the continuous and vigorous back- and-forth exchange of moisture between land and the atmosphere plays a significant role in the overall water cycle.

The land surface is the primary sink of atmospheric momentum. Friction due to natural and manmade obstacles such as hills, trees, and buildings slow down the wind. During this process kinetic energy in the wind is converted to heat. Friction due to land surface is typically expressed as a function of the surface roughness. Surfaces with tall, rigid densely packed obstacles, such as tropical forests and urban areas, have higher roughness and hence generate stronger friction than relatively smooth surfaces such as sandy deserts.

The Earth's surface is also a source and a sink of atmospheric gases such as carbon dioxide, oxygen, and nitrogen. Carbon dioxide is absorbed by plants during photosynthesis and released from the surface back into the atmosphere during respiration by plants and animals. This natural carbon cycle has been significantly modified by humans since the Industrial Revolution due to the large amount of carbon dioxide released by fossil fuel burning without a corresponding natural sink. Oxygen is absorbed by plants and animals during respiration and released back into the atmosphere by plants as a byproduct of photosynthesis. Atmospheric nitrogen is absorbed by bacteria in the soil and nitrogen-fixing plants and released back into the atmosphere during decomposition of biomass.

Natural Phenomena

The land surface is spatially heterogeneous due to variability in topography and vegetation cover. Consequently, the magnitudes of the exchange processes also vary in space. For example, forests absorb more heat but also release more water during evapotranspiration than adjacent grasslands. Such spatial heterogeneity in land–atmosphere inter-actions result in various natural phenomena such as slope-valley winds, sea breezes, and monsoons.

Slope and Valley Winds

Slope and valley winds are localized wind circulations generated by topography (Figure 13.2). They are part of the same circulations system but formed by distinctly different mechanisms. A fundamental difference is that slope winds flow across the valleys, whereas valley winds flow along the valleys. Slope winds are generated by temperature difference between mountain slopes and adjacent valleys. Air near the slope is warmer during the daytime compared with the air at the same altitude away from the slope. This causes air from the valley to move upward along the mountain slope creating the upslope or anabatic wind. The process is reversed in the night when winds flow downward leading to the downslope or katabatic winds.

Valley winds are formed due to temperature difference between narrow valleys and adjacent plains. Due to the topographic constraints, valleys contain a smaller volume of air than open plains. Consequently during the daytime a valley heats up faster leading to a localized low-pressure region over the valley floor. Upslope winds that develop in the morning transport air upward further reducing air pressure in the valley. By early afternoon this low-pressure region becomes strong enough so that air from the adjacent plains flows into the valley in response to the pressure differential. In the evening the valley cools fast and downslope winds bring cold air down creating a high pressure region over the valley floor. In response, valley winds start flowing out into the plains. Valley winds are typically stronger than slope winds and play an important role in the transport of heat, moisture, and pollutants in mountainous regions all over the world.

Sea Breezes

Sea breezes are circulations generated due to land–water temperature contrasts in coastal areas (Figure 13.3). Because land has a lower heat capacity than water, the land surface warms up faster than the sea surface. After sunrise, as the land heats up, it heats the air above it. Since warm air is lighter, it rises up, creating a low-pressure region over the land. In response to this pressure gradient, cold air from the sea flows in towards the land resulting in a sea breeze. In the evening, the land cools faster than the sea causing the sea breezes to weaken and then finally die out. If temperature over the land falls below the sea surface temperature, a low pressure is created over the sea, which results in a land breeze that flows from land to sea. The strength of sea/land breeze depends on the temperature contrast between the land and the sea. Strong sea breezes can often generate thunderstorms, especially during the summer in tropical regions. Sea breezes are observed in coastal regions all over the world. An interesting example is the Florida peninsula where sea breezes develop both along the Gulf of Mexico coast in the west and

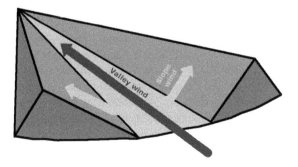

FIGURE 13.2 Typical slope and valley winds during the afternoon.

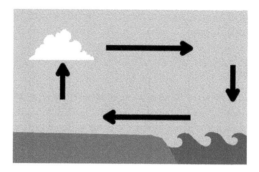

FIGURE 13.3 Circulation patterns and clouds associated with a typical sea breeze.

the Atlantic coast in the east. Circulations forced by land–water temperature gradients similar to sea breezes are also observed at the boundaries of large lakes such as the Great lakes in North America and large rivers such as the Amazon where they are known as lake and river breezes, respectively.

Asian Monsoon

Monsoons are atmospheric processes characterized by seasonal reversal of winds. It is widely acknowledged that the Asian monsoon acts like a giant sea breeze. The Asian continental land mass quickly heats up in the summer while the ocean is still cold. This temperature difference causes moist winds to flow inland resulting in significant amount of rainfall over south and east Asia. This rainfall is a significant source of freshwater for farmers in that region. The landmass starts to cool down in the Fall but the ocean is still warming up. Consequently, dry winds flow out into the ocean from the land. Scientists are currently investigating if similar land-atmosphere coupling exists in other recently discovered monsoon systems such as the North American Monsoon and the West African Monsoon.

Manmade Activities

Rapid population growth and increasing demand for food, housing, energy, and other natural resources is leading to an increase in tropical deforestation, agriculture, and urbanization. These land use changes are changing the magnitudes of land-atmosphere interactions, thereby significantly affecting regional climate, energy, and water resources.

Tropical Deforestation

Deforestation is an acute problem particularly for tropical regions where forests are being cleared to meet the growing demand for food, fuel, timber, and other agricultural products such as textiles and biofuels. By most estimates about half of the world's tropical rainforests have disappeared while the rest are under threat. About 80% of tropical deforestation is driven by commercial and subsistence agriculture with logging for timber also making a significant but smaller contribution. Large-scale conversion of natural forests into farmlands and pastures has severe impacts on the climate and hydrology at local, regional, and even global scales. Perhaps the most important impact of tropical deforestation is that on the global carbon cycle. Forests absorb carbon dioxide during photosynthesis and store it in their roots, stems, and leaves. Forest degradation releases this stored carbon back to the atmosphere thereby contributing to global climate change. Tropical deforestation released approximately 1 PgC/yr during 2000s, representing 6–17% of the total anthropogenic carbon emissions. Deforestation also leads to a suite of changes in land–atmosphere exchange of heat and moisture that act competitively to alter local and regional climate. Deforested lands have higher albedo or reflectivity and usually reflect more sunlight causing a cooling effect. On the other hand, deforestation reduces transpiration resulting in a strong warming that more than offsets the albedo-induced cooling effect. Deforestation also has a strong impact on the water cycle. Trees can access groundwater through their deep roots and release water into the atmosphere through transpiration. Deforestation cuts off this process thereby reducing atmospheric moisture content. Consequently, deforested regions tend to be warmer and drier than nearby forests. In spite of community and government efforts to control tropical deforestation, it is still continuing, albeit at a slower rate. Hence, the environmental impacts of deforestation are likely to persist in the near future.

Agriculture

Agriculture is the dominant driver of manmade land use/land cover change. As a result of continuous agricultural expansion throughout human history, farms have replaced almost 50% of global forests and grasslands. Conversion of forests to farmlands increases albedo leading to a cooling effect. This effect is masked in tropics and leads to a net warming as discussed in the previous section. However, clearing temperate forests for agriculture has a cooling effect because the albedo-induced cooling is much stronger in those regions. Converting grasslands to farmlands does not significantly affect the exchange of heat and momentum because crop albedo and roughness are similar to that of grasses. However, this conversion leads to a drying of the soil and more evapotranspiration due to increased water uptake by crops. Intensive agricultural practices also release stored soil organic carbon into the atmosphere thereby contributing to global warming. Increasing demand for food implies agricultural expansion is not going to slow down in the near future. However, improved agricultural practices and efficient irrigation techniques can reduce its impact on climate to a certain extent.

Urbanization

Driven by economic, political, and social factors, more and more people are moving from rural to urban areas. According to the United Nations, the proportion of human population living in cities has grown from 13% in 1900 to more than 50% today and is likely to go up to 70% by 2050. The expansion of cities and suburbs leads to the conversion of natural vegetation into residential, commercial, and industrial complexes. These changes affect the exchange of mass, momentum, and energy between the surface and the atmosphere. The most well-known impact of urbanization is the urban heat island (UHI) effect where urban areas are significantly warmer than the surrounding regions (Figure 13.4). Brick, concrete, and other materials used in building construction and asphalt in the roads absorb more solar radiation than the natural vegetation. Additionally, human activity such as industries and vehicular transport

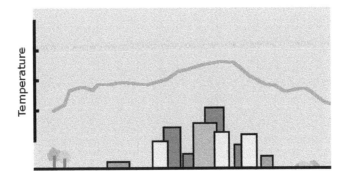

FIGURE 13.4 Typical urban heat island temperature profile showing significant warming over urban areas compared with the surroundings.

generate more heat than natural ecosystems. The presence of buildings prevents this extra heat from getting radiated back into the atmosphere. Also, due to relative lack of vegetation there is less cooling due to evapotranspiration. All these processes cause urban areas to be warmer than their surroundings by several degrees. According to a recent NASA study, large cities in the northeastern United States are 13–16°F warmer than the surrounding area during the daytime in the summer due to the UHI effect. Urbanization also affects the surface water balance. Buildings, parking lots, and roads reduce the infiltration of rainwater into the ground resulting in increased surface runoff and reduced ground water recharge. Although urbanization significantly affects the environment in cities and suburbs, it tends to reduce the impact of human land use in rural areas that may revert back to their original natural state. The projected growth of urbanization in the future implies that environmental impacts will continue to intensify.

Bibliography

Laurance, W.F. Reflections on the tropical deforestation crisis, Biol. Conserv. **1999**, *91*, 109–117.

Memon, R.A.; Leung, D.Y.C.; Chunho, L. A review on the generation, determination and mitigation of urban heat island, J. Environ. Sci. **2008**, *20*, 120–128.

Miller, S.T.K.; Keim, B.D.; Talbot, R.W.; Mao, H. Sea breeze: Structure, forecasting and impacts, Rev. Geophys. **2003**, *41*, 1011.

Moutinho, P.; Schwartzman, S. (Eds.), *Tropical Deforestation and Climate Change;* Amazon Institute for Environmental Research: Washington, DC, 2005; 132 pp.

Schmidli, J.; Rotunno, R. Mechanisms of along-valley winds and heat exchange over mountainous terrain, J. Atmospheric Sci. **2010**, *67*, 3033–3047.

14

Ozone and Ozone Depletion

Jason Yang
Ball State University

Introduction

Ozone is a colorless gas that is naturally present in the atmosphere. Each ozone molecule contains three atoms of oxygen and is denoted chemically as O_3. The word "ozone" is derived from the Greek word ózein (*ozein* in Latin), meaning "to smell" [1]. Ozone mainly exists in two layers of the atmosphere. Approximately 10% of the total atmospheric ozone resides in the troposphere, a layer of the atmosphere from ground up to 15 km altitude. The ozone in this layer is called tropospheric ozone or commonly known as the ground-level ozone. The remaining 90% of the atmospheric ozone is in the stratosphere, a layer between 17 km and up to 50 km above the Earth's surface. The ozone in this layer is called stratospheric ozone or commonly known as the ozone layer (Figure 14.1) [2].

Ozone molecules in the troposphere and the stratosphere are chemically identical, but they play different roles and have different effects on human beings and other living organisms. Ground-level ozone is a key component of photochemical smog, an air pollutant of many cities around the world, and plays a destructive role in the environment and ecosystem. Many studies have documented the harmful effects of ground-level ozone on crop production, forest growth, and human health [3–5]. Stratospheric ozone, on the other hand, plays a beneficial role by absorbing most of the biologically damaging ultraviolet-B (UV-B) radiation ranging from 280 to 320 nm. Without the absorption by the ozone layer, UV-B radiation would penetrate the atmosphere and reach the Earth's surface. Many experimental studies of plants and animals and clinical studies of humans have shown the harmful effects of excessive exposure to UV-B radiation [6,7]. A simple statement summarizing the different effects of atmospheric ozone is that it is "good up high, bad nearby" [8].

The dual role of atmospheric ozone links it to two separate environmental issues. One issue relates to the increased "bad" ozone in the troposphere. Although ground-level ozone is not emitted directly from car engines or by industrial operations, motor vehicle exhaust, industrial emissions, and chemical solvents are major anthropogenic sources to form ozone near ground [9]. Many countries around the world have developed programs to regulate the emission of ozone-forming substances (OFSs) that cause excess ground-level ozone by setting up air quality standards for ozone. Emission control programs, including

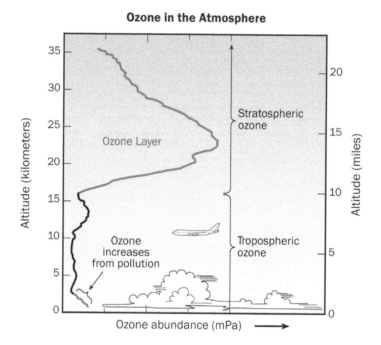

Ozone in the Atmosphere

FIGURE 14.1 Ozone presents throughout the troposphere and the stratosphere. Most ozone resides in the strato-spheric "ozone layer" above the Earth's surface. (NOAA Earth System Research Laboratory, www.esrl.noaa.gov/csd/assessments/ozone/2014/twentyquestions/.)

removal of highway tolls to reduce traffic congestions, use of vapor recovery devices for gasoline pumps, and enforcement of vehicle emission inspections, have been implemented in the United States since the 1980s and other countries to control the increase in ground-level ozone [10].

The second issue of atmospheric ozone on the environment relates to the depleted "good" ozone in the stratosphere. Ground-based and satellite instruments have measured decreases in the amount of ozone in the stratosphere since the 1980s [11]. The most extreme case occurs over some parts of Antarctica, where up to 60% of the total overhead amount of ozone disappeared during some periods of the Southern Hemisphere spring (August–October). Similar processes occur in the Arctic region that have also led to significant ozone depletion during late winter and spring in many recent years. It is found that the primary cause for the ozone depletion in the stratosphere is the presence of some ozone-depleting substances (ODSs) such as chlorine- and bromine-containing gases [12]. These gases can dissociate and release chlorine atoms in the presence of UV-B radiation, and chlorine atoms then go on to catalyze ozone destruction. In 1987, many countries have joined an effort to stop the ozone deple-tion in the stratosphere by signing an international agreement known as the "Montreal Protocol" [13]. In this agreement, governments have decided to develop more "ozone-friendly" substitutes and eventually discontinue those productions that emit ozone-depleting gases.

Atmospheric Ozone

Tropospheric Ozone

In the troposphere, ozone is formed by a chemical reaction involving NO_x and volatile organic com-pounds (VOCs) subjected to favorable weather conditions such as low humidity, plenty of sunlight, and moderate to low wind speed [14]. Emissions from industrial facilities and electric utilities, motor vehicle exhaust, gasoline vapors, and chemical solvents are major anthropogenic sources for NOx and VOCs.

As a major component of urban smog, ozone is harmful to human health by causing sore and watery eyes, soreness in the throat and sinuses, and difficulty in breathing [5]. A statistical study of 95 large urban communities in the United States showed a significant positive association between ground-level ozone levels and premature death [15]. It is estimated that a one-third reduction in urban ozone concentrations would save roughly 4,000 lives per year. Ground-level ozone also damages vegetation such as forests and crops. In the United States alone, ground-level ozone is responsible for an estimated $500 million in reduced crop production each year [16]. To address the issue of ground-level ozone, U.S. Environmental Protection Agency (EPA) has set up air quality standards for ozone since 1971 and kept revising it in 1979, 1997, 2008, and 2011, respectively, to provide requisite protection of public health and welfare [17].

Tropospheric ozone is being monitored, predicted, and declared in advance in many countries today. For example, U.S. EPA coordinates the collection of real-time ground-level ozone measurements and posts the current air quality information, including combined PM and ozone, to the public through the AirNow website [18]. Therefore, people can avoid outdoor activities on those days with a higher ozone level; meanwhile, they take some simple steps in their daily life (e.g., limiting vehicle usage) to reduce the production of OFSs.

Stratospheric Ozone

In the stratosphere, ozone is constructed and destructed naturally by chemical reactions involving oxygen molecules and UV-B radiation, which is illustrated in Figure 14.2. First, one diatomic oxygen (O_2) splits into two atomic oxygen (O) by absorbing UV-B radiation. The free unstable atomic oxygen then combines with other diatomic oxygen in the air to form ozone (O_3). The O_3 absorbs UV-B radiation and splits into two diatomic oxygen and two atomic oxygen. The atomic oxygen then joins up with diatomic oxygen to form ozone again. This process is repeated over and over, thereby absorbing a large amount of UV-B radiation that would otherwise reach the Earth's surface.

The stratospheric ozone can also be destructed by ODSs from natural processes and human activities, including large sources of chlorine- and bromine-containing gases. First, the chlorine (Cl) takes away an atomic oxygen from ozone to form chlorine monoxide (ClO) and leave a normal diatomic oxygen (Equation 14.1). The newly formed chlorine monoxide can destroy a second ozone molecule to yield another chlorine (Cl) (Equation 14.2). This process is repeated (Equation 14.1) and continues to destroy ozone [19].

$$Cl + O_3 \rightarrow ClO + O_2 \tag{14.1}$$

$$ClO + O_3 \rightarrow Cl + 2O_2 \tag{14.2}$$

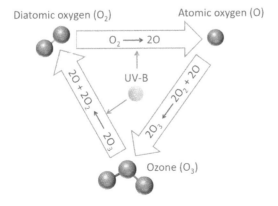

FIGURE 14.2 Natural formation and destruction of ozone molecule (O_3) in the stratosphere.

Ozone Depletion

The Ozone Hole

Ozone depletion in the stratosphere was first noticed by British Antarctica Survey based on ground-based observations and reported in the scientific journal *Nature* in 1985 that there was a substantial thinning of the ozone layer over Antarctica in early spring [11]. Soon after, measurements from NASA satellites confirmed the ozone depletion over the Antarctic and further showed that in each Southern Hemisphere spring, the ozone layer depletion extended over a large region centered near the south pole [20].

Stratospheric ozone depletion varies strongly with latitude over the globe. There is no significant trend found in the tropics, and there is only a very small amount of declines (3%–6%) for middle latitudes. However, the ozone decrease can be up to 30% in the Arctic and even 60% over the Antarctica. Ozone depletion above the Antarctic is a seasonal phenomenon, occurring primarily between August and November with peak depletion in early October. This severe ozone depletion is commonly called "ozone hole," which is defined as an area of the stratosphere in which the ozone levels have dropped to as low as 33% of their pre-1980 values. Therefore, the ozone hole is not technically a "hole" where no ozone presents, but a region in the stratosphere with exceptionally depleted ozone. The size of the ozone hole is usually determined by the geographic region contained within the 220-Dobson unit (DU) contour in the total ozone map. Based on NASA Earth Observatory Image presented in Figure 14.3, the maximum size of the ozone hole in most years after 1985 far exceeds the size of the Antarctic continent.

Reasons and Consequence of Ozone Depletion

It is now widely accepted that the stratospheric ozone is gradually being destroyed by man-made chemicals referred to as ODSs, including chlorofluorocarbons (CFCs), hydrochlorofluorocarbons (HCFCs), halons, and other gases. These substances were formerly used and sometimes are still used in coolants, foaming agents, fire extinguishers, solvents, pesticides, and aerosol propellants. Once released into the atmosphere, these gases degrade very slowly until they reach the stratosphere where they are broken down by the intensive UV radiation and release chlorine and bromine molecules, which deplete the ozone layer there (Equations 14.1 and 14.2).

The role of sunlight in ozone depletion is the reason why the Antarctic ozone depletion is the greatest during the early Southern Hemisphere spring. Even though nacreous or polar stratospheric clouds (PSCs) are at their most abundant during winter, there is no light over the Antarctic to drive the chemical reactions. During the spring, however, the increasing solar radiation raises the stratospheric temperature and removes the PSCs via sublimation, thereby releasing the trapped compounds that can deplete ozone. At the end of Southern Hemisphere spring, warm temperatures can break up the vortex and ozone-rich air flows in from lower latitudes, so that the PSCs are destroyed; as a result, the ozone depletion process shuts down and the ozone hole closes [21].

Ozone depletion is expected to increase the UV-B radiation on the Earth's surface, which could lead to damage to crops and human health. Studies have shown that an increase in UV-B radiation would be expected to reduce crop yields for a number of economically important species such as rice and soybeans [7,22]. Some scientists suggest that marine phytoplankton, which are the base of the ocean food chain, are already under stress from UV-B radiation. This could have adverse consequences for human food supplies from the oceans [23]. Another main public concern regarding the ozone depletion is the effects of increased UV-B radiation on human health. Research has shown that UV-B radiation is the main cause of sunburn and tanning, as well as the formation of vitamin D3 in the skin, which has negative influences on the immune system of human beings. UV-B radiation contributes significantly to the aging of the skin and eyes, even causing skin cancer [24].

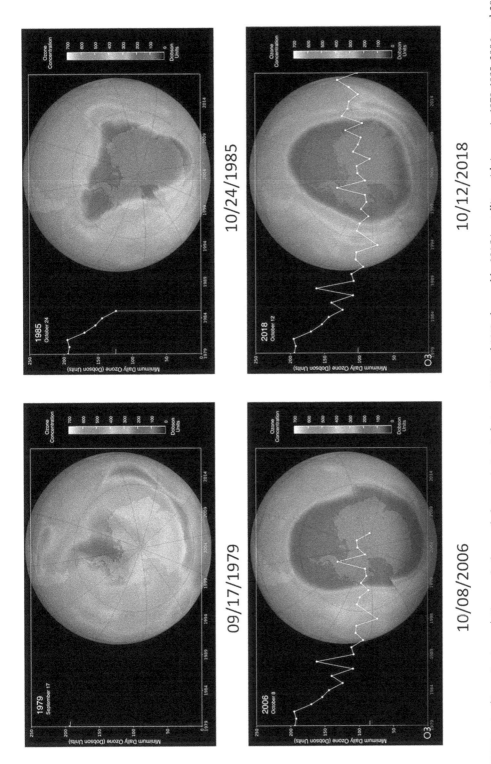

FIGURE 14.3 (See color insert.) The trend of ozone hole over Antarctic between 1979 and 2018 observed by NASA satellites with images in 1979, 1985, 2006, and 2018. (NASA's Goddard Space Flight Center, https://svs.gsfc.nasa.gov/12816.)

Trends of Ozone Depletion

Research on ozone depletion advanced rapidly after it was found in the 1980s, leading to the identification of CFCs and other halocarbons as the cause. Since then, industries have developed more ozone-friendly substitutes for the CFCs and other ODSs. The 1987 "Montreal Protocol on Substances that Deplete the Ozone Layer," ratified now by over 190 countries, established legally binding controls on the production and consumption of ODSs. As a result, the total abundance of ozone-depleting gases in the atmosphere has begun to decrease in recent years, which has been observed using instruments such as Total Ozone Mapping Spectrometer (TOMS) on NASA satellites [25]. A report in 2007 showed that the ozone hole over the Antarctic was closing and the smallest it had been for about a decade [26]. Another report in 2010 found that global ozone and ozone in the polar regions are no longer decreasing but are not yet increasing, and a recent NASA study stated that for the first time, scientists found the direct proof of ozone hole recovery due to chemicals ban of Montreal Protocol (Figure 14.3) [27].

However, significant depletion of the ozone layer in the Arctic stratosphere has drawn more attention recently. A record ozone-layer loss was observed on March 15, 2011, with about 50% of the ozone over the Arctic having been depleted [28]. A study published in the journal *Nature* in October 2011 stated that between December 2010 and March 2011, up to 80% of the ozone in the Arctic region was depleted [29]. The level of ozone depletion over the Arctic was severe enough that scientists said it could be compared to the "ozone hole" over Antarctica every October. And for the first time, sufficient ozone loss occurred that can be reasonably described as an Arctic ozone hole [30].

Conclusion

Ozone can be either bad or good to the environment and ecosystem depending on its location in the atmosphere. Both the tasks to reduce the tropospheric "bad" ozone and to recover the stratospheric "good" ozone are challenging. Efforts have been made to reduce the emission of those OFSs such as NO_x and VOCs that form ozone on the ground. Great efforts have also been made to reduce the emission of those ODSs such as CFCs and halons, which will eventually reach the stratosphere and deplete ozone. The interaction between science in identifying the problem, technology in developing alternatives, and governments in devising new policies is thus an environmental "success story in the making." Indeed, the Montreal Protocol and its subsequent Amendments and Adjustments have served as a model in international cooperation for other environmental issues now facing the global community. If the nations of the world continue to follow the provisions of the Montreal Protocol, the decrease in ODSs will continue throughout the 21st century and the stratospheric ozone is expected to recover to its pre-1980 level around the year 2070.

References

1. Rubin, M.B. The history of ozone. The schönbein period, 1839–1868. *Bulletin for the History of Chemistry*, 2001. 26(1): p. 16.
2. WMO. Scientific Assessment of Ozone Depletion: 2002 in Global Ozone Research and Monitoring Project - Report No. 47. 2003. p. 498.
3. Fishman, J.; Creilson, J.K.; Parker, P.A.; Ainsworth, E.A.; Vining, G.G.; Szarka, J.; Booker, F.L.; Xu, X. An investigation of widespread ozone damage to the soybean crop in the upper Midwest determined from ground-based and satellite measurements. *Atmospheric Environment*, 2010. 44(18): pp. 2248–2256.
4. Emberson, L.D.; Büker, P.; Ashmore, M.R. Assessing the risk caused by ground level ozone to European forest trees: A case study in pine, beech and oak across different climate regions. *Environmental Pollution*, 2007. 147(3): pp. 454–466.

5. Stedman, J.R.; Kent, A.J. An analysis of the spatial patterns of human health related surface ozone metrics across the UK in 1995, 2003 and 2005. *Atmospheric Environment*, 2008. 42(8): pp. 1702–1716.

6. Van Der Leun, J.C. Effects of Increased UV-B on Human Health. In *Studies in Environmental Science*; Lee, S.D., Wolters, G.J.R., Schneider, T., Grant, L.D., Editors. 1989, Elsevier: Amsterdam. pp. 803–812.

7. Kumagai, T.; Hidema, J.; Kang, H.S.; Sato, T. Effects of supplemental UV-B radiation on the growth and yield of two cultivars of Japanese lowland rice (Oryza sativa L.) under the field in a cool rice-growing region of Japan. *Agriculture, Ecosystems & Environment*, 2001. 83(1–2): pp. 201–208.

8. Ozone - Good Up High Bad Nearby. U.S. Environmental Protection Agency (EPA). September, 2014 [accessed 2018 November 29]; Available from: https://cfpub.epa.gov/airnow/index.cfm?action=gooduphigh.index.

9. WMO. Scientific Assessment of Ozone Depletion: 2006, in Global Ozone Research and Monitoring Project - Report No. 50. 2007. p. 572.

10. Yang, J.; Miller, D.R. Trends and variability of ground-level ozone in Connecticut over the period 1981–1997. *Journal of Air & Waste Management Association*, 2002. 52: p. 8.

11. Lubinska, A. Ozone depletion: Europe takes a cheerful view. *Nature*, 1985. 313(6005): pp. 727.

12. Hegglin, M.I.; Fahey, D.W.; McFarland, M.; Montzka, S.A.; Nash, E.R. Twenty Questions and Answers About the Ozone Layer: 2014 Update, Scientific Assessment of Ozone Depletion: 2014, 88 pp., World Meteorological Organization: Geneva, Switzerland, 2015.

13. Morrisette, P.M. The evolution of policy responses to stratospheric ozone depletion. *Natural Resources Journal*, 1989. 29: p. 27.

14. Brimblecombe, P. *Air Composition and Chemistry*. 2nd ed., Cambridge University Press: New York, 1996. p. 253.

15. Bell, M.L.; McDermott, A.; Zeger, S.L.; Samet, J. M.; Dominici, F. Ozone and short-term mortality in 95 US urban communities, 1987–2000. *Journal of the American Medical Association*, 2004. 292: p. 6.

16. Ground-level Ozone. U.S. Environmental Protection Agency (EPA). Last updated on November 7, 2018 [accessed 2019 April 9]; Available from: www.epa.gov/ground-level-ozone-pollution.

17. Ozone (O3) Standards - Table of Historical Ozone NAAQS. U.S. Environmental Protection Agency (EPA) Last updated on February 20, 2018 [accessed 2019 March 28]; Available from: www.epa.gov/ground-level-ozone-pollution/table-historical-ozone-national-ambient-air-quality-standards-naaqs.

18. AirNow. EPA/NOAA/NPS Last updated on April 15, 2010 [accessed 2019 April 15]; Available from: http://airnow.gov/.

19. Gabler, R.E.; Petersen, J.F.; Trapasso, L.M.; Sack, D. *The Atmosphere, Temperature, and the Heat Budget, in Physical Geography*, Brooks/Cole, Cengage Learning: Belmont, CA, 2009. p. 89.

20. Ozone, NASA. Last updated on August 3, 2017 [accessed 2019 March 8]; Available from: www.nasa.gov/ozone.

21. Ozone Facts: What is the Ozone Hole? U.S. National Aeronautics and Space Administration (NASA) Last updated on October 18, 2018 [accessed 2019 January 7]; Available from: http://ozonewatch.gsfc.nasa.gov/facts/hole.html.

22. Koti, S.; Reddy, K.R.; Kakani, V.G.; Zhao, D.; Gao, W. Effects of carbon dioxide, temperature and ultraviolet-B radiation and their interactions on soybean (Glycine max L.) growth and development. *Environmental and Experimental Botany*, 2007. 60(1): pp. 1–10.

23. Nielsen, T.; Ekelund, N.G.A. Influence of solar ultraviolet radiation on photosynthesis and motility of marine phytoplankton. *FEMS Microbiology Ecology*, 1995. 18(4): pp. 281–288.

24. Tevini, M. Editor. *UV-B Radiation and Ozone Depletion: Effects on Humans, Animals, Plants, Microorganisms, and Materials*, Lewis Publishers: Boca Raton, FL, 1993.

25. Total Ozone Mapping Spectrometer (TOMS). Last updated on August 11, 2018 [accessed 2019 March 2]; Available from: https://en.wikipedia.org/wiki/Total_Ozone_Mapping_Spectrometer.

26. Ozone Hole Closing Up, Research Shows. ABC News. Last updated on November 15, 2007 [accessed 2011 December 2]; Available from: www.abc.net.au/news/2007-11-16/ozone-hole-closing-up-research-shows/727460.

27. NASA Study: First Direct Proof of Ozone Hole Recovery Due to Chemicals Ban, Last updated on Jan. 4, 2018 [accessed 2019 April 18]; Available from www.nasa.gov/feature/goddard/2018/nasa-study-first-direct-proof-of-ozone-hole-recovery-due-to-chemicals-ban.

28. Dell'Amore, C. First North Pole Ozone Hole Forming? *National Geographic News*. Last updated on March 23, 2011; [accessed 2019 March 2]; Available from: http://news.nationalgeographic.com/news/2011/03/110321-ozone-layer-hole-arctic-north-pole-science-environment-uv-sunscreen/.

29. Manney, G.L.; Santee, M.L.; Rex, M.; Livesey, N.J.; Pitts, M.C.; Veefkind, P.; Nash, E.R.; Wohltmann, I.; Lehmann, R.; Froidevaux, L.; Poole, L.R. Unprecedented Arctic ozone loss in 2011. *Nature*, 2011. 478(7370): pp. 469–475.

30. Arctic ozone loss at record level. *BBC News Online*. Last updated on October 2, 2011; [accessed 2019 Feb. 2]; Available from: www.webcitation.org/6297yWsLm.

15

Transpiration and Physical Evaporation: United States Variability

Toshihisa Matsui
University of Maryland

David M. Mocko
National Aeronautics and Space Administration (NASA)

Introduction

Terrestrial transpiration and physical evaporation are Earth's breath that transports water and heat from the soil or plants to the atmosphere. Transpiration is the water vapor exchange from plant leaves to the surrounding air through numerous tiny pores in the leaf, called *stomata*. The water that transpires from the leaves has its origin in the soil and travels from the plant's root structure through the stems to the leaves. Simultaneously, plants exchange other gas species (e.g., CO_2 and O_2) for photosynthesis, and uptake minerals from the soils. Thus, transpiration is the critical indicator for plant productivity. Physical evaporation is vaporization of the standing water on the soil or on plants to the overlying atmosphere, or the diffusion/ turbulent transport of water vapor of soil moisture from the soil surface to the overlying air.

The seasonal and regional variability of transpiration and evaporation is important in the context of the surface water budget and energy budget. Over large areas and an annual time scale, the surface water budget equation is expressed as

$$P = Et + R + dW \tag{15.1}$$

where P is surface precipitation; Et is transpiration and physical evaporation; R is surface runoff; and dW is the change in the surface moisture storage (such as soil moisture, snow cover, or reservoir). Note that ground-water movement is negligible in comparison with other terms for this time and spatial scale. Variability of water inputs from the atmosphere to the surface (P) will be balanced by the water outputs ($Et + R + dW$). On the same scale, the surface energy budget equation is expressed as

$$Q_{SW}(1-\alpha) + (1-\varepsilon)Q_{LW} = Q_L + Q_H + Q_{LW}^{emit} \tag{15.2}$$

where Q_{SW} is the incoming surface shortwave solar flux; a is the surface albedo; Q_{LW} is the downward atmospheric thermal flux; e is the thermal emissivity; Q_L is the turbulent latent heat flux (i.e., physical evaporation and transpiration); Q_H is the turbulent sensible heat flux; and Q_{LW}^{emit} is the thermal flux emitted from the surface. Thus, variability of energy inputs from the atmosphere to the surface $[Q_{SW} (1 - \alpha) + (1 - \varepsilon) Q_{LW}]$ will be balanced by the energy outputs from the surface to the atmosphere $(Q_l + Q_H + Q_{LW}^{emit})$.

Equations 15.1 and 15.2 are coupled through Et and Q_L. For example, if Et is in the unit of mm/day and Q_L is in the unit of W/m²,

$$Et = Q_L / (\rho_{liq}\lambda) = aQ_L$$

where ρ_{liq} is liquid water density (in kg/m³) and λ is the latent heat of vaporization (in J/kg); the density and λ vary slightly with temperature and pressure, but for typical conditions, $a = 0.0337$ (m²/W·mm/day). Plant transpiration is limited to the availability of root-zone soil moisture, and soil evaporation is limited to the top-soil moisture. Thus, Q_L is also coupled to dW. With respect to the energy and water balance equations, this entry examines seasonal and regional variability of Et through 30-year observations and simulations of a land-surface model (LSM).

Seasonal and Regional Variability of *Et* over the Conterminous United States

According to Equations 15.1 and 15.2, variability of Et or Q_L is extracted from the seasonal and regional variability of input terms: P and $Q_{SW} (1 - a)$. In this section, we discuss the roles and variability of net solar radiation, precipitation, land-cover type, and Et over the Conterminous United States (CONUS) through reviewing state-of-the-art observational datasets and numerical model simulation results from the North American Land Data Assimilation System (NLDAS).[1] The NLDAS project constructs atmospheric forcing and LSM datasets from the best available observations and model outputs. Unlike in situ or satellite data, NLDAS can provide spatially and temporally complete datasets that satisfy energy and budget balances in Equations 15.1 and 15.2. The use of separate observational platforms for datasets of P, Et, and other terms can lead to inconsistencies and imbalances in the closure of the water and energy cycle equations. For this section, we have compiled a 30-year seasonal climatology (March/April/May (MAM); June/July/August (JJA); September/October/November (SON); December/January/February (DJF)) from NLDAS Phase 2[2] datasets using precipitation observations and output from the Mosaic LSM from 1979 to 2009. It is important to note that NLDAS-2 precipitation is a state-of-the-art dataset over CONUS for such a long period, based primarily on a 1/8th-degree daily gridded analysis of rain gauge observations. This precipitation (and other NLDAS-2 surface meteorology, such as Q_{SW} and Q_{Lw}) is used as input to the LSM. The Mosaic LSM (like all LSMs) contains several parameterizations and assumptions in its model formulations to calculate Et, R, dW, Q_L, and Q_H. In the following sub-sections, we discuss the seasonal and regional variability of the key water and energy terms.

Precipitation and Solar Radiation

Surface precipitation, P, varies seasonally and regionally, characterized primarily by synoptic weather (i.e., large-scale circulation), mesoscale circulation (local disturbance), and thermodynamics (daytime heating). When precipitation reaches the surface, some water may be trapped on the canopy, stored on the surface of the soil (including as snow), be infiltrated into the soil, or run off to a river. Infiltrated moisture to the soil becomes the storage term, dW (Equation 15.1), that is the direct source of Et, through transpiration and/or soil evaporation.

Figure 15.1 shows a 30-year seasonal climatology of the precipitation from NLDAS-2. In the MAM, DJF, and SON periods, the coastal regions of the North West show a large amount of the precipitation generated by extratropical cyclones from the Pacific Ocean. The moisture flux from the north Pacific also generates snowfall or orographic rainfall over the mountains in the west. In the eastern domains, precipitation is peaked around the lower Mississippi Basin due to moisture flux from the Gulf of Mexico. Precipitation is higher along a narrow range of the northeast coastal region in the MAM and the SON periods. In the JJA period, the largest precipitation appears in the Deep South due to effects of sea breezes. Excepting winter, the South West tends to be drier for all seasons. The Great Plains become very dry during winter, and have light amounts of precipitation during other seasons.

Surface incoming solar radiation (Q_{SW}) varies along with seasonal cycles. Its seasonal variability is, at the first order, governed by the mean solar zenith angle due to tilt of earth's rotation axis relative to the sun. Its regional variability is affected by the presence of clouds and aerosols. Surface albedo (a: defined as a fraction of the incoming shortwave flux that is reflected by the surface) also affects the regional variability of the surface net solar radiation. Albedo differs widely depending on the land-cover type and vegetation *senescence,* ranging from 10% with dense forest to 50% over desert, but it could be up to 95% with the presence of fresh snow.

Figure 15.2 shows a 30-year seasonal climatology of the surface net solar radiation, $Q_{SW} (1 - a)$, from NLDAS-2 Mosaic. It is obvious that the largest net solar radiation is in the JJA period, whereas the lowest is in the DJF period, and is generally stratified in the latitudinal zones. This confirms that the largest factor for determining the net surface solar radiation is the mean solar zenith angle. Even in the same latitudinal zone, net solar radiation shows regional (longitudinal) variability at some extent, especially highlighted in the MAM period. For example, net solar radiation is ~40 W/m² higher in California than in Nevada. In Colorado, the Rocky Mountain region has spotty regions with low net solar radiation. These types of regional variability are attributed to differences in albedo associated with the land-cover type and the presence of cloud/aerosols and surface snow.

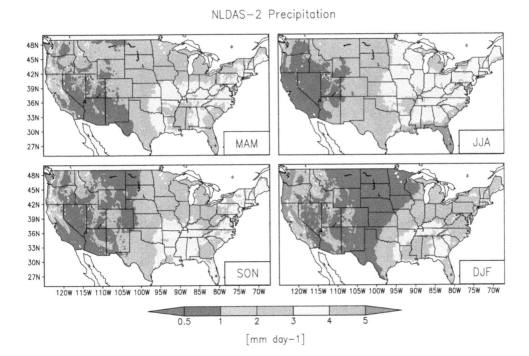

FIGURE 15.1 **(See color insert.)** Thirty-year seasonal climatology (Mar 1979–Feb 2009) of surface precipitation in NLDAS Phase 2.

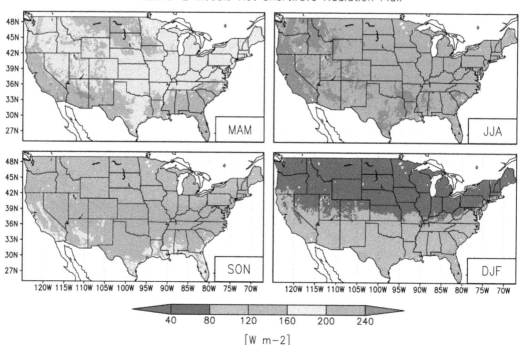

FIGURE 15.2   (See color insert.) Thirty-year seasonal climatology (Mar 1979–Feb 2009) of surface net solar radiation from the Mosaic LSM in NLDAS Phase 2.

Natural vegetation competes and evolves by taking the best advantage of solar radiation and rainfall patterns. Figure 15.3 shows the satellite-derived land-cover map used in NLDAS. Forest types extend over much of the eastern domain of the CONUS, coastal regions of the North West, and spotted over the Mountain region, where enough orographic rainfall sustains their life. Trees with a large canopy require more soil moisture to maintain their photosynthesis, transpiration, and mineral uptake. Some tree species adapt to the region where mean temperature significantly changes seasonally by becoming dormant during the cold seasons (e.g., deciduous forest). For areas with moderate climate throughout the seasons, tree species tend to be evergreen (e.g., evergreen forest). Grass and shrub species have smaller bodies, which require less number of resources to maintain. They can be dormant for long periods during drought and/or cold. Thus, they are well adapted for the dry regions, such as in the Great Plains, South West, and some of the Mountain regions. Cropland exists elsewhere in the CONUS, especially extended over the Midwest, lower Mississippi river basin, and the Great Plains. Agricultural practices have modified crop species, soil, and moisture sources for their climate and economy through a long-time history. Thus the distribution of cropland is not strictly regulated by the rainfall and net radiation patterns.

Transpiration and Physical Evaporation

For a given solar insolation and downwelling thermal radiation, surface total available energy is determined, balanced by energy release from the turbulent sensible heat flux (Q_H), the turbulent latent heat flux (Q_L), and the thermal flux emitted from the surface (Q_{LW}^{emit}). For further detail, these three terms in Equation 15.2 can be expressed as follows.

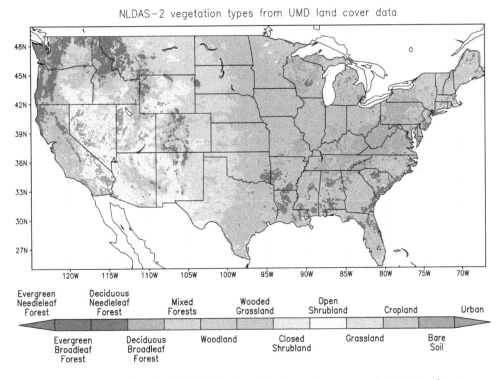

NLDAS−2 vegetation types from UMD land cover data

Evergreen Needleleaf Forest · Deciduous Needleleaf Forest · Mixed Forests · Wooded Grassland · Open Shrubland · Cropland · Urban

Evergreen Broadleaf Forest · Deciduous Broadleaf Forest · Woodland · Closed Shrubland · Grassland · Bare Soil

FIGURE 15.3   University of Maryland (UMD) dominant land-cover type map used in NLDAS Phase 2.

For a plant-leaf level,

$$Q_{net} = \rho_{atm}\lambda\left[q_{air} - q^{sat}(T_{leaf})\right]g_L + \rho_{atm}C_P(T_{air} - T_{leaf})g_H + \varepsilon\sigma T_{leaf}^4, \tag{15.3a}$$

and for soil level,

$$Q_{net} = \rho_{atm}\lambda(q_{air} - q_{soil})g_L + \rho_{atm}C_P(T_{air} - T_{soil})g_H + \varepsilon\sigma T_{soil}^4 \tag{15.3b}$$

where ρ_{atm} is dry air density; λ is latent heat of vaporization; q_{air} is water vapor mixing ratio of surrounding air; $q^{sat}(T_{leaf})$ is saturated water vapor mixing ratio at leaf temperature (T_{leaf}); g_L is water vapor conductance (stomatal and leaf aerodynamic conductance for plant leaf, soil aerodynamic conductance for soil); g_H is heat conductance (leaf or soil aerodynamic conductance); C_p is the specific heat of dry air, T_{air} is surrounding air temperature; ε is emissivity; σ is Stefan–Boltzmann constant; q_{soil} is water vapor mixing ratio of soil surface; T_{soil} is soil surface temperature.

Et plays a primal role to determine the energy balance between Q_H, Q_L, and Q_{LW}^{emit}. For example, strong wind and high stomatal conductance provide the best environment for plants to transpire more water vapor from leaves to the surrounding air, and consequently less energy is available for sensible heat flux and thermal emission, which is strongly coupled with leaf temperature. Consequently, leaf temperature must be reduced until it balances the net radiation through Equation 15.3a. Large *Et* indicates larger net solar radiation and available soil moisture from frequent precipitation, thus linking to larger exchange of CO_2 for photosynthesis (plant production) and larger uptake of root-zone soil moisture associated with various minerals required for their growth. Alternatively, a plant may become highly stressed due to lack of available soil moisture or a lack of wind, and transpiration and associated latent heat flux will be restricted, which will quickly increase leaf temperature, enhancing sensible heat flux and thermal emission. If such condition is extended for a long term, plants start wilting and become dormant to survive in a severe environment.

Soil evaporation performs a similar way to transpiration; it controls soil skin temperature and sensible heat flux (Equation 15.3b). A significant difference between soil evaporation and plant transpiration is the depths of soil moisture. Soil evaporation is linked to top-soil moisture, whereas plant transpiration is linked to root-zone soil moisture (up to several meters in depth). During short-term drought situation, deep-root tree species can maintain high Et and Q_L with deep soil moisture, even if physical evaporation is limited from the soil surface. It should be also noted that Equation 15.3a will be scaled to the leaf area index. So Equation 15.3a dominates the Equation 15.3b for dense forests, and vice versa for a less vegetated area.

Figure 15.4 shows the seasonal Et from the 30-year NLDAS-2 Mosaic climatology. Overall patterns of seasonal and regional variability of Et can be extracted from those of precipitation (Figure 15.1) and net solar radiation (Figure 15.2). From MAM to DJF, the trend of CONUS-mean Et is well explained by that of net solar radiation; e.g., Et is largest in JJA, decreases from MAM to SON, and lowest in DJF. The same trend is shown in the surface net solar radiation. For a given net solar radiation, seasonal precipitation and landcover type explains the regional variability. In MAM, light Et (~1–2 mm/day) is spread over the Great Plains and West in general. Moderate Et (2–4 mm/day) appears in eastern domains of the CONUS and the North West Coast. The similar regional pattern continues in the JJA period, but Et over the eastern domains is ~1 mm/day larger than the MAM period. Also, from MAM to JJA, the peak of Et shifts from the South to Midwest cropland regions. In the SON period, minimal Et (<0.5 mm/day) is shown across the South West, and Et decreases approximately 2 mm/day over eastern domains. In the DJF period, the minimal Et region is spread widely over the West and Midwest, and small Et (0.5–2 mm/ day) remains in the Southeast and the West Coast.

Figure 15.5 shows a scatter plot between NLDAS-2 precipitation and Mosaic Et. All values are a 30-year seasonal climatology, each mark represents a single grid-point within the 1/8th-degree NLDAS grid

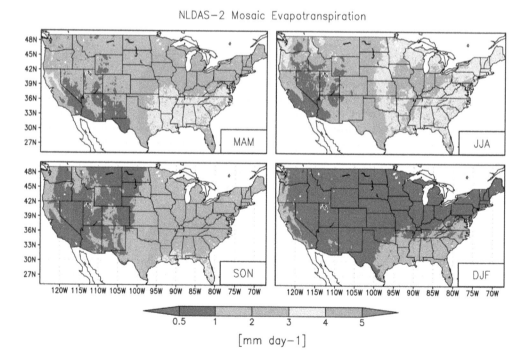

FIGURE 15.4 Thirty-year seasonal climatology (Mar 1979–Feb 2009) of Et from the Mosaic LSM in NLDAS Phase 2.

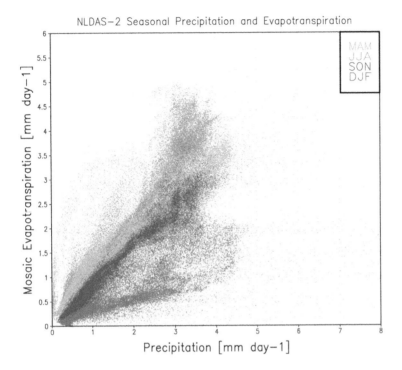

FIGURE 15.5 **(See color insert.)** Scatter plot between surface precipitation and Mosaic *Et* for the MAM, JJA, SON, and DJF periods in NLDAS Phase 2. Each scattered point represents a grid point over CONUS, and all values are based on a 30-year seasonal climatology (Mar 1979–Feb 2009).

(52,476 marks for each season); it therefore correlates Figures 15.1–15.4. For all seasons, precipitation and *Et* are well correlated with each other; it suggests that precipitation variability characterizes the regional *Et* variability. Slopes between *Et* and precipitation (defined as Et/precipitation) become steeper in the order from DJF (~0.25), SON (~0.6), MAM (~1.0), to JJA (~1.25), which is in the same order to the seasonal variability of surface net solar radiation over the CONUS (Figure 15.2). This means that JJA (DJF) has the largest (smallest) energy that is used to recycle surface precipitation back to the atmosphere by *Et*. Also of note is the number of points where the *Et* is larger than precipitation, especially during JJA, as shown in Figure 15.5 and comparing Figures 15.1 and 15.4. This effect is caused by changes in the *dW* storage term, showing evidence of soil moisture and snow cover that fell as precipitation in previous seasons being converted to *Et* during a season when more energy is available. Overall, Figure 15.5 suggests that, at first order, variability of mean surface net solar radiation determines the seasonal variability of *Et*, whereas precipitation variability determines the regional variability of *Et*. Variability of land-cover (vegetation) type (Figure 15.3) and regional variability of net surface radiation (Figure 15.2), as well as soil moisture and snow cover storage also affect the spread of the scatter plots.

Conclusion

Terrestrial transpiration and physical evaporation (Et) are important parameters for the surface energy and water budget equations. Variability of *Et* also feeds back to weather and climate through modulating planetary boundary layer, lower-troposphere thermodynamic profile, mesoscale circulation, and potentially general circulation.[3] It is critical to understand and monitor regional and seasonal variability of *Et* together with variability of vegetation senescence, land-cover, precipitation, and surface net radiation.

This entry examined *Et* variability from state-of-the-art NLDAS-2 data, and concluded that net surface radiation strongly controls seasonal variability of *Et*, whereas precipitation distribution strongly affects regional variability of *Et* over the CONUS. However, due to potential errors in observations and in numerical simulations, NLDAS (as well as other similar *Et* and precipitation datasets) and the conclusion could be biased to some extent. Currently, eddy turbulent covariance measurements at numerous locations globally have created an extensive surface network (FLUXNET).[4] It is the most direct and accurate *Et* in situ observational database, and in addition it is temporally continuous. Nevertheless, FLUXNET lacks spatial continuity, and its historical record of observations is somewhat limited.

Satellite- or aircraft-based visible–infrared imager sensors can measure various aspects of surface information, and can be translated into *Et* data. One of the notable measurements is the MOD 16 *Et* product, which adapts the Penman–Monteith equation to the various surface information derived from the MODerate Resolution Imaging Spectroradiometer (MODIS) sensor in Terra and Aqua satellites.[5] MOD16 *Et* can estimate *Et* at high spatial resolution (1km) over the globe, whereas satellite orbital patterns and algorithm limit the *Et* estimates at instantaneous (a.m. or p.m.) times to only once, possibly twice, a day. Technically, MOD16 *Et* product is available from an 8-day average; average *Et* of up to 8-day instantaneous *Et* estimate. It also involves uncertainties of *Et* retrieval with the Penman–Monteith algorithm.

An LSM, such as used within NLDAS, is composed of a set of surface boundary conditions and soil–vegetation–atmosphere–transfer parameterizations; thus, it can estimate *Et* spatially and temporally complete at satellite-resolvable resolution. Uncertainties of LSMs attribute to the boundary conditions, atmospheric forcing, and model physics. Thus none of the three is the best possible method to estimate *Et*; in other words, *Et* estimates from in situ, satellite, and LSMs are always necessarily to facilitate monitoring the regional and seasonal variability of *Et*. More comprehensive intercomparison and reviewing are available in Jiménez et al.[6] Continuous maintenance and development of an in situ *Et* network, satellite *Et* estimates, and LSMs will be required in the future, and simultaneously, a better assimilation method between the three datasets must be developed.

Acknowledgments

This work was supported by the NASA Modeling Analysis and Prediction project. The authors are grateful to Dr. D. Considine at NASA HQ. The NLDAS project is supported by NOAA's Climate Prediction Program for the Americas (CPPA) and by the NOAA Climate Program Office's Modeling, Analysis, Predictions, and Projections (MAPP) program. The data used in this effort were acquired as part of the activities of NASA's Science Mission Directorate, and are archived and distributed by the Goddard Earth Sciences (GES) Data and Information Services Center (DISC).

References

1. Mitchell, K.E.; Lohmann, D.; Houser, P.R.; Wood, E.F.; Schaake, J.C.; Robock, A.; Cosgrove, B.A.; Sheffield, J.; Duan, Q.; Luo, L.; Higgins, R.W.; Pinker, R.T.; Tarpley, J.D.; Lettenmaier, D.P.; Marshall, C.H.; Entin, J.K.; Pan, M.; Shi, W.; Koren, V.; Meng, J.; Ramsay, B.H.; Bailey, A.A. The multi-institution North American Land Data Assimilation System (NLDAS): Utilizing multiple GCIP products and partners in a continental distributed hydrological modeling system. J. Geophys. Res. **2004**, *109*, D07S90. doi:10.1029/ 2003JD003823.
2. Xia, Y.; Mitchell, K.; Ek, M.; Sheffield, J.; Cosgrove, B.; Wood, E.; Luo, L.; Alonge, C.; Wei, H.; Meng, J.; Livneh, B.; Lettenmaier, D.; Koren, V.; Duan, Q.; Mo, K.; Fan, Y.; Mocko, D. Continental-scale water and energy flux analysis and validation for the North American Land Data Assimilation System project phase 2 (NLDAS-2): 1. Intercomparison and application of model products. J. Geophys. Res. **2012**, *117*, D03109, 27 pp. doi:10.1029/2011JD016048.

3. Pielke Sr., R.A. Influence of the spatial distribution of vegetation and soils on the prediction of cumulus convective rainfall. Rev. Geophys. **2001**, *39*, 151–177.

4. Baldocchi, D.; Falge, E.; Gu, L.; Olson, R.; Hollinger, D.; Running, S.; Anthoni, P.; Bernhofer, Ch.; Davis, K.; Evans, R.; Fuentes, J.; Goldstein, A.; Katul, G.; Law, B.; Lee, X.; Malhi, Y.; Meyers, T.; Munger, W.; Oechel, W.; Paw, K.T.; Pilegaard, K.; Schmid, H.P.; Valentini, R.; Verma, S.; Vesala, T.; Wilson, K.; Wofsy, S. FLUXNET: A new tool to study the temporal and spatial variability of ecosystem-scale carbon dioxide, water vapor, and energy flux densities. Bull. Am. Meteorol. Soc. **2001**, *82* (11), 2415–2434.

5. Mu, Q.; Zhao, M.; Running, S.W. Improvements to a MODIS Global Terrestrial Evapotranspiration Algorithm. R. Sensing Environ. **2011**, *115*, 1781–1800.

6. Jiménez, C.; Prigent, C.; Mueller, B.; Seneviratne, S.I.; McCabe, M.F.; Wood, E.F.; Rossow, W.B.; Balsamo, G.; Betts, A.K.; Dirmeyer, P.A.; Fisher, J.B.; Jung, M.; Kanamitsu, M.; Reichle, R.H.; Reichstein, M.; Rodell, M.; Sheffield, J.; Tu, K.; Wang, K. Global intercomparison of 12 land surface heat flux estimates. *J. Geophys. Res.* **2011**, *116*, D02102. doi:10.1029/2010JD014545.

16

Water Storage: Atmospheric

Forrest M. Mims III
*Geronimo Creek
Observatory*

Introduction

This entry examines the variability, global distribution, and abundance of water vapor stored in the atmosphere. The hydrologic cycle is summarized, and some of the environmental effects of water vapor are mentioned. Also discussed are various methods employed to measure water vapor, including ground-based Global Positioning System (GPS) receivers and satellite instruments that have provided detailed knowledge about the global distribution of water vapor. Water vapor is the principal greenhouse gas, and concerns are addressed about the need to improve the understanding of its role in the climate system and the spatial and temporal accuracy of global total water vapor measurements.

Water Vapor

The properties of water are unique among Earth's ingredients that comprise the climate system. Only water is present as a gas, liquid, and solid. Water vapor is the most variable of the major gases that form Earth's atmosphere. Its concentration in a given parcel of air, which can vary from virtually none to several percent, is significantly affected by temperature, geography, and elevation and is subject to dramatic changes over the course of seasons and during the passage of weather systems. Because of the variability of water vapor, it is customarily ignored in lists of the fractional composition of the atmosphere's major gases. Thus, while a perfectly dry atmosphere is 78.08% nitrogen, 20.95% oxygen, 0.93% argon,[1] 0.039% carbon dioxide,[2] and an assortment of trace gases, accounting for the presence of water vapor reduces these percentages accordingly. Of the several major gases that comprise the bulk of the atmosphere, only

139

water vapor makes its presence known by condensing into visible clouds of liquid droplets or frozen particles. It is also the most variable of the atmosphere's major gases.

Despite its relatively modest fraction of the atmosphere, water vapor is essential for the hydrologic cycle and is responsible for tropospheric weather and maintaining Earth's average global temperature above the freezing point. The vapor phase of water eventually leads to precipitation that profoundly influences agriculture and the transformation of the landscape by erosion. Water vapor also enhances haze, particularly when it condenses on sulfate and other hygroscopic aerosols and significantly reduces visibility by increasing the aerosol optical thickness of the atmosphere.[3] Because people are cooled by the evaporation of perspiration, water vapor in the ambient air has a significant impact on a person's comfort level on warm days. Water vapor is also the most significant of the greenhouse gases.[4] Calculations made well over a century ago showed that the oceans would be frozen were it not for the warming effect of the atmosphere. Tyndall, who discovered that water vapor and carbon dioxide absorb infrared radiation, expressed it best when he wrote that water vapor "is a blanket more necessary to the vegetable life of England than clothing is to man. Remove for a single summer-night the aqueous vapor from the air... and the sun would rise upon an island held fast in the iron grip of frost."[5]

The Hydrologic Cycle

The environmental factors that lead to evaporation and condensation explain and control the hydrologic cycle, the continuous cycling of water between Earth and its atmosphere. Water molecules are constantly being evaporated from the oceans and other surface water unless the adjacent air is saturated with water vapor. Additional vapor arises from the evaporation of small droplets in the spray of waves. The rate of evaporation is increased appreciably when surface water is warmed by sunlight. Some of this heat is transferred to the overlying air, and this gives rise to convection as the warm air expands and its reduced density causes it to rise, along with its water vapor. Cool air can contain less water vapor than warm air, and as the warm air rises, it may cool to a point where it becomes fully saturated with water vapor. The water vapor will then condense into droplets or deposit as ice crystals onto microscopic particles of dust, sea salt, bacteria, and other aerosols, which are collectively described as condensation or ice nuclei, respectively.

Visible clouds are formed when the water vapor in large volumes of moist air is transformed into liquid droplets or ice crystals. While clouds contain considerable amounts of liquid or frozen water, they remain suspended in the air because their density is slightly less than or equal to the volume of dry air. Cloud droplets or ice crystals are initially only a fraction of a micrometer in diameter. They grow in size as they merge into drops a few millimeters in diameter that fall from the cloud when the tug of gravity exceeds the force of convective updrafts or as ice crystals that grow by deposition of water vapor and by the collision and collection of ice crystals to make snowflakes. Falling snowflakes can melt into raindrops if the lower atmosphere is above the freezing point.

The interconnected bodies of water known as oceans cover 71% of Earth's surface. Most water vapor originates from the oceans, where most precipitation falls and where it may be stored for many years before returning to the atmosphere. Precipitation that falls over land may be stored in soil or in aquifers, or it may run off into streams and rivers and be captured in reservoirs or returned to the ocean. Some precipitation is stored in glaciers. The hydrologic cycle is completed when the portion of the atmosphere depleted of its moisture by precipitation is restocked with fresh evaporation from bodies of water, with lesser contributions provided by transpiration from plants, exhalation and evaporated perspiration from animals and insects and sublimation from snow and ice. Volcanic eruptions, natural steam vents, gas wells, steam-powered and water-cooled electricity generators, and other processes also inject small amounts of water vapor into the atmosphere. Some of the water vapor in the upper atmosphere is a byproduct of the oxidation of methane.[6]

Total Abundance of Global Water Vapor

The total amount of water stored in the atmosphere can be calculated from the liquid equivalent of the globally averaged water vapor between the surface and the top of the atmosphere, a meteorological parameter with many names, including precipitable water (PW), integrated precipitable water (IPW), and total column water vapor (TCWV). TCWV is expressed as the depth of liquid water that will result if the total column is brought to the surface at standard temperature and pressure. Although column water vapor generally refers to the total precipitable water over a specific location, it may also apply to specific columns anywhere within the atmosphere.

Although TCWV over a specific site can be remotely inferred by a ground-based instrument to an accuracy of approximately 2–10%, expanding the measurement to the entire globe and providing regular updates is a challenging task that has yet to be satisfactorily completed. Important questions have been raised about some of the methods and differences in the findings of the major projects that have attempted to provide a detailed understanding of the global distribution of TCWV and its regional and global trends.[7] These studies blended data from various satellite instruments and upper air soundings by balloon-borne radiosondes. The National Centers for Environmental Prediction (NCEP)-National Center for Atmospheric Research (NCAR) reanalysis-2 project found a global average TCWV of 24.68 mm from 1988 to 1999.[8] The NASA Water Vapor Project (NVAP) found a global average TCWV of 24.46 mm from 1988 to 2001.[9] Despite problems with both studies, their global TCWV findings were within a few percent of an earlier estimate of 25 mm.[10]

The liquid equivalent of the water stored in the atmosphere can be estimated by subtracting the volume of Earth calculated from its volumetric mean radius of 6371 km[11] from the same radius to which the thickness of the atmosphere's TCWV has been added. The 1988–2001 global TCWV[9] gives an average total volume of 12,476 km³ of water stored in the atmosphere. While this is about five times the volume of all Earth's rivers,[12] it is slightly less than 0.001% of Earth's oceans.[13]

Global Distribution of Water Vapor

The total water vapor over both hemispheres has a distinctive annual cycle, with average TCWV significantly higher in summer than in winter, especially in the northern hemisphere. This is clearly shown in the summer and winter satellite images of the United States in Figures 16.1 and 16.2. Because the

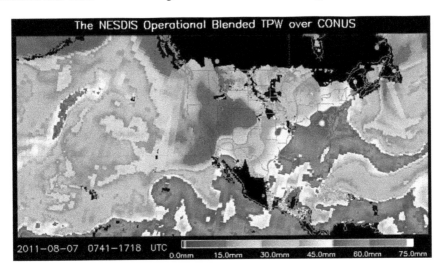

FIGURE 16.1   (See color insert.) Enhanced total water vapor over the continental United States on a summer day (7 August 2011). NESDIS Operational Blended TPW Products.

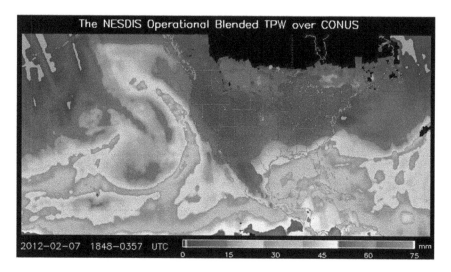

FIGURE 16.2 (See color insert.) Reduced total water vapor over the continental United States on a winter day (7 February 2012). NESDIS Operational Blended TPW Products.

hemispherical variations are asymmetrical, the global average TCWV also has an annual cycle, with one study giving 23 mm from November to January and 26.5 mm during July.[14] Spatial variations are substantial, with the polar regions and the Himalaya mountain range having a minimum TCWV less than 1 mm and the equatorial zone having a maximum TCWV of 70 mm or more. Annual TCWV measured by a GPS receiver at Hawaii's Mauna Loa Observatory, an alpine site (3.4 km above the mean sea level) surrounded by ocean, ranges from less than 1 to 15 mm, whereas TCWV at nearby sea level sites ranges from 20 to 50 mm.[15]

The greatest abundance of water vapor is stored in a band that encircles Earth known as the Intertropical Convergence Zone (ITCZ). This especially moist region swings from generally north of the Equator in July to generally south of the Equator in January. The ITCZ is obvious in the satellite image in Figure 16.3.

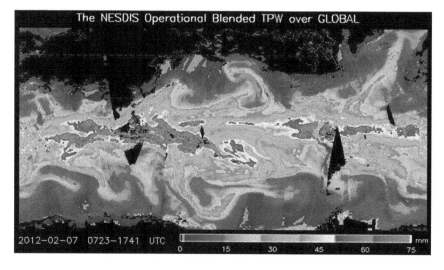

FIGURE 16.3 (See color insert.) Expanded view of Figure 16.2 showing global total water vapor. Note the significantly enhanced moisture over the tropics in the Intertropical Convergence Zone. Dark areas, mainly over land, indicate an absence of data (7 February 2012). NESDIS Operational Blended TPW Products.

Heating from extended hours of direct solar radiation transports considerable moist air in the ITCZ up into the middle and upper troposphere, where its water vapor condenses into heavy downpours. The atmospheric water vapor lost to precipitation is replenished by evaporation the next day.

Vertical Distribution of Water Vapor

While water vapor is present from the surface to the stratosphere, more than half is usually stored in the first few kilometers nearest to the surface. If Earth could be reduced to the size of a basketball, most of the planet's water vapor would be found in a film of air around the ball having an average thickness of several sheets of office paper. The amount of water vapor in a given parcel of air can be expressed as the ratio of the mass of the water vapor to that of an equivalent volume of completely dry air. This mixing ratio declines exponentially with elevation[16] until the water vapor mixing ratio in the stratosphere between 16 and 28 km declines to about 1/1000 of that at sea level.[17]

Residence Time in the Atmosphere

Because annual precipitation matches the sum of evaporation and the other processes that inject water vapor into the atmosphere, the mean residence time of a water molecule in the atmosphere can be estimated by dividing the average of global TCWV by the average global precipitation of 2.5 mm/day.[18] A global TCWV of 24.46 mm[9] gives a typical residence time of nearly 10 days. Some water vapor may return to a liquid state almost immediately, whereas vapor that reaches the stratosphere may remain there considerably longer than 10 days. Over geological time a given molecule of water spends minuscule time in the atmosphere. It is much more likely to be found immersed in an ocean, stored in an aquifer or locked in a glacier.

Measuring Atmospheric Water Vapor

The measurement of water vapor on a global scale is of increasing interest in view of its significance as a greenhouse gas and its role in climate models. TCWV is measured from the surface by various kinds of sun photometers that monitor changes in near-infrared sunlight caused by the absorption of water vapor[19,20] and by microwave radiometers that receive frequencies emitted by water molecules.[21] TCWV can also be measured by an inexpensive infrared thermometer pointed at a cloud-free zenith sky.[22] Water vapor delays the microwave signals from GPS satellites, and this provides a high-quality method for inferring TCWV.[23] The U.S. National Oceanic and Atmospheric Administration's (NOAA) Ground-Based GPS-IPW Network processes data from a vast array of more than 550 GPS receivers across the United States and a number of other countries and islands. The network provides automatic TCWV measurements that are posted online every half hour.[15] This GPS network is arguably the most precise source of site-specific TCWV data, and its expansion around the world would greatly improve the understanding of water vapor and its cycles and trends over land.

Since 1930 the vertical distribution of water vapor has been measured by humidity sensors flown from upper air sounding balloons.[24] Integrating the balloon humidity measurements provides the TCWV. Studies of the global distribution of water vapor based only on upper air sounding balloon measurements were biased because most sounding balloons are launched from land, which comprises only 29% of Earth's surface. Additional biases were caused by different humidity sensor designs and inconsistent performance, especially at reduced temperatures.

Earth satellites equipped with water vapor sensing instruments have dramatically improved the understanding of the global distribution of water vapor. Most satellite instruments monitor the infrared or microwave wavelengths emitted by water vapor.[25] Selective monitoring of these wavelengths permits TCWV and the water vapor at various levels in the atmosphere to be measured day and night. Some satellite instruments monitor TCWV by measuring changes in the water vapor absorbing near-infrared sunlight reflected from both oceans and land.[26] Near real-time water vapor imagery from various satellite sensors is available online, including the National Environmental Satellite, Data, and Information Service (NESDIS) Operational Blended TPW Products.[27] As shown in Figures 16.1 and 16.2, NESDIS imagery is particularly informative over the continental United States due to the inclusion of TCWV measurements by NOAA's GPS network.

Conclusion

Water vapor, the most variable of the major gases that form Earth's atmosphere, profoundly influences agriculture, weather, climate, and the transformation of the landscape by erosion. Because of its direct and indirect roles in weather and climate, water vapor is a subject of intense scientific scrutiny.[28] Satellite observations of the water vapor stored in the atmosphere have dramatically improved our understanding of its global distribution, abundance and seasonal cycles. Yet the identification and analysis of regional and global trends are made difficult by uncertainties in both the accuracy of instrumental measurements and how best to merge data from ground, upper air, and satellite instruments. A routine understanding of the global distribution of water vapor and its trends will require years of additional observations by a mix of carefully planned and calibrated ground- and space-based instruments. Long-range planning and closer collaboration among the various international institutions and agencies that monitor water vapor and analyze its trends will enhance this process.

Acknowledgments

The author is grateful to the atmospheric scientists who provided advice about water vapor and its measurement that led to his TCWV measurements since 1990 and this entry, including Seth Gutman, Brent Holben, Yoram Kaufman, Roger Pielke Sr., Robert Roosen, Glenn Shaw, and Frederick Volz.

References

1. Lide, D.R., Ed.; Atmospheric composition. In *CRC Handbook of Chemistry and Physics 2006–2007*, 87th Ed.; Taylor and Francis: Boca Raton, FL, 2007; 14–3 pp.
2. http://scrippsco2.ucsd.edu/data/in_situ_co2/monthly_mlo. csv (accessed 6 February 2012).
3. Tang, I.N. Chemical and size effects of hygroscopic aerosols on light scattering coefficients. J. Geophy. Res. **1996**, *101* (D14), 19245–19250.
4. Meckler, S.B. *Special Report: Water Vapor in the Climate System*; American Geophysical Union: Washington, DC., 1995.
5. Tyndall, J. On radiation through the Earth's atmosphere. Philosophical Mag. **1863**, *4* (25), 200–206.
6. McCormick, M.P.; Chou, E.W.; McMaster, L.R.; Chu, W.P.; Larsen, J.C.; Rind D.; Outmans, S. Annual variations of water vapor in the stratosphere and upper troposphere observed by the stratospheric aerosol and gas experiment II. J. Geophy. Res. **1993**, *98* (D3), 4867–4874.
7. Trenberth, K.E.; Fusillo, J.; Smith, L. Trends and variability in column-integrated water vapor. Climate Dynamics **2005**, *24* (7–8), 741–758.
8. Amen, G.G.; Kumar, P. NVAP and Reanalysis-2 global precipitable water products: intercomparison and variability studies. Bull. Amer. Meteor. Soc. **2005**, *86*, 245–256.
9. Vonder Haar, T.H.; Forsythe, J.M.; McKague, D.; Randel, D.L.; Ruston, B.C.; Woo, S. *Continuation of the NVAP Global Water Vapor Data Sets for Pathfinder Science Analysis*, Science and Technology Corporation Technical Report 3333, October 2003, Fig. 24, 26 http://eosweb.larc.nasa.gov/PRODOCS/nvap/sci_tech_report_3333.pdf (accessed at 06 February 2012).
10. Trenberth, K.E.; Guillemot, C.J. The total mass of the atmosphere. J. Geophys. Res. **1994**, *99* (D11), 23,079–23,088.
11. Moritz, H. Geodetic Reference System 1980. J. Geodesy **2000**, *74* (1), 128–162.

12. Claudio Cassardo, C.; Jones, J.A.A. Managing water in a changing world. Water **2011**, *3*, 618–628 http://www.mdpi.com/2073-4441/3/2/618/pdf (accessed 6 Feb. 2012).

13. Trenberth, K.E.; Smith, L.; Qian, T.; Dai, A.; Fusillo, J. Estimates of the global water budget and its annual cycle using observational and model data. J. Hydrometeorol. **2007**, *8* (4), 758–769.

14. Randel, D.L.; Vonder Haar, D.H.; Ringerud, M. A.; Stephens, G.L.; Greenwald, TJ.; Combs, C.L. A new global water vapor dataset. Bull. Amer. Meteor. Soc. **1996**, *77*, 1233–1246.

15. http://gpsmet.noaa.gov (accessed 10 Feb 2012).

16. U.S. Standard Atmosphere, 1976. *National Oceanic and Atmospheric Administration, National Aeronautics and Space Administration, U.S. Air Force*; U.S. Government Printing Office: Washington, DC, 1976; Table 20, 44.

17. Oltmans, S.J.; Vömel, H.; Hofmann, D.J.; Rosenlof, K.H.; Kley, D. The increase in stratospheric water vapor from balloonborne, frostpoint hygrometer measurements at Washington, D.C., and Boulder, Colorado. Geophys. Res. Lett. **2000**, *27* (21), 3453–3456.

18. Huffman, G.J.; Adler, R.F.; Arkin, P.; Chang, A.; Ferraro, R.; Gruber, A.; Janowiak, J.; McNab, A.; Rudolf, B.; Schneider, U. The Global Precipitation Climatology Project (GPCP) combined precipitation dataset. Bull. Amer. Meteor. Soc. **1997**, *78* (1), 5–20.

19. Fowle, F.E. The spectroscopic determination of aqueous vapor. Astrophys. J. **1912**, *35*, 149–162.

20. Mims III, F. M. An inexpensive and stable LED Sun photometer for measuring the water vapor column over South Texas from 1990 to 2001. Geophys. Res. Lett. **2002**, *29* (13), 20–21 to 20–24.

21. Liljegren, J.C. Two-channel microwave radiometer for observations of total column precipitable water vapor and cloud liquid water path. In *Fifth Symp. on Global Change Studies*; Nashville, T.N., Ed., American Meteorological Society: Boston, MA, 1994; 262–269 pp.

22. Mims, F.M.; Chambers, L.H.; Brooks, D.R. Measuring total column water vapor by pointing an infrared thermometer at the sky. Bull. Amer. Meteor. Soc. **2011**, *92* (10), 1311–1320.

23. Gutman, S.; Benjamin, S. The role of ground-based GPS meteorological observations in numerical weather modeling. GPS Solutions **2001**, *4* (4), 16–24.

24. Pettifer, R. From observations to forecasts—Part 2. The development of in situ upper air measurements. Weather **2009**, *64* (11), 302–308.

25. Schmit, T.J.; Feltz, W.F.; Menzel, W.P.; Jung, J.; Noel, A.P.; Heil, J.N.; Nelson, J.P.; Wade, G.S. Validation and use of GOES sounder moisture information. Wea. Forecasting **2002**, *17* (1), 139–154.

26. Kaufman, Y.J.; Gao, B.C. Remote sensing of water vapor in the near IR from EOS/MODIS. IEEE Trans. Geosci. Remote Sens. **1992**, *30*, 871–884.

27. http://www.osdpd.noaa.gov/bTPW/index.html (accessed 6 February 2012).

28. Vonder Haar, T.H.; Bytheway, J.L.; Forsythe, J.M. Weather and climate analysis using improved global water vapor observations. Geophys. Res. Letts. **2012**, *39* (15), L15802, doi:10.1029/2012GL052094.

II

Weather and Climate

Agroclimatology

Dev Niyogi and
Olivia Kellner
Purdue University

Introduction

Agroclimatology can be considered the study of local climate (determined by water, soil, and radiation in a given area, along with biomass and daily weather) and that local climate's interaction with agriculture and crop production for food, fiber, fuel, and availability of feed for livestock. It can help answer questions such as: how do temperature and precipitation patterns affect agricultural productivity? Agroclimatology can have many foci: climatic influences on crop production, availability of feed for livestock, crop modification to help plants withstand climate extremes, resilience of pests, sustaining yields in the face of an increasing world population, the influence of the microscale environment on crop yield, and agroclimatic modeling.[1,2] An emerging feature of the agroclimatic sciences has seen studies completed to understand the effect of climate change on agriculture, as well as the impact of agriculture on regional climates.[3–6]

Agroclimatology in the Beginning

The relation between world agricultural regions and global climatic patterns is well appreciated. It is also linked to a region's culture with seasonal festivals and events that celebrate the beginning and end of growing seasons. Agroclimatology is a relatively young science rooted in systematic weather observations of temperature and precipitation undertaken by many regional and national weather agencies such as, but not limited to, the U.S.A.'s National Weather Service, the Met Office of the United Kingdom, and the India Meteorological Department. Large-scale weather and climate events such as the Dust Bowl (~1920s, 1930s, and 1950s) that struck the U.S.A. have also contributed to the growth and evolution of agroclimatology as a practical science.[1,2,7] Agroclimatology can, however, be traced back to the domestication of native plants in societies that began thousands of years ago with a series of trial and error with local flora. This primitive science of plant domestication sometimes led to the collapse of societies due to weather and climate shifts that led to prolonged drought and crop failures (i.e., Anasazi of North America, the collapse of Polynesian society on Easter Island, and Mesopotamia that spanned modern day Turkey, Syria, Iran, and Iraq)[8] showing the need for a detailed understanding of weather, climate, and food production. Successes in crop domestication through the recognition of weather and climate patterns led to the establishment of distinctive crop-growing regions such as the Fertile Crescent in the Mediterranean region due to its mild climate and the U.S. Corn Belt due to its suitable glacial till soils and adequate rains. These two locations are examples of the many suitable crop-growing regions around the world where agricultural economies are well rooted.

Regular weather observations began with the development of the telegraph and railroad networks around the world during the Industrial Revolution (mid to late 1800s), which eventually led to the development of the national weather observation organizations across developed countries that kept record of daily weather data across the country. Daily weather observations were then shared via telegraph allowing for the development of daily weather maps that have grown through the century into the modern weather maps that meteorologists and climatologists use today.[9] As meteorology grew in the late 1800s, the development of regional weather climatologies from archived weather data of observed temperature and precipitation around the world began as well.[2,10] Building from these climatic data sets, climate systems were developed as a practical guide for sustaining flora and fauna. The Köppen climate system (1918) and the Holdridge life zones system (1947) are examples of two climate classification systems that characterize specified regions/climate zones based on temperature and precipitation (Köppen) or on precipitation, potential evapotranspiration, and humidity (Holdridge). These climate systems provide a quick overview of the region's agricultural potential.

Agroclimatology Today

In the early 1920s, the initial step of implementing agroclimatology as a practicing science was made by the United Kingdom Royal Meteorological Society with an expression of the need to study the effect of weather phenomena on crop growth and total yield.[7] Research investigating the impacts between climatic variables and crop yields began with simple regression analysis. These analyses eventually expanded over subsequent decades to simple statistical and empirical crop models by the 1970s and 1980s.[1,2,7] The development of remote-sensing technologies after World War II and into the 1970s and 1980s resulted in research programs and instruments capable of direct measurements of soil and moisture fluxes in the field (in situ) without human interference (remotely sensed data). Microsensors, automated weather stations, and remote-sensing technologies opened up a new dimension of detail in agroclimatology. The collection of meteorological, biophysical, and biogeochemical variables is now available in an unprecedented manner and this data is used to study and develop plant–climatic relationships. Some examples include assessing the effect of nitrogen levels on plant productivity or the effect of leaf area index on evapotranspiration and water demand in local areas, each of which is used to understand crop sustenance. This has allowed for much more detailed research and discoveries between crops and locally observed weather/climate patterns to take place.[1,2,11] Through the 1980s, crop modification, decision support systems and tools, and more detailed research have energized the development of crop models. Availability of observed weather patterns and observed surface data at micro to county-level scales has led to a broader understanding of hydrologic cycles within crop-climate analysis. Today's crop production models can simulate crop physiology and productivity reasonably well and have advanced to incorporate climate change scenarios providing insight into the possible impacts of varying weather and climate conditions on crop yields. In other words, while simple and statistical relations between climatic patterns and crop response continue to be important, there is a growing ability to develop more sophisticated, predictive tools that appear to adequately mimic complex agroclimatic interactions at multiple scales.

Current efforts to improve crop production amidst shifting climate patterns and dynamic economies include the efforts of international research scientists and crop models working in sync on projects such as the Agricultural Model Intercomparison and Improvement Project (AgMIP) and the Decision Support System for Agrotechnology Transfer (DSSAT).[12,13] Community efforts such as AgMIP consist of specialized teams comprising experts in agronomy, economics, and climate, devoted to the detailed study of specific crops, and collaborating and comparing more than multiple crop-specific (such as maize, wheat, rice, and sugarcane) models for crop response to carbon dioxide (CO_2), temperature, and other environmental factors. The research framework model for such collaborative assessments is to conduct studies at different locations to represent variation in crop production around the world and seek to find answers to, e.g., the following questions: Are models similar when responding to climate forcing parameters such as increased CO_2? Is the accuracy of ensemble model prediction better than individual model prediction? Does the detail of input data into the model affect how the model responds? The end goals of such efforts are to increase capacity through new adaptation strategies and methods for major agricultural regions in the developing and developed world. A number of crop models such as DSSAT, which is a compilation of crop simulation models founded on soil–plant–atmosphere dynamics for more than 28 crops, are now routinely available. The models simulate crop growth when given soil and meteorological input parameters to help assess the impacts of climate variability and change from farm-level to regional agricultural areas (Figure 17.1).

Importance of Agroclimatology

Agroclimatology is important for numerous reasons. Understanding weather and climatic impacts to crops and soils is important in order to feed increasing world populations. It is also important in turn to economic/commodity markets, livestock production (a large majority of livestock feed is developed from corn), and climatic risk assessment. Agroclimatology also becomes important in cost-benefit analysis when considering irrigation, fertilizers, tile drainage systems, cropping systems, and yield.[14]

The U.S.A. is the world's largest producer of corn. Extreme climatic events such as the 2012 drought can greatly influence crop production/yield, drastically impact commodity prices, and impact the agricultural sector of the economy. The 2012 drought across a large portion of the U.S.A. is a recent example of an extreme climatic event that dropped yield estimates of corn to 123.4 bushels per acre, the lowest since 1995. While not immediately felt in food prices by consumers, the effect of low yields can take a

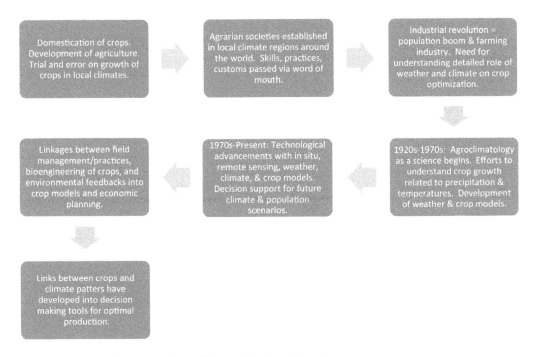

FIGURE 17.1 General narrative chart of the growth of agroclimatology.

year to trickle through production and increase the cost of consumables. Feedlots paid lower prices for feed cattle in 2013 as a result of the higher cost of feed in 2013 caused by reduced availability of pasture and decreased yield from the 2012 drought. The decrease in available pasture results in cattle being fed over a longer term with feed, but at lower weights because of the higher cost of feed in 2013. It is speculated that this will lead to greater production declines by 2014, which will increase cattle prices for almost 2 years after the drought occurred.[15]

Climatic risk assessment determines the risk (i.e., potential) of crop failure or success due to weather and climate variability over a growing season or over several growing seasons. Risk is often determined from the analysis of past growing seasons and potential future weather and/or climate scenarios as computed by crop models and expert projections. Usually, once a climatic risk assessment has been completed, a cost-benefit analysis is undertaken by producers based on the assessment of global vulnerabilities. Decisions include choice of when to plant amidst the balance of late frost risk at the beginning of the season, pest issues during the growing season, and having enough radiation/temperature (degree days) to complete maturity through harvest. Decisions related to crop insurance, fertilizers, and pesticide purchases are agronomically as well as economically important, and weather plays a dominant role in ensuring profitability. Thus, agroclimatology plays an important role in helping manage weather and climate risks for producers and stakeholders, and in turn better mitigating and adapting to climate change and variability.

Agroclimatology Today: Some Important Variables

Agroclimatology in the Twenty-First Century

Agroclimatology today focuses more readily on providing guidance in the form of weather and climate products, aiming to develop resilience in crops to climatic extremes, pests, diseases, and weeds to improve crop growth and production at local and regional scales.[1] Although agroclimatology is specifically focused on studying the impacts of weather/climate patterns on crop production and soils, it is a multidisciplinary field comprised of, but not limited to, knowledge from agronomists, soil scientists, biologists, hydrologists, meteorologists, climatologists, physical geographers, human geographers, sociologists, and economists.[2] Embedded within the agroclimatic notion is the societal need to provide decision support for current agriculture practices and weather catastrophes such as droughts or late frosts.[16]

Agroclimatology is connected to several sub-sciences: soils, hydrology, weather and climate, and agronomy. Different soil types possess distinctive hydraulic and thermodynamic properties that influence the growth rate of plants and the movement of water and nutrients through the soil and on the land surface. Surface and subsurface hydrology are affected not only by the type of soil but the type and amount of vegetation present over that soil. The density of a plant biomass on the surface influences the amount of rainfall reaching soils and the amount of radiation reaching the land surface. Plant roots can also affect infiltration and runoff rates. Weather and climate affect agronomic productivity mainly through precipitation, temperature, and radiation for plant photosynthesis and growth.

Multi-scale coupled soil–vegetation–atmosphere transfer (SVAT) processes regulate the hydrologic, energy, and nutrient transfer balance in agricultural landscapes (Figure 17.2). These transfer processes cascade across different scales that ultimately impact crop yield and profitability. Crop transpiration and photosynthesis regulate the nutrients used by the plant via the gross primary production (GPP) and the net primary productivity (NPP) of the agroclimatic system. The NPP, or yield, is linked with transpiration through the plant canopy and evaporation from the soil surface. The soil surface is the fundamental level where the carbon/nutrient and water link is established, but can be scaled beyond the leaf, plant, and to larger regional scales of total biomass/vegetation in a field. Thus, a detailed understanding of SVAT processes often becomes essential in providing agroclimatic guidance (Table 17.1).[17]

Changes in regional agricultural crop cover and greenness/phenology lead to changes in regional dew points and temperatures from the evapotranspiration from plants and/or irrigation. The spatial

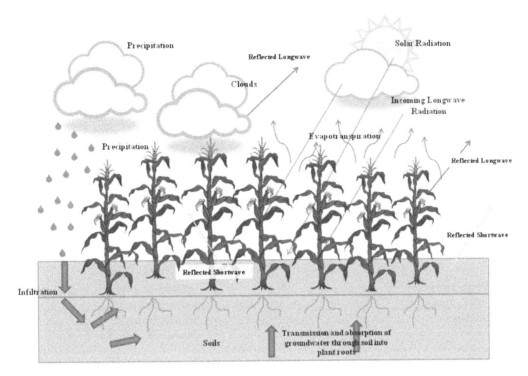

FIGURE 17.2 The soil–vegetation–atmosphere–transfer model (SVAT) represents the continuous feedback of radiative energy between the atmosphere, biosphere (crop), lithosphere (soil/land surface), and water in all its physical states: solid-ice crystals of clouds; liquid raindrops; and gas–water vapor/evapotranspiration. Primary components and associated parameters include the following: 1) soil—water retention, soil hydraulic conductivity, bulk density, water content, water potential, temperatures, infiltration, and evaporation; 2) canopy (vegetation/biomass)—leaf area index (LAI), physiology of the plant, leaf and plant density, seasonality, and evapotranspiration; 3) atmosphere—air temperature, humidity, wind speed, and solar radiation; and 4) hydrological cycle—evaporation, condensation, precipitation, infiltration, groundwater flow, surface runoff, and ponding.

extent and greenness of crop cover can result in increased moisture in the atmosphere and lowering of the surface air temperature.[18,19] This in turn can lead to changes in regional convective potential, cloud cover, and in some cases rainfall (Figure 17.3). Understanding these linkages is difficult, yet this is where agroclimatological assessments tend to help by providing a framework and understanding how a particular region would be expected to behave in a statistical/climatological sense. The fundamentals of such statistical/climatological relationships lie in the understanding of variables that are important for agricultural principles. Some of the climatic variables that are analyzed include temperature (maximum, minimum, average), rainfall characteristics (distribution, intensity), and solar radiation. Additional variables such as sunshine hours, humidity, winds, soil temperature, soil moisture, and evaporation are also needed but are generally difficult to measure with high spatiotemporal resolution or fidelity and are estimated from different models.[20,21]

Primary Agroclimatological Parameters

Weather and Climate

Weather can be defined by daily precipitation, temperature, and other dynamic (wind) and thermodynamic (humidity) weather patterns. Weather characterization ranges from temporal and spatial scales of seconds to decades and millimeters to thousands of kilometers. The most common temporal and spatial

TABLE 17.1   Common Soil and Vegetation Parameters in Land Surface Models Embedded within Some Crop Models and Weather/Climate Models

Primary Parameters: Dominant Types of Vegetation and/or Land Cover (USGS/NLCD Classifications)	Secondary Parameters (Estimated or Prescribed)
1) Vegetation type: cultivated or hay/pasture land (more detailed in crop models), forest type: deciduous, evergreen, mixed, shrubs	Saturated volumetric moisture content
	Wilting point volumetric water content
2) Dominant type of soil texture (USDA textural classification):	Soil thermal coefficient at saturation
Sand	Depth of the soil column
Loamy sand	Fraction of vegetation
Sandy loam	Minimum surface resistance
Loam	Leaf area index (LAI)
Sandy clay loam	Roughness length
Silty clay loam	Albedo
Clay loam	Emissivity
Sandy clay	Soil thermal conductivity
Silty clay	Soil thermal diffusivity
Clay	Soil hydraulic conductivity
3) Meteorological parameters:	Soil hydraulic diffusivity
Daily T_{max} and T_{min}	Soil heat capacity
Net solar radiation	Soil bulk density
Precipitation	Moisture flux resistance
Relative humidity	
Evapotranspiration	
Soil moisture	

Source: Adapted from Noilhan & Planton.[21]

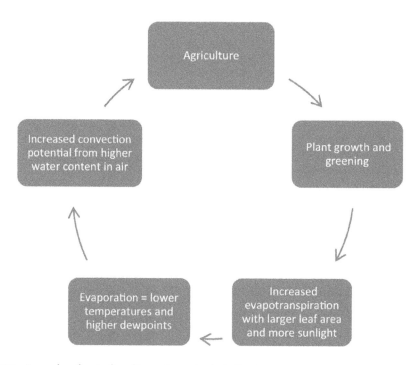

FIGURE 17.3   Examples of agricultural impacts on physical climate.

scales for daily weather are microscale (areas less than 1 km, seconds to minutes), mesoscale (1–100 km, several hours to a day), synoptic scale (100–1000 s km, days to a week), and planetary scale (1000 s km and lasts up to several weeks).[22,23] The average weather and its variability over a period of time delineate a specific area's climate.[10] The climate of a region is typically defined by the average and variance of temperature and precipitation, which is influenced by topography, proximity to water bodies, size of the given landmass, or any geographical feature including urban areas. Mountain ranges can lead to regional climate zones with higher precipitation climates on the windward side, while more arid and temperate climates are found downwind.[10,24] Coastal climates are moderated daily and seasonally by ocean waters and currents due to the higher specific heat capacity of water and resulting land/sea breezes.[10,22] Urban climates impact local temperatures, wind patterns, and boundary layer depths because of differences in surface layer energy and heat balances. A lack of vegetation and prolific extent of impervious, highly reflective, and/or highly absorbent land-cover types generate this type of localized climate.[22,23,25]

The average (and variation) over 30 years of recorded weather patterns such as daily rainfall, high and low temperature, and mean temperature is generally used to define a region's (climate zone, state, or multi-state region) set of climate "normals."[26] These climate normals are used extensively in agriculture to determine the baseline and the ensuing anomaly for any given season. In the U.S.A., climate divisions (typically nine per state) are determined to help the broader applications community be aware of temperature and precipitation patterns/and shifts at regional and state scales. Sometimes, additional information on frost depth, soil temperatures, and average soil moisture are available through the 30-year climatological summaries (Figure 17.4). For many regions within the U.S.A., the climate divisions are also aligned with the United States Department of Agriculture's crop reporting districts and assist climate–crop yield assessments.

Soils

The basic soil classifications are based on particle size. Soils are typically classified as gravel (greater than 2.00 mm), sand (0.05–2.00 mm), silt (0.002–0.05 mm), or clay (less than 0.002 mm). The amount of quartz and carbon content of a given soil type is also included as a characteristic in some models to clarify the surface layer soil type from the underlying soil layers. Each basic soil type has specific thermal and hydraulic characteristics that influence the temperature and moisture content of that specific soil. Soil heat capacity (amount of heat required to produce a given change of the temperature of the body that is largely influenced by the presence of water), thermal conductivity (how fast heat is transferred through soil), and hydraulic conductivity (how fast water is transferred through soil) are just a few of the many parameters that govern how suitable a subsurface climate may be for a specific crop.[27]

The land-surface (soil and vegetative surface) response is governed by solar radiation. Shortwave (SW) radiation is absorbed and continually reemitted by the land surface, with peak absorption and emission during the afternoon, and a dominant outward flux of longwave (LW) radiation in the form of sensible and ground heat flux back to space in the night.[20,22,23] Land surface radiation absorption or reflection is characterized by a ratio of absorbed to emitted SW radiation called albedo (a). Typical values of albedo are 0.05–0.40 for dark and wet to light and dry soils, 0.18–0.25 for agricultural crops, 0.15–0.20 for deciduous forests, 0.05–0.15 for coniferous forests, 0.0–0.10 (small zenith angle) to 0.10–1.0 (large zenith angle) for water, and 0.40 (old) to 0.95 (fresh) for snow.[23]

The values of incident net radiation (NR) are quantified through the radiation balance equation, which is the sum of incoming shortwave radiation (SW↓), outgoing longwave radiation (LW↑), sensible heat (H) (thermal heat transfer from the ground to overlying air via conduction and convection), latent heat (E) (heat released into the air as water vapor condenses back into water), and ground heat (G) (thermal heat energy transferred from the ground).[20,22,23] All together, these variables help quantify the surface radiation budget in relation to surface albedo that is primarily governed by surface characteristics:

$$SW \downarrow (1-\alpha) + LW \uparrow = NR$$

$$H + E + G = NR$$

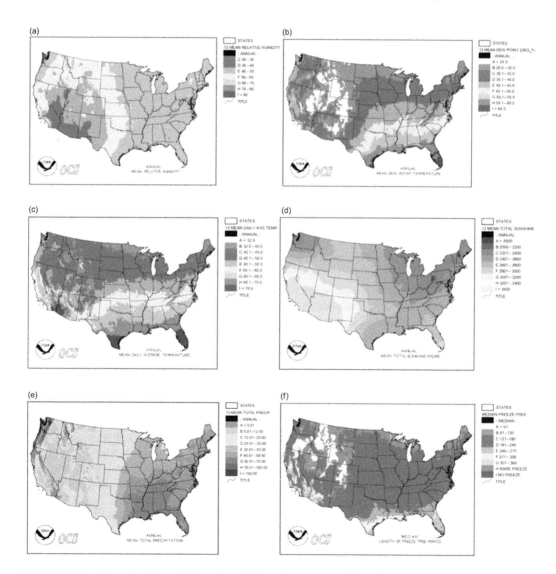

FIGURE 17.4 (See color insert.) Examples of climatology maps collected from the National Climatic Data Center's Climate Maps of the United States database. Maps are developed from the 1961 to 1990 period of record from official weather and climate station sites unless otherwise noted. **(a)** Annual mean relative humidity in percent, **(b)** annual mean dew point temperature in degrees Fahrenheit, **(c)** annual mean daily average temperature in degrees Fahrenheit, **(d)** annual mean total sunshine hours, **(e)** annual mean total precipitation in inches, and **(f)** median length of freeze-free period in days.

Data to understand energy fluxes is typically available from research field sites (e.g., Fluxnet, fluxnet. ornl.gov[28]). Data availability is limited because specialized measurements of surface sensible heat flux, latent heat flux, ground heat flux, carbon dioxide flux, moisture flux, evaporation/transpiration, surface temperature, soil moisture, and soil temperatures are challenging to obtain and require sophisticated equipment that needs ample care and maintenance. These data are typically used to develop and test newer plant-yield, environmental response relationships. Crop modelers develop empirical equations that can be applied in models and regional-scale analysis. In recent years, satellite platforms such as the NASA Moderate Resolution Imaging Spectroradiometer (MODIS) aboard Terra and Aqua, Atmospheric Infrared Sounder (AIRS), and the Advanced Microwave Scanning Radiometer-Earth

Observing System (AMSR-E) have also greatly contributed to the understanding of the Earth's surface and atmospheric radiation balances.[29–31] Near real-time analysis of leaf area index (LAI), normalized difference vegetation index (NDVI), enhanced vegetation index, evapotranspiration, surface temperature, and moisture stress index are some of the measurements these satellite platforms provide for research and application into crop models, along with near real-time information pertinent to irrigation and harvest.[32] Surface flux data along with satellite data has been assimilated into weather forecast, climate, and crop growth models, achieving a higher level of understanding of weather/ climate and crop production forecasts.[33,34]

Water

Agroclimatic decisions are acutely linked to hydroclimatology. Surface and ground water are readily coupled with the Earth's surface because they impact crop growth and evapotranspiration into the overlying atmosphere (which determines surface moisture flux). The moisture flux of the land surface is controlled by the saturation and temperature of the overlying air, wind speeds, turbulent eddies, and intensity of sunlight reaching the surface in the form of SW radiation.[8,11,20–22,35] Radiation reaching the Earth's soil surface contributes to the evaporation rate of water from the soil and is further impacted by vegetation. Plant physiology plays a large role in surface moisture flux: stomata (microscopic pores or openings on the leaves of plants) opens and closes based on environmental conditions such as carbon dioxide availability, soil moisture, atmospheric saturation, wind, and sunlight, releasing moisture into the atmosphere along with water vapor and oxygen exchange.[20,22,35] Moisture flux is a quantitative measure of soil moisture that moves up through the soil, through the plant, and evaporates out of the soil and transpires from the plant.[22,23,27,35] Soil moisture excess or deficits in the form of floods or droughts (along with temperature) have a direct impact on plant health and yield. The timing of water stress is also important in determining the crop response to the impact of stress. Crops 2–3 weeks after planting are typically more vulnerable than 2 months into the growing season.

Crop Management (vegetation/biomass)

Climate variables such as base temperature, growing degree days (GDDs), average frost days, first frost day, and last frost day help determine crop progress. Temperature becomes an important driver of crop development and is typically used as a resource available to the plant. This is quantified as GDDs, which are used to determine phenology such as when the seeds of row crops will germinate and grow in conjunction with determining pest development rates within the soil:

$$GDD = ((T_{max} - T_{min})/2) - T_{base}$$

The base temperature for corn is commonly 10°C, and typically after 10 consecutive GDDs (base time for corn), seeds germinate and grow.[20,35] Base temperature will vary by crop.

The last spring frost consideration is also crucial to seed germination and plant growth. If a seed is planted before the last frost, farmers risk plant loss because the seed will fail to germinate. As a result, climatologies of important crop management variables have been developed: 1) the first and last frost dates (e.g., in the Midwest U.S.A. in states such as Indiana, temperatures below 32°F and 28°F are used as baseline temperatures); 2) snowfall amounts; 3) temperatures above 95°F; 4) rainfall; 5) evapotranspiration loss; etc. Climatologies also exist for high-impact weather events such as high winds that can impact spray operations, hail that can cause crop losses, and heavy rain that can result in erosion and/or flood and surface ponding in fields. Developing a climatology of a specific variable includes taking the actual values and anomalies and averaging them with quality control procedures such as duration of measurement, time of day the measurement is observed, and site-specific location information such as proximity to trees, buildings, concrete, and elevation to develop the desired climatology. These climatologies are taken either at a station level (like the U.S. Cooperative Observer Network) if observations exist, or at climate division levels over a 30-year period (e.g., 1981–2010). Additional weather networks exist across different spatial and temporal scales that contribute to the development of climatologies and provide real-time weather data to producers and forecasters. Many states have what are called "mesonetworks" that comprise weather observation stations in counties across a given state. These stations can provide hourly readings of temperature, precipitation, winds, evapotranspiration estimates, and pressure in addition to daily maximum and minimum temperatures, and 24-hour rainfall/ precipitation totals. In the U.S.A., these data are typically available from respective state climate offices.

Extreme climatic events such as floods or droughts during the 1930s, 1950s, 1980s, and most recently 2012 contribute greatly to reduced yield, crop profitability, and world market stability. Forecasting climate extremes such as floods and drought still remains highly uncertain despite advancements in weather and climate modeling. Global features such as the El Nino Southern Oscillation (ENSO) are a large determining factor in North American weather regimes.[10] When compounded across seasons or even years, ENSO-related weather patterns can manifest and lead to anomalous events such as the 2012 drought.[36]

Current agroclimatic research is aligned toward characterizing droughts and their impacts on crop yields, including linking the interplay between agriculture–droughts–economic decisions and impacts. Drought is more than a lack of rainfall, and a number of indices have been applied to assess it. Drought impacts are characterized on the basis of timing (e.g., middle of the growing season well after planting), intensity, duration, spatial extent, and location (e.g., urban versus rural regions). Drought indices include basic analysis such as percentage of normal and quantitative meteorological measures, e.g., the Standardized Precipitation Index (SPI) and Palmer Drought Severity Index (PDSI).[37]

Development of climatologies include weather and climate variables such as evapotranspiration, annual precipitation, total sunshine hours, plant hardiness zones, and mean freeze-free period for agricultural applications (Figure 17.4). These climatologies provide a baseline value to compare to current weather and climate patterns, letting agricultural producers and stakeholders know if seasonal weather patterns are trending above or below normal. This could allow producers to better prepare for a change in their projected yield for the season, and hopefully mitigate loss if sufficient time is still present to act accordingly. For example, the Useful to Usable (U2U) climate information project sponsored by the U.S. Department of Agriculture (USDA) is developing an agricultural climatology for the U.S. Corn Belt region that includes the ENSO phase at the climate division level for cereal crop producers.[38] This climatology can provide guidance to farmers related to irrigation, fertilizer application, planting, and harvest-reducing crop vulnerability to changing weather patterns.[14]

Agroclimatology in the Future

An important question going forward is how can societies, economies, and countries sustain and continue to expand food supplies for a continually growing world population. This problem is further compounded with shifts in climate patterns, particularly temperature and precipitation across the world.[1,2,39,40] Agronomic decisions and productivity is thus linked to environmental and economic sustainability. Climate projections indicate a high probability of the subtropics becoming drier while the midlatitudes will have continued shifts in temperature, rainfall, cloud cover, and related climatic patterns.[41] The agricultural community has been working with adaptation approaches, which at a macroscale will need to address the current distribution of arable land for agriculture and the increased demand for more hybrid crops to withstand wetter/drier climates. New soil diseases can pose a threat in regions experiencing changes in seasonal temperatures as growing seasons become longer and the time a field lays fallow shortens.[42,43] Understanding these climate–pathogen relationships in a changing climate regime will likely become increasingly important.

In addition to the demand for hardier crops to surpass the challenges of a changing climate, the demand for biofuels will likely continue to rise as governments continue to mandate the reduction of greenhouse gas emissions. Current research in this area includes determining the best crop rotation cycles to protect soil nutrients while also maximizing the use of crops and crop waste such as corn and corn stalks, oil seeds, crop residue, and woody biomass for the production of bioenergy and ethanol. However, growing crops for bioenergy and ethanol instead of food for populations leads to societal questions that are also at the forefront of the climate change debate. In essence, future challenges facing agroclimatology include better decision-making tools, data acquisition, availability and uniformity (spatial and temporal), links between economic and regional decision makers, the development of more accurate and detailed crop and forecast models, and a more detailed understanding of climate change. These challenges require collaboration between different disciplines.

Conclusion

Agroclimatology is at an interesting juncture, becoming central to today's most challenging questions in a world of continued population growth, increasing food demand, and climate change. With technological advancements such as multiscale remote-sensing data, vegetation and moisture stress mapping, land data assimilation systems linked with crop models, and high resolution crop, weather and climate datasets, models will likely play an important role in the understanding and evolution of agroclimatology. Weather and climate data collection, assessment and mitigation of extremes such as floods and droughts, and climate and crop modeling will continue into the future as agroclimatology grows as a science. In addition, the feedback of agricultural climates at field and regional scales needs to be further researched to grasp the full understanding of heat and moisture fluxes between soils, the overlying boundary layer, vegetation, and the feedback into crop yields.

Despite large technological advancements in data collection and analysis, and a more detailed understanding of the agroclimatic system than ever before, questions regarding the future vulnerability of agro-climate systems still remain unanswered. Collaborative research and efforts to mitigate the effects of climate change need to be aggressively supported and addressed to allow for continued food production, development of bioenergies, and human survival in a world currently experiencing climate variability and change.[2] Changes in temperature and precipitation as projected using climate change studies will impact crops and non-crop species. As summarized by a recent USDA synthesis report,[43] the projected variability in precipitation and location shifts will require changes in water management practices (which will further feedback into the local climate system). Projected changes in temperature indicate the northward advance of frost-free days, opening the doors to additional regions for crop growth but also creating an environment that would be potentially conducive for invasive weeds and pests.[43]

There are at least two enduring challenges that impact current monitoring and modeling efforts in agroclimate. The first is due to scale disparity: 1) field measurements are often "point" data while effects are often regional scale in nature and not well captured; and 2) if remotely sensed/ satellite data are used for assessing the regional view, the dominant impact of local-scale decisions and micro features that can influence crop yields are not captured. Combination approaches involving the assimilation of multiscale products into a gridded assessment are currently underway and may likely alleviate the uncertainty due to this problem.[44,45] The second challenge can be linked to capturing the diversity in agronomic practices. For example, most studies consider the relationship between climate and crop patterns for a "typical" crop and the variations between hybrids are poorly assessed. Similarly, variability in farm-scale practices such as planting date, presence of tilling, crop cover, distances between crop rows, fertilizer use, and pest risk are also poorly captured and cause uncertainty in current assessments. Additional challenges due to socioeconomic choices and economic tradeoffs are also difficult to capture.

Glossary

Anomaly: The deviation of a measurable unit, (e.g., temperature or precipitation) in a given region over a specified period from the long-term average, often the 30-year mean, for the same region.[46]

Base temperature: The temperature below which plant growth is zero. The base temperature varies by crop.[47]

Biochemistry (biochemical): The study of chemical processes within (or pertaining to the physical processes of) and related to living organisms that explain the processes of life such as cell biology and signaling, development and disease, energy and metabolism, genetics, molecular biology, and plant biology.[48]

Bioenergy: Renewable energy developed from biological sources such as plant material to be used for heat, electricity, or fuel for vehicles.[49]

Biogeochemistry (biogeochemical): The study of the physical (or pertaining to the physical processes of), chemical, biological, and geological processes and reactions that govern the composition of and changes to the natural environment. It mainly consists on studying the cycles and interaction of elements carbon, nitrogen, sulfur, and phosphorous as they move through the atmosphere, hydrosphere, and lithosphere.[50]

Boundary layer: In general, a layer of air adjacent to a bounding surface. Specifically, the term most often refers to the **planetary boundary layer**, which is the layer within which the effects of friction are significant. For the Earth, this layer is considered to be roughly the lowest one or two kilometers of the atmosphere. It is within this layer that atmospheric turbulence dominates the exchanges, and temperatures are most strongly affected by daytime insolation and nighttime radiational cooling, and winds are affected by friction with the Earth's surface. The effects of friction die out gradually with increasing height, so the "top" of this layer cannot be exactly defined.[51]

Climate: The average weather over a 30-year period.[46]

Climate change: A non-random change in climate that is measured over several decades or longer. The change may be due to natural or human-induced causes.[46]

Climate system: The system consisting of the atmosphere (gases), hydrosphere (water), lithosphere (solid rocky part of the Earth), and biosphere (living) that determine the Earth's climate.[46]

Climate variability: Variation in a climatic parameter such as temperature or precipitation that varies from its long-term mean over a shorter time scale than climate change. Variability occurs on times scales of daily, seasonal, annual, inter-annual, to several years and is inclusive of changes observed with teleconnections such as the El Nino Southern Oscillation.[52]

Climatologies: A quantitative description of climate showing the characteristic values of climate variables over a specific region.[46]

Climatology: The description and scientific study of climate.[46]

Condensation: The physical process by which water vapor in the atmosphere changes to liquid in the form of dew, fog, or cloud; the opposite of evaporation.[46]

Convection: Transfer of heat by fluid motion between two areas with different temperatures. In meteorology, the rising and descending air motion is caused by differential heat or density/pressure. Atmospheric convection is almost always turbulent and is the dominant vertical transport process over tropical oceans and during sunny days over continents. The terms "convection" and "thunderstorms" are often used interchangeably, although thunderstorms are only one form of convection. In the ocean, convection is prominent in regions of high heat loss to the atmosphere and is the main mechanism for deep water formation.[46]

Degree day: For any individual day, degree days indicate how far that day's average temperature departed from 65°F. Heating degree days (HDD) measure heating energy demand. It is a measure to indicate how far the average temperature fell below 65°F. Similarly, cooling degree days (CDD), which measure cooling energy demand, indicate how far the temperature averaged above 65°F. In both cases, smaller values represent less fuel demand, but values below 0 are set equal to 0, because energy demand cannot be negative. Furthermore, since energy demand is cumulative, degree day totals for periods exceeding 1 day are simply the sum of each individual day's degree day total. For example, if some location had a mean temperature of 60°F on day 1 and 80°F on day 2, there would be 5 HDDs for day 1 (65 minus 60) and 0 for day 2 (65 minus 80, set to 0). For the day 1 + day 2 period, the HDD total would be 5 + 0 = 5. In contrast, there would be 0 CDDs for day 1 (60 minus 65, reset to 0), 15 CDDs for day 2 (80 minus 65), resulting in a 2-day CDD total of $0 + 15 = 15$.[46]

Dew point: The point at which the air at a certain temperature contains all the moisture possible without precipitation occurring. When the dew point is 65°F, one typically begins to feel the humidity. The higher the temperature associated with the dew point, the more uncomfortable one feels.[46]

Drought: Drought is a deficiency of moisture that results in adverse impacts on people, animals, or vegetation over a sizeable area. NOAA and USDA together with its partners provide short-and long-term drought assessments through the U.S. Drought Monitor.[46]

Eddy: Swirling currents of air at variance with the main current.[51]

Enhanced vegetation index (EVI): A satellite product designed to monitor the health of vegetation. It is considered to be an improvement upon NDVI (see below) correcting for distortions in reflected light caused by particulate matter and corrects for ground cover below vegetation. EVI is calculated in a similar manner to NDVI.[53]

Ensemble forecast: Multiple predictions from an ensemble of slightly different initial conditions and/ or various versions of models. The objective is to improve the accuracy of the forecast through averaging the various forecasts, which eliminates non-predictable components, and to provide reliable information on forecast uncertainties from the diversity amongst ensemble members. Forecasters use this tool to measure the fidelity of a forecast.[46]

ENSO (El Nino-Southern Oscillation): Originally, ENSO referred to El Niño/Southern Oscillation, or the combined atmosphere/ocean system during an El Niño warm event. The ENSO cycle includes La Niña and El Nino phases as well as neutral phases, or ENSO cycle, of the coupled atmosphere/ocean system though sometimes it is still used as originally defined. The Southern Oscillation is quantified by the Southern Oscillation Index (SOI).[46]

Evaporation: The physical process by which a liquid or solid is changed to a gas; the opposite of condensation.[46]

Evapotranspiration: Combination of evaporation from free water surfaces and transpiration of water from plant surfaces to the atmosphere.[51]

Extratropical: In meteorology, the area north of the Tropic of Cancer and the area south of the Tropic of Capricorn. In other words, the area outside the tropics.[46]

Flux: The rate of transfer of fluids, particles, or energy per unit area across a given surface (amount of flow per unit of time).[51]

Gridded assessment: The process of spatially interpolating observed weather and/or climate data across spatial grids or a specified resolution for the development of a continuous data set for analysis.

Gross primary production: The total amount of energy or nutrients created by an ecological unit/ organism in a given length of time. Some of this energy is used for cellular respiration and maintenance of the organism tissue. The remaining energy not used for this process is **net primary production**.[54]

Gross primary productivity: The rate at which photosynthesis or chemosynthesis (gross primary production) occurs in an organism/ecosystem.[54]

Growing degree days: A heat index that relates the development of plants, insects, and disease organisms to environmental air temperature. The index varies depending on whether it is a cool, warm, or very warm season plant. For example, a corn growing degree day (GDD) is an index used to express crop maturity. The index is computed by subtracting a base temperature of 50°F from the average of the maximum and minimum temperatures for the day. Minimum temperatures less than 50°F are set to 50, and maximum temperatures greater than 86°F are set to 86. These substitutions indicate that no appreciable growth is detected with temperatures lower than 50 or greater than 86. If the maximum and minimum temperatures were 85 and 52, you would calculate the GDD by $((85 + 52/2) - 50) = 18.5$ GDD.[46]

Hardiness zone: Eleven zones in the U.S.A. based on average annual minimum temperatures. Zones are used by horticulturalists and nurseries to characterize plants by their hardiness; the hardiness zone maps can then be used to determine the likely survivability of particular plant species and varieties according to one's local growing area.[47]

Holdridge life zone system: A classification of ecosystems/climate zones developed by American botanist, Leslie Holdridge. The life zone system is based on the principle understanding that climate and plants have an intertwined relationship. This model relates climates and plants based on three properties: temperature, humidity, and precipitation. The classification of a life zone is determined though a triangular diagram that quantifies annual precipitation on the right side, potential evapotranspiration ratios on the left side, and humidity provinces along the base of the triangle. Tiers of the triangle designate latitudinal regions and altitudinal belts.[56]

Hybrid: A heterozygous (different alleles that code for the same gene or trait) offspring of two genetically distinct parents. An example would be a mule, offspring of a female horse and a male donkey. Sweet corn varieties are hybrids selectively bred for color(s) and sweetness.[55]

Insolation hours measure the duration of direct solar radiation for a given location on the Earth.[60]

Invasive species: Non-native or exotic species of plants, animals, and pests whose introduction to an ecosystem causes or is likely to cause economic or environmental harm or harm to human health.[47]

Köppen climate system: A widely used climate classification system developed by Vladimir Köppen. The Köppen classification system is based on a subdivision of terrestrial climates into five major types with four climates defined by temperature criteria and one type defined by dryness/aridity. Although Köppen's classification did not consider highland climate regions, the highland climate category, or H climate, is sometimes added to climate classification systems to account for elevations above 1,500 m (about 4,900 ft).[57]

Leaf area index (LAI): A dimensionless value that is used to characterize plant canopies and is defined as the one-sided green leaf area per unit ground surface area. It ranges from 0 to 10, with lower values representing less dense plant canopies and higher values more dense canopies. It is used as an eco-physiological measure of the photosynthetic and transpirational surface within a canopy, and/or as a remote-sensing measure of the leaf reflective surface within a canopy.[58]

Moisture/water stress: Most commonly expressed via discoloration and wilting of the plant due to the plant having to use more energy to remove water from the soil. The amount of force needed for a plant to remove water from the soil is known as the matric potential. When soil moisture is low, plants have to use more energy to remove water from the soil, thus the matric potential is greater. When the soil is dry and the matric potential is strong, plants show symptoms of stress.[61]

Normalized Difference Vegetation Index (NDVI): A satellite product that shows the health of vegetation. It is a ratio of the difference between near-infrared radiation and visible radiation detected by the satellite sensor to the sum of near-infrared radiation and visible radiation detected by the satellite sensor. Healthy vegetation absorbs most of the visible spectrum and reflects a large portion of the near-infrared spectrum. Unhealthy or sparse vegetation reflects more visible radiation and less near-infrared radiation. NDVI values range from –1 to 1. A zero means no vegetation and a positive value close to 1 indicates a high density of green leaves/healthy vegetation.[53]

Palmer Drought Severity Index (PDSI): An index that compares the actual amount of precipitation received in an area during a specified period with the normal or average amount expected during that same period. It was developed to measure the lack of moisture over a relatively long period of time and is based on the supply and demand concept of a water balance equation. Included in the equation are evaporation, soil recharge, runoff, temperature, and precipitation data.[46]

Phenotype: The outward appearance of the individual or organism. The phenotype results from interactions between genes, and between the genotype and the environment.[55]

Relative humidity: An estimate of the amount of moisture in the air relative to the amount of moisture that the air can hold at a specific temperature. For example, if it is 70°F near dawn on a foggy summer morning, the relative humidity is near 100%. In the afternoon, the temperature soars to 95°F and the fog disappears. The moisture in the atmosphere has not changed appreciably, but the relative humidity drops to 44% because the air has the capacity to hold much more moisture at a temperature of 95°F than it does at 70°F. But even when the relative humidity is "low" at 44%, it is a very humid day when the temperature is 95°F. For this reason, a better measure of comfort is dew point.[46]

Remote sensing: The science/process of collecting data through the detection of energy by sensors that is reflected from earth. Sensors are commonly onboard satellites or attached to aircraft. The sensors are not in the immediate environment where data is being collected. Sensors are either active or passive. Passive sensors detect only reflected radiation. Active sensors emit energy that is reflected back to the sensor to detect features on the Earth.[59]

Spectral instrument: An in situ (within the immediate environment) or remotely placed sensor that detects and records electromagnetic radiation from the sun that is reflected, transmitted, or absorbed (and subsequently emitted) from a surface or element. All surfaces or elements (such as trees, crops, soils, etc.) have a distinct spectral signature based on wavelength(s).

Standardized Precipitation Index (SPI): A probability-based way of measuring drought as a measure of precipitation across different time scales. This index is negative for drought and positive for wet conditions.[62]

Subtropical: A climate zone adjacent to the tropics with warm temperatures and little rainfall.[46]

Sunshine hours: A climatological indicator of cloudiness—measures the duration of sunshine in a given period (commonly a day or year) for a given location on the Earth. **Transpiration**: Water discharged into the atmosphere from plant surfaces.[51]

Tropics: Areas of the Earth within 20° north and south of the equator.[51]

Acknowledgments

NIFA/USDA—Useful to Usable (U2U): Transforming Climate Variability and Change Information for Crop Producers: Agriculture and Food Research Initiative Competitive Grant no. 2011–68002–30220, NSF INTEROP dri-NET, USDA NIFA Drought Trigger projects at Purdue through Texas A&M, and NSF COSIEN project at Purdue through UC Berkeley.

References

1. Decker, W.L. Developments in agricultural meteorology as a guide to its potential for the twenty-first century. Agric. Forest Meteorol. **1994**, *69* (1), 9–25.
2. Steiner, J.L.; Hatfield, J.L. Winds of change: A century of agroclimate research. Agron. J. **2008**, *100* (Suppl. 3), S-132-S-152. doi: 10.2134/agronj2006.0372c.
3. Pielke, R.A., Sr.; Pitman, A.; Niyogi, D.; Mahmood, R.; McAlpine, C.; Hossain, F.; Goldewijk, K.K.; Nair, U.; Betts, R.; Fall, S.; Reichstein, M.; Kabat, P.; de Noblet-Ducoudré, N. Land use/land cover changes and climate: Modeling analysis and observational evidence. WIREs Clim. Change **2011**, *2* (6), 828–850. doi: 10.1002/wcc.144.
4. Dirmeyer, P.A.; Niyogi, D.; de Noblet-Ducoudré, N.; Dickinson, R.E.; Snyder, P.K. Impacts of land use change on climate. Int. J. Climatol. **2010**, *30* (13), 1905–1907. doi: 10.1002/joc.2157.
5. Niyogi, D.; Kishtawal, C.; Tripathi, S.; Govindaraju, R.S. Observational evidence that agricultural intensification and land use change may be reducing the Indian summer monsoon rainfall. Water Resour. Res. **2010**, *46* (3), W03533. doi: 10.1029/2008WR007082.

6. Mishra, V.; Cherkauer, K.A.; Niyogi, D.; Lei, M.; Pijanowski, B.C.; Ray, D.K.; Bowling, L.C.; Yang, G. A regional scale assessment of land use/land cover and climatic changes on water and energy cycle in the upper midwest United States. Int. J. Climatol. **2010**, *30* (13), 2025–2044. doi: 10.1002/joc.2095.

7. Monteith, J.L. Agricultural meteorology: Evolution and application. Agric. Forest Meteorol. **2000**, *103* (1), 5–9.

8. Diamond, J.M. *Collapse: How Societies Choose to Fail or Succeed;* The Viking Press (Penguin Group USA): New York, **2004**; 592 pp.

9. Cox, J.D. Storm Watchers: The Turbulent History of Weather Prediction from Franklin's Kite to El Nino; Wiley & Sons: Hoboken, 2002; 252 pp.

10. Robinson, P.J.; Henderson-Sellers, A. *Contemporary Climatology;* Pearson Prentice Hall: Essex, **1999**; 317 pp.

11. Lenschow, D.H. Boundary layer processes and flux measurements. In *Handbook of Weather, Climate, and Water: Atmospheric Chemistry, Hydrology, and Societal Impacts;* John Wiley & Sons, Inc.: Hoboken, 2003; 966 pp.

12. Rosenzweig, C.; Jones, J.W.; Hatfield, J.L.; Mutter, C.Z.; Adiku, S.G.K.; Ahmad, A.; Beletse, Y.; Gangwar, B.; Guntuku, D.; Kihara, J.; Masikati, P.; Paramasivan, P.; Rao, K.P.C.; Zubair, L. The Agricultural Model Intercomparison and Improvement Project (AgMIP): Integrated regional assessment projects. In *Handbook of Climate Change and Agroecosystems: Global and Regional Aspects and Implications;* Hillel, D.; Rosenzweig, C., Ed.; Imperial College Press, 2012; Vol. 2, 263–280.

13. Jones, J.W.; Hoogenboom, G.; Porter, C.H.; Boote, K.J.; Batchelor, W.D.; Hunt, L.A.; Wilkens, P.W.; Singh, U.; Gijsman, A.J.; Ritchie, J.T. DSSAT cropping system model. Eur. J. Agron. **2003**, *18* (3), 235–265.

14. Takle, E.S.; Anderson, C.J.; Anderson, J.; Angel, J.; Elmore, R.; Gramig, B.M.; Guinan, P.; Hilberg, S.; Kluck, D.; Massey, R.; Niyogi, D.; Schneider, J.; Shulski, M.; Todey, D.; Widhalm, M. Climate forecasts for corn producer decision-making. Earth Interact. **2013**, doi: 10.1175/2013EI000541.1, *in press.*

15. Crutchfield, S. U.S. Drought 2012: Farm and Food Impacts. United States Department of Agriculture, Economic Research Service, http://www.ers.usda.gov/topics/in-the-news/us-drought-2012-farm-and-food-impacts.aspx#. UnEd8HDqhyI (Last updated on 26 July, 2013).

16. Prokopy, L.S.; Haigh, T.; Mase, A.S.; Angel, J.; Hart, C.; Knutson, C.; Lemus, M.C.; Lo, Y.; McGuire, J.; Morton, L.W.; Perron, J.; Todey, D.; Widhalm, M. Agricultural advisors: A receptive audience for weather and climate information? Weather Clim. Soc. **2013**, *5* (2), 162–167.

17. Campbell, G.S.; Norman, J.M. *An Introduction to Environmental Biophysics,* 2nd Ed.; Springer Science + Business Media: New York, 1998; 286 pp.

18. Fall, S.; Niyogi, D.; Gluhovsky, A.; Pielke Sr., R.A.; Kalnay, E.; Rochon, G. Impacts of land use land cover on temperature trends over the continental United States: Assessment using the North American regional reanalysis. Int. J. Climatol. **2010**, *30* (13), 1980–1993. doi: 10.1002/ joc.1996.

19. Niyogi, D.; Mahmood, R.; Adegoke, J.O. Land-use/landcover change and its impacts on weather and climate. Boundary-Layer Meteorol. **2009**, *133* (3), 297–298. doi: 10.1007/s10546–009-9437-8 (Editorial).

20. Griffiths, J.F., Ed. *Handbook of Agricultural Meteorology;* Oxford University Press: New York, 1994; 320 pp.

21. Noilhan, J.; Planton, S. A simple parameterization of land surface processes for meteorological models. Mon. Weather Rev. **1989**, *117* (3), 536–549.

22. Stull, R.B. *An Introduction to Boundary Layer Meteorology;* Springer LLC: New York, 1988; 683 pp.

23. Oke, T.R. *Boundary Layer Climates,* 2nd Ed.; Routledge: New York, 1988; 464 pp.

24. Holton, J.R. *An Introduction to Dynamic Meteorology,* 4th Ed.; Elsevier Academic Press: Waltham, 2004; 535 pp.

25. Landsberg, H.E. *The Urban Climate;* Academic Press: New York, 1981; 275 pp.
26. Arguez, A.NOAA's1980–2010 Climate Normals. 14 February 2013, http://lwf.ncdc.noaa.gov/oa/climate/normals/usnormals.html, 2013.
27. Houser, P.R. Infiltration and soil moisture processes. In *Handbook of Weather, Climate, and Water: Atmospheric Chemistry, Hydrology, and Societal Impacts;* John Wiley & Sons, Inc.: Hoboken, 2003; 966 pp.
28. Baldocchi, D.; Falge, E.;Gu, L.; Olson, R.; Hollinger, D.; Running, S.; Anthoni,P.; Bernhofer, C.; Davis, K.; Evans, R.; Fuentes, J.; Goldstein, A.; Katul, G.; Law, B.; Lee, X.; Malhi, Y.; Meyers, T.; Munger, W.; Oechel, W.; Paw, K.T.; Pilegaard, K.; Schmid, H.P.; Valentini, R.; Verma, S.; Vesala, T.; Wilson, K.; Wofsy, S. FLUXNET: A new tool to study the temporal and spatial variability of ecosystem-scale carbon dioxide, water vapor, and energy flux densities. Bull. Am. Meteorol. Soc. **2001**, *82* (11), 2415–2434.
29. Maccherone, B.; Frazier, S. About MODIS, http://modis.gsfc.nasa.gov/about/, 2013.
30. Graham, S.; Parkinson, C. AMSR-E, http://aqua.nasa.gov/about/instrument_amsr.php, 2011.
31. Ray, S. How AIRS Works, http://airs.jpl.nasa.gov/instrument/how_AIRS_works/, 2013.
32. Frolking, S.; Xiao, X.; Zhuang, Y.; Salas, W.; Li, C. Agricultural land-use in China. A comparison of area estimates from ground-based census and satellite-borne remote sensing. Global Ecol. Biogeogr. **1999**, *8* (5), 407–416.
33. Olioso, A.; Chauki, H.; Courault, K.; Wingeron, J. Estimation of evapotranspiration and photosynthesis by assimilation of remote sensing data into SVAT models. Remote Sens. Environ. **1999**, *68* (3), 341–356.
34. Kumar, A.; Niyogi, D.; Chen, F.; Barlage, M.; Ek, M.B.; Peters-Lidard, C.D. Assessing impacts of integrating MODIS vegetation data in the Weather Research Forecasting (WRF) model coupled to two different canopy-resistance approaches. Conference papers and proceedings, 90th American Meteorological Society Annual Meeting, Atlanta, Georgia, Jan 17–21, 2010.
35. Monteith, J.L. Climatic variation and the growth of crops. Q. J. R. Meteorol. Soc. **1981**, *107* (454), 749–774.
36. Hoerling, M.; Schumert, S.; Mo, K. *An Interpretation of the Origins of the 2012 Central Great Plains Drought,* NOAA Drought Task Force, Narrative Team, Silver Spring, Maryland, 2013;1–50, http://www.drought.gov/media/pgfiles/2012-Drought-Interpretation-final.web-041013_V4.0.pdf
37. Guttman, N.B. Comparing the Palmer Drought Index and the Standardized Precipitation Index. J. Am. Water Resour. Assoc. **1998**, *34* (1), 113–121.
38. Prokopy, L. *Useful to Useable: Transforming Climate Variability and Change Information for Cereal Crop Producers.* Agriculture and Food Research Initiative Competitive Grant no. 2011–68002–30220, https://drinet.hubzero.org/groups/u2u, 2013.
39. Downing, T.E.; Stowell, Y. Household food security and coping with climatic variability in developing countries. In *Handbook of Weather, Climate, and Water: Atmospheric Chemistry, Hydrology, and Societal Impacts;* John Wiley & Sons, Inc.: Hoboken, 2003; 966 pp.
40. Glantz, M.H. *Currents of Change El Nino Impact on Climate and Society;* Cambridge University Press: Cambridge, 1996, 194 pp.
41. Bates, B.C.; Kundzewicz, Z.W.; Wu, S.; Palutikof, J.P., Eds. *Climate Change and Water.* IPCC Secretariat: Geneva, 2008, 210 pp.
42. Southworth, J.; Pfeifer, R.A.; Habeck, M.; Randolph, J.C.; Doering, O.C.; Gangadhar Rao, D. Sensitivity of winter wheat yields in the midwestern United States to future changes in climate, climate variability and CO_2 fertilization. Clim. Res. **2002**, 22 (1), 73–86.
43. Walthall, C.L.; Hatfield, J.; Backlund, P.; Lengnick, L; Marshall, E.; Walsh, M.; Adkins, S.; Aillery, M.; Ainsworth, E.A.; Ammann, C.; Anderson, C.J.; Bartomeus, I.; Baumgard, L.H.; Booker, F.; Bradley, B.; Blumenthal, D.M.; Bunce, J.; Burkey, K.; Dabney, S.M.; Delgado, J.A.; Dukes, J.; Funk, A.; Garrett, K.; Glenn, M.; Grantz, D.A.; Goodrich, D.; Hu, S.; Izaurralde, R.C.; Jones, R.A.C.; Kim, S.-H.; Leaky, A.D.B.; Lewers, K.; Mader, T.L.; McClung, A.; Morgan, J.; Muth, D.J.; Nearing, M.;

Oosterhuis, D.M.; Ort, D.; Parmesan, C.; Pettigrew, W.T.; Polley, W.; Rader, R.; Rice, C.; Rivington, M.; Rosskopf, E.; Salas, W.A.; Sollenberger, L.E.; Srygley, R.; Stöckle, C.; Takle, E.S.; Timlin, D.; White, J.W.; Winfree, R.; Wright-Morton, L.; Ziska, L.H. Climate change and agriculture in the United States: Effects and adaptation. In *USDA Technical Bulletin*; U.S. Department of Agriculture: Washington, DC, 2013, http://www.usda.gov/.oce/climate_change/effects_2012/CC%20and%20 Agriculture%20Report%20(02–04–2013)b.pdf

44. Liu, X.; Niyogi, D.; Charusombat, U. Estimating Corn Yields Regionally across Midwest Using the Hybrid Maize Model with a Land Data Assimilation System, Abstract GC21H-07 of Presentation, American Geophysical Union Fall Meeting, San Francisco, California, Dec 3–7, 2012, http://198.61.161.98/abstracts/meetings/2012/FM/sections/GC/sessions/GC21H/abstracts/ GC21H-07.html.

45. Niyogi, D.; Liu, X. Adaptability of the Hybrid-Maize Model and the Development of a Gridded Crop Modeling System for the Midwest U.S. Presentation, ASA, CSSA, and SSSA International Annual Meetings, Cincinnati, Ohio, Oct 23, 2012. https://scisoc.confex.com/crops/2012am/ webprogram/Paper70435.html.

46. Climate Prediction Center Online Climate Glossary, http://www.cpc.ncep.noaa.gov/products/ outreach/glossary.shtml#C, 2013.

47. Womach, J. *Agriculture: A Glossary of Terms, Programs, and Laws, 2005 Ed.*, CRS Report for Congress; The Library of Congress: Washington D.C., http://www.cnie.org/NLE/CRSreports/05jun/97-905. pdf, 2005. (29 October 2013).

48. What is biochemistry? Biochemical Society, http://www.biochemistry.org/?TabId=456 (accessed November 2013).

49. Bioenergy. United States Department of Agriculture, Economic Research Service, http://www.ers. usda.gov/topics/farm-economy/bioenergy.aspx#.Um_cl3DqhyI (accessed November 2013).

50. Biogeochemistry. Woods Hole Oceanographic Institution, https://www.whoi.edu/main/topic/ biogeochemistry (accessed November 2013).

51. National Weather Service Online Glossary, http://w1.weather.gov/glossary/ (accessed November 2013).

52. Ramamasy, S.; Baas, S. *Climate Variability and Change*: Adaptation to Drought in Bangladesh; Asian Disaster Preparedness Center, Food and Agriculture Organization of the United Nations, ftp://ftp.fao.org/docrep/fao/010/a1247e/a1247e00.pdf, 2007. (29 October 2013).

53. Weier, J.; Herring, D. Measuring Vegetation (NDVI and EVI). NASA Earth Observatory, http:// earthobservatory.nasa.gov/Features/MeasuringVegetation/measuring_vegetation_1.php, 2000. (29 October 2013).

54. Amthor, J.S.; Baldocchi, D.D. Terrestrial higher plant respiration and net primary production. In *Terrestrial Global Productivity*; Academic Press: San Diego, CA, 2001; 33–59.

55. Agricultural Thesaurus and Glossary. United States Department of Agriculture, National Agricultural Library, http://agclass.nal.usda.gov/glossary.shtml (accessed November 2013).

56. Meier, B.L.; Osborn, J.; Knight, C. Life Zones Reflect Climate: Climate Change Demands Future Planning. NOAA Earth System Research Laboratory, Global Systems Division, http://www.esrl. noaa.gov/gsd/outreach/education/poet/LifeZones.pdf, 2013.

57. Arnfield, A.J. Köppen Climate Classification. *Encyclopedia Britannica*, http://www.britannica. com/EBchecked/topic/322068/Koppen-climate-classification (accessed October 2013).

58. Chen, J.M.; Rich, P.M.; Gower, S.T.; Norman, J.M.; Plummer, S. Leaf area index of Boreal forests: Theory, techniques, and measurements. J. Geophys. Res. **1997**, *102* (D24), 29429–29443.

59. Remote Sensing. National Oceanic and Atmospheric Administration http://oceanservice.noaa. gov/facts/remote-sensing.html (accessed October 2013).

60. Jarraud, M. *Guide to Meteorological Instruments and Methods of Observation, WMO-No. 8*, 7th Ed.; World Meteorological Organization: Geneva, Switzerland, 2008, ISBN 978-92-63-10008-5.

61. Pearson, K.; Bauder, J. How and When Does Water Stress Impact Plant Growth and Development? Presentation, American Society of Agronomy, Crop Science Society of America, and Soil Science Society of America 2003 Annual Meeting, Denver, Colorado, http://waterquality.montana.edu/docs/irrigation/a9_bauder.shtml, 2005.

62. Heim, R. Climate of 2013 – April: U.S. Standardized Precipitation Index. National Climatic Data Center, http://www.ncdc.noaa.gov/oa/climate/research/prelim/drought/spi.html (accessed November 2013).

18

Climate: Classification

Long S. Chiu
George Mason University

Introduction

Weather is a snapshot of the state of the atmosphere. The state variables include air pressure, temperature, moisture, cloud conditions, wind, and other environmental parameters. In a broad sense, climate is a "long-term" average of the weather.[1] According to the Intergovernmental Panel on Climate Change (IPCC), "Climate [is the] statistical description in terms of the mean and variability of relevant quantities over a period ranging from months to thousands or millions of years. The period is usually taken to be 30 years, as defined by the World Meteorological Organization (WMO)" while other periods are also used, depending on the available data and specific application.[2] Climate classification is the division of the Earth's climate into contiguous regions of relatively homogeneous statistics of climate elements or according to the dominant physical processes at work. This entry describes three basic approaches to climate classification with an example each, discusses state-of-the-art observations, and argues for integrating the observations for refinement and further development of climate classes to facilitate the assessment, monitoring, change detection of the Earth's climate, and sustainable exploitation of the Earth's natural resources.

Climate Classification Schemes

Brief History

The earliest climate classification was by the Greeks who divided the Earth according to the location relative to the Sun's inclination. Contemporary works in climate are mostly shaped by German meteorologists. For example, in 1817, Alexander von Humboldt (1769–1859) drew annual mean temperatures on a world map. Seasonal temperature ranges with improvements were included by Wladimir Köppen (1846–1940) in 1884. He later collaborated with Rudolf Geiger (1894–1981) in establishing his system of climate classification.[3,4] Tor Bergeron (1891–1971), the Germany-educated Swedish meteorologist, developed a classification scheme in terms of causes of climate in 1928. Alisov[5] classified climate regimes according to the physical causes such as circulation type, air mass formation, and frontal systems.[1] American climatologist and geographer, C. W. Thornthwaite (1899–1963) developed a hierarchical classification in 1931, essentially in terms of the annual pattern of soil-moisture conditions. His scheme was made more explicit with the introduction of potential evaporation in 1948, which, however, was superseded by the physical-based formulation of evaporation by Penman.[6]

Broadly, we can define three approaches to climate classification: genetic, empirical, and applied classification schemes. The genetic approach focuses on the cause of climate. An example is the air mass classification developed by Bergeron that later developed into the spatial synoptic classification (SSC) system often employed in synoptic climatology. An empirical approach is data driven and is focused on the effect of climate on a specific region. An example is the Köppen–Geiger classification scheme, which utilizes frequently observed atmospheric variables such as temperature and precipitation that are related to the land biomes. An application classification is created for, or is an outgrowth of, specific applications. An example is the Thornthwaite classification, which considers the role of vegetation on the soil-water availability.

Mathematical Climate Classification

Climate comes from the Greek *klima* which means inclination. The earliest Greek climate classification divides the hemispheres into three zones based on their latitudes: the "summerless," "intermediate," and "winterless" zones, labeled the torrid, temperate, and frigid zones, which are bounded by the Arctic Circle, Antarctic Circle, Tropic of Cancer, and Tropic of Capricorn, in that order.[1] Major improvements by Supan[7] take account of the actual observations of temperature, precipitation, wind, and orography. The Earth is divided into a hot belt, two temperate belts, and two cold caps, and regions are classified as polar, temperate, tropical, continental, marine, and mountain climates, and other types, with variations. He further classified the Earth into 34 climate provinces. However, no attempts were made to relate climate types to different regions.[1]

Bergeron and Spatial Synoptic Classification

The Bergeron classification is the most widely accepted form of air mass classification, focused primarily on the origin of the air masses. Three letters in the air mass classification specify the air mass types. The first letter describes its moisture properties; c is used for continental air masses (dry) and m for maritime air masses (moist). The second letter describes the thermal characteristic of its source region: T for tropical, P for polar, A for Arctic or Antarctic, M for monsoon, E for equatorial, and S for subsidence air (dry air formed by significant downward motion in the atmosphere). The third letter designates the stability of the atmosphere. If the air mass is colder (warmer) than the ground below it, it is labeled k (w).[1] During the 1950s, air mass identification was used in weather forecasting that later developed into synoptic climatology.[8]

Based upon the Bergeron classification scheme, the SSC includes six categories of air masses: Dry Polar (similar to continental polar), Dry Moderate (similar to maritime superior), Dry Tropical (similar to continental tropical), Moist Polar (similar to maritime polar), Moist Moderate (a hybrid between maritime polar and maritime tropical), and Moist Tropical (similar to maritime tropical, maritime monsoon, or maritime equatorial).

Figure 18.1 shows the world distribution of air masses according to the SSC system. Note that the SSC classifies air masses over land as well as over ocean. More refined techniques such as cluster analyses have been used for SSC based on weather observations in the United States.[9]

Köppen–Geiger Classification

The objective of Köppen's classification is the association of climate elements that determine the land biomes. His scheme is based on annual and seasonal temperature and precipitation values and is represented by three letters. The first letter recognizes the five primary types and are labeled A through E, namely A for tropical; B, dry; C, mild midlatitude; D, cold midlatitude; and E, polar. The second letter

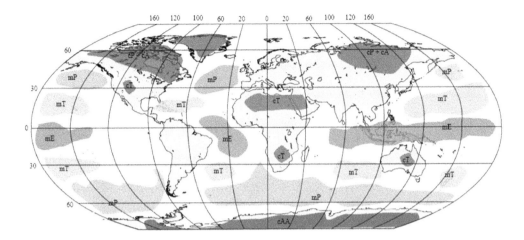

FIGURE 18.1 World distribution of air mass classification; m and c denote maritime and continental, respectively, and P, T, E, M, A, and S denote polar, tropical, equatorial, monsoon, arctic (or Antarctic), and subsiding air masses, in that order. A third letter k or w indicates whether the air is colder (k) or warmer (w) than the underlying surfaces to indicate the stability of the atmosphere.
Source: http://en.wikipedia.org/wiki/Air_mass, cited March 16, 2012.

shows the seasonal precipitation pattern, with w, s, and f, where w represents a dry winter—the driest winter month has at the most 1/10th of the precipitation found in the wettest summer month; s is a dry summer—the driest summer month has at the most 30 mm (1.18 in.) of rainfall and has at the most one-third the precipitation of the wettest winter month, and f, when the condition of neither s nor w is satisfied. The third letter represents the seasonal temperature conditions and is designated a when the warmest month averages above 22°C, and b, when conditions do not meet the requirements for a, but there still are at least four months above 10°C.

The five primary classifications can be further divided into secondary classifications such as rain forest, monsoon, tropical savanna, humid subtropical, humid continental, oceanic, Mediterranean, steppe, subarctic, tundra, polar ice cap, and desert climates.

Figure 18.2 shows a distribution of the climate classification according to the Köppen's scheme developed by the United Nations Food and Agricultural Organization (FAO).[10,11] The temperature data are based on the Climate Research Unit (CRU)[12] analysis and the rainfall data are from the Global Precipitation Climatology Centre (GPCC)[13] data. The climate classes are described as follows.

Rain forests are characterized by heavy rainfall, with minimum normal annual rainfall between 1750 and 2000 mm. Mean monthly temperatures exceed 18°C during all months of the year.

A monsoon is characterized by a seasonal prevailing wind, which lasts for several months before the rainy season in the region. Regions within North America, South America, sub-Saharan Africa, Australia, and East Asia are monsoon regimes.

A tropical savanna is a grassland biome located in semiarid to semihumid climate regions of subtropical and tropical latitudes, with average temperatures remaining at or above 18°C year round and rainfall between 750 and 1270 mm a year. They are widespread in Africa, and are found in India, the northern parts of South America, Malaysia, and Australia.

In the humid subtropical climate zone, winter rainfall (and sometimes snowfall) is associated with large storms that are steered by westerlies (winds from west to east). Most summer rainfall is accompanied by thunderstorms and occasionally associated with tropical cyclones. Humid subtropical climates lie on the east side continents, roughly between latitudes 20° and 40° away from the equator.

A humid continental climate is marked by variable weather patterns and a large seasonal temperature variance. Places with more than 3 months of average daily temperatures above 10°C, a coldest month

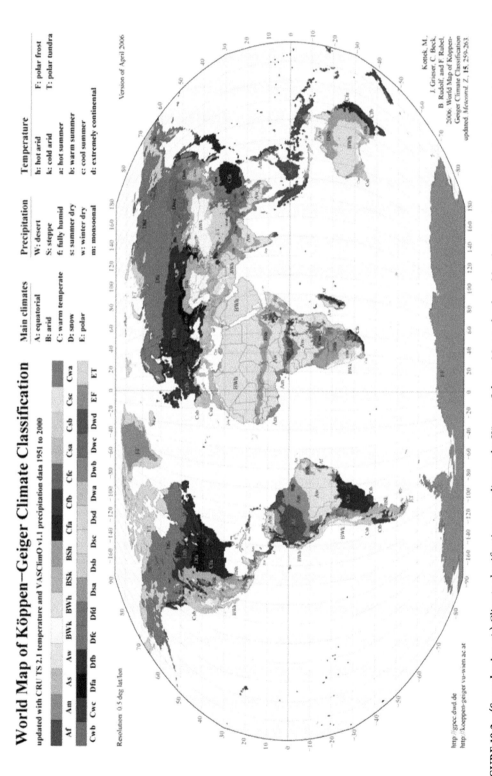

FIGURE 18.2 (See color insert.) Climate classification according to the Köppen Scheme Map is developed by the United Nations Food and Agricultural Organization (FAO). Temperature and rainfall data are based on the Climate Research Unit (CRU) and the Global Precipitation Climatology Centre (GPCC) data, respectively.
Source: Grieser et al.,[11] Mitchell et al.,[12] and Beck et al.[13]

temperature of below $-3°C$, but which do not meet the criteria for an arid or semiarid climate, are classified as continental.

An oceanic climate is typically found along the west coasts at the middle latitudes of all the world's continents, and in southeastern Australia, and is accompanied by plentiful precipitation year round. It is characterized by warm, but not hot summers and cool, but not cold winters, and a relatively narrow annual temperature range and an adequate and evenly distributed precipitation throughout the year.

The Mediterranean climate regime resembles the climate of the lands in the Mediterranean Basin, parts of western North America, parts of West and South Australia, southwestern South Africa and parts of central Chile. The climate is characterized by hot, dry summers and cool, wet winters.

A steppe is a dry grassland with an annual temperature range in the summer of up to $40°C$ and during the winter down to $-40°C$.

A subarctic climate has little precipitation, with monthly temperatures above $10°C$ for 1 to 3 months of the year, and permafrost in large parts of the area due to the cold winters. Winters within subarctic climates usually include up to 6 months of average below-freezing temperatures ($0°C$).

Tundra occurs in the far Northern Hemisphere, north of the Taiga Belt, including vast areas of northern Russia and Canada, characterized by a temperature range of $10°C$ to $0°C$ for the warm months, annual rainfall 150–250 mm, and absence of trees.

A polar ice cap, or polar ice sheet, is a high-latitude region that is covered in ice. Subfreezing temperatures occur most of the year due to a net radiative deficit of incoming solar and outgoing longwave radiation.

A desert is a landscape form or region that receives very little precipitation. Deserts usually have a large diurnal and seasonal temperature range. Depending on location, daytime high temperatures can reach up to $45°C$ in summer, and low night-time temperatures in winter can go down to below $0°C$ due to extremely low humidity. Many deserts are formed in rain shadows, as mountains block the moisture on the windward side and leave the leeward side as rain shadows.

The Thornthwaite Scheme

The Thornthwaite[14] scheme is a hierarchical classification scheme, explained essentially in terms of the annual pattern of soil-moisture conditions. The soil-moisture conditions are determined by the monthly rainfall as input and evaporation as an output. Evaporation is indicated by temperature conditions. Thornthwaite[15] quantified his earlier scheme and introduced potential evapotranspiration (PE) to describe thermal efficiency (TE). PE is the maximum evaporation available, which is not limited by the available moisture of the underlying surface. The TE is compared to precipitation to form a moisture index (Im) to show amounts and periods of soil-water surplus (deficit) when precipitation is larger (smaller) than PE. Definite break points in the moisture regimes are used to define climatic boundaries. Table 18.1 shows the moisture provinces corresponding to the Im computed from monthly PE and precipitation. The thermal provinces are defined by the annual PE.

TABLE 18.1 Moisture and Thermal Indices Used in the Thornthwaite Scheme Classification

	Moisture Province	Moisture Index (Im)	Thermal Province	Annual PE (cm)
A	Perhumid	100 and above	Megathermal	>114.0
B4	Humid	80 to 99.9	Mesothermal	99.8 to 114.0
B3	Humid	60 to 79 9	Mesothermal	85.6 to 99.7
B2	Humid	40 to 59.9	Mesothermal	71.3 to 85.5
B1	Humid	20 to 39.9	Mesothermal	57.1 to 71.2
C2	Moist subhumid	0 to 19.9	Mesothermal	42.8 to 57.0
C1	Dry subhumid	−19.9 to 0	Mesothermal	28.6 to 42.7
D	Semiarid	−39.9 to −20	Tundra	14.3 to 28.5
E	Arid	−60 to −40	Frost	0 to 14.2

The moisture descriptors are hyperhumid (A), humid (B1,...,B4), moist and dry subhumid (C_1 and C_2), semiarid (D), and arid (E) conditions. The moist conditions (A, B, and C_1) are further subdivided into the following subclasses: r showing little or no water deficiency, s: moderate summer water deficiency, w: moderate winter water deficiency, s_2: large summer water deficiency, and w_2: large winter water deficiency. Similarly, dry conditions (C_2, D, and E) have the following subclasses; d shows little or no water surplus, s: moderate winter water surplus, w: moderate summer water surplus, s_2: large winter water surplus, and w_2: large summer water surplus.

Thermal descriptors represents the water need and include frost (denoted E'), tundra (D'), microthermal (C'_1 and C'_2), mesothermal ($B'_1..B'_4$), and megathermal (A') regimes. A microthermal climate is one of low annual mean temperatures, generally between 0°C and 14°C, which experiences short summers and has a potential evaporation between 14 and 43 cm. A mesothermal climate lacks persistent heat or persistent cold, with potential evaporation between 57 and 114 cm. A megathermal climate is one with persistent high temperatures and abundant rainfall, with potential annual evaporation in excess of 114 cm. The seasonality is represented as a' for regions with summer precipitation contributing to less than 48% of annual total, b' (between 48% and 68%), c'_1 and c'_2 between 68% and 88%, and d' higher than 88%.

To illustrate this scheme, San Francisco, California, is classified as $C_1B'_1s_2a'$, i.e., dry subhumid, first mesothermal, with large winter water surplus, and a temperature efficiency regime of normal to megathermal.

More Recent Developments

With advances in measurement techniques, collection methodology, analysis procedures, and improved understanding of the working of the Earth systems, more sophisticated classification schemes are available. For example, better quality-controlled rainfall data are collected by the Global Precipitation Climatology Center as part of the WMO Global Precipitation Climatology Project, which are used in Figure 18.2. Budyko's climate classification is based on the soil-energy requirement. Population growth and man's activities on the Earth system prompted Brooks's[16] classification scheme. The emergence of megacities with their own climate leads to the development of urban climate by Landsberg.[17] The Thornthwaite scheme is modified and is the basis of the Palmer Drought Indices used in monitoring drought and crop conditions.

Climate data were collected on the Earth's surface with ground-based instruments. The advance of space technology and information acquired through remote sensing has increased our knowledge of the Earth systems. Atmospheric variables, such as air pressure, wind, temperature and humidity profile, precipitation, soil moisture, evapotranspiration, atmospheric composition, chemical constituents, and vegetation, are now available from remote-sensing techniques.[18] For example, improved precipitation data are collected by the international Tropical Rainfall Measuring Mission (TRMM) and its follow-on Global Precipitation Mission (GPM). International collaborations have also resulted in co-ordinated measurements of atmospheric trace gases, biogeochemical substances, and flux measurements. These observations cover the large expense of the earth's surface that has not been surveyed before. Advances in information technology facilitate data integration and data mining. More sophisticated statistical procedures, such as eigenvector analysis and cluster analysis, have been used to group the climates of places and to define areas of "similar" climates. Climate classifications are suited for monitoring climate change and impact on society. Generally, they provide more robust change detection criteria for monitoring climate change due to human activities.

Conclusion

The climate at each location on the Earth is dependent on the interactions of the components of the Earth systems: the atmosphere, hydrosphere, land surface, and biosphere and are shaped by human activities (or the anthrosphere). The progress on climate classification is dependent on the available data and their targeted applications. We described three approaches to climate classification, the genetic, empirical, and application classifications with examples.

The Köppen–Geiger classification scheme describes the integrated effect of climate on vegetation, which is related to the biomes. The Thornthwaite scheme includes the effect of vegetation on regulating climate through evapotranspiration. As photosynthetic process ultimately determines the habitability of the Earth, land-based classification schemes can be extended to include the ocean biota and primary productivity. This will include oceanic variables, such as sea-surface temperature, salinity, and biogeochemical variables not used before in climate classification. The integration of terrestrial and aquatic biomes will allow a better assessment, monitoring, and sustainable exploitation of the natural resources of the Earth on which our lives depend.

Acknowledgments

The author acknowledges support for his research on Earth system science by the U.S. National Aeronautics and Space Administration and National Oceanic and Atmospheric Administration.

References

1. American Meteorological Society (AMS). *Glossary of Meteorology,* 2nd Ed.; AMS, 45 Beacon Street: Boston, MA 02108–3693, 2000. http://glossary.ametsoc.org/.

2. Solomon, S.; Qin, D.; Manning, M.; Chen, Z.; Marquis, M.; Averyt, K.B; Tignor, M.; Miller, H.L., Eds.; *Contribution of Working Group I to the Fourth Assessment Report of the Intergovernmental Panel on Climate Change*; Cambridge University Press: Cambridge, United Kingdom and New York, NY, USA, 2007. IPCC; http://www.ipcc.ch/.

3. Köppen, W.P. *Grundriss der Klimakunde,* 2d Ed.; Walter de Gruyter: Berlin, 1931.

4. Köppen, W.P.; Geiger, R. *Handbuch der Klimatologie;* Gebruder Borntraeger: Berlin, 1930–1939; 6 Vol.

5. Alisov, B.P. *Die Klimate der Erde;* Deutscher Verlag der Wissenschaften: Berlin, 1954; 277 pp.

6. Penman, H.L. Natural evaporation from open water, bare soil and grass. Proc. Roy. Soc. Lond. **1948**, *A* (194), S. 120–145.

7. Supan, A. Die Temperaturzonen der Erde. Petermanns Geog. Mitt. **1879**, 25, 349–358.

8. Schwartz, M.D. Detecting structural climate change: an air mass-based approach in the North Central United States, 1958–1992. Ann. Assoc. Am. Geographers **1995**, *85* (3), 553–568.

9. Sheridan, S.C. The redevelopment of a weather-type classification scheme for North America. Intl. J. Climatol. **2002**, 22, 51–68. (http://sheridan.geog.kent.edu/ssc.html, cited May 31, 2012).

10. Kottek. M.; Geisier, J.; Beck, C.; Rudolf, B; Rubel, F. Map of Koppen-Geiger climate classification update. Meteorol. Z. **2006**, *15,* 259–263;http://koeppen-geiger.vu-wien.ac.at/pdf/kottek_et_al_2006_A4.pdf. Retrieved January 20, 2012.

11. Jürgen G.; René, G.; Stephen, C.; Michele, B. New gridded map of the Koeppen's Climate classification, 2006. http://www.fao.org/nr/climpag/globgrids/KC_classification_en.asp (cited May, 2012).

12. Mitchell, T.L Jones, P. An improved method of constructing a database of monthly climate observations and associated high-resolution grids. Int. J. Climatol. **2005**, *25* 693–712. http://www.cru.uea.ac.uk/.

13. Beck, C.; Grieser J.; Rudolf, B. A new monthly precipitation climatology for the global land areas for the period 1951 to 2000, Klimastatusbericht. 2004; 181–190, DWD.

14. Thornthwaite, C. The climates of North America according to a new classification. Geogr. Rev. **1931**, *21,* 633–655.

15. Thornthwaite, C.W. An approach toward a rational classification of climate. Geogr. Rev. **1948**, *38, 55–94.*

16. Brooks, C.E.P. *Climate in Everyday Life;* Greenwood Press, London, 1950; 314 pp.

17. Landsberg, H.; *Urban Climate;* International Geophysics Series. Academic Press: New York, 1981; Vol. 28, 277 p.
18. Chiu, L.S. Earth observations. In *Advanced Geoinformation Science;* Yang, C., Miao, Q., Wong, D., Yang, R., Eds.; CRC Press: Taylor and Francis, 2011; 486.

Bibliography

Pidwirny, M. Climate Classification and Climatic Regions of the World. *Fundamentals of Physical Geography, 2nd Ed.* 2006, http://www.physicalgeography.net/fundamentals/7v.html, Viewed January 20, 2012.
Ritter, Michael E. *The Physical Environment: An Introduction to Physical Geography.* 2006. http://www.uwsp.edu/geo/faculty/ritter/geog101/textbook/title_page.html (cited May 31, 2–12, 2012).
Rohli, R.V.; Vega, A.J. *Climatology,* 2nd Ed; Jones and Barlett Learning, 2010; 426 pp.
Trewartha, G.T. *An Introduction to Climate,* 3d Ed.; Appendix A, 1954; 223–238.

19

Climate: Extreme Events

Philip Sura
Florida State University

Introduction

Extreme events in weather and climate are by definition scarce, but they can have a significant physical and socioeconomic impact on people and countries in the affected regions. Alongside the intuitive knowledge that hurricanes, tornados, severe storms and rain, droughts, floods, and extreme temperatures might qualify, how can we define extreme events more quantitatively? There are several approaches used in climate research. One often-used [e.g., by the Intergovernmental Panel on Climate Change (IPCC)] definition of an extreme event of the variable under consideration is based on the tails of its climatological (i.e., reference) probability density function (PDF) at a particular geographical location. Extreme events would normally be as rare as or rarer than the, for example, 5th and 95th percentiles. The specific percentile values are not rigorously defined, so often other, more or less stringent, ranges are used (e.g., the 1st and 99th or 10th and 90th percentiles), depending on the particular application. Another routinely used definition of extreme events is that of block maxima or minima. Block maxima/minima are the highest/lowest values attained at a specified location during a given time interval (e.g., daily, monthly, seasonal, or annual). If the interval is the whole period for which observations are available, a block maxi- mum/minimum is called the "absolute extreme." Note that both definitions do not depend on the particular shape (e.g., Gaussian or non-Gaussian) of the PDF. Yet the Gaussian distribution is very often used to estimate the odds of extreme events, neglecting the non-Gaussianity of real world observations. Therefore, an extreme event can also be defined as the non-Gaussian tail of the data's PDF. This definition implies that a high amplitude event does not qualify as extreme if it is described by Gaussian statistics. In a nutshell, because there are different definitions, it is important to be aware of the one being used in a particular application.

Whichever definition is used, understanding extremes has become an important objective in weather and climate research because weather and climate risk assessment depends on knowing and understanding the statistics of the tails of PDFs. At this point it is essential to define weather and climate. As described in almost all fundamental meteorological textbooks,[1,2] weather is varying on timescales of hours, days, to a few weeks, whereas climate varies on longer timescales of months, years, and decades. There is, of course, a certain overlap and the terms weather and climate are typically used in a loose way, specifying the timescales as needed for particular applications. It should be noted that nonlinear multi-scale interactions render a strict separation of timescales unfeasible, making it all put impossible to attribute an individual extreme event to a changing climate.

There is broad consensus that some of the most hazardous effects of climate change are related to a potential increase (in frequency and/or intensity) of extreme weather and climate events. (A notable exception is gradual sea level rise, which will result in the inundation of large densely populated regions; by our definition this is not an extreme event, but it is surely catastrophic.) The overarching goal of studying extremes is, therefore, to understand and then manage the risks of extreme events and related disasters to advance strategies for efficient climate change adaption. Although numerous important studies have focused on changes in mean values under global warming, such as mean global temperature (one of the key variables in almost every discussion of climate change; see, for example, reports from the IPCC available at http://www. ipcc.ch), the interest in how extreme values are altered by a changing climate is a relatively new topic in climate research. The reasons for that are primarily two-fold. First, for a comprehensive statistical analysis of extreme events, high-quality and high-resolution (in space and time) observational data sets are needed. It is only recently that global high-quality daily observations became available to the international research community. Second, we need extensive simulations of high-resolution climate models to (hopefully) simulate realistic climate variability. Again, only recently long enough high-resolution numerical simulations of climate variability became feasible to study extreme events in some detail.

Sampling

The general problem of understanding extremes is, of course, their scarcity: it is very hard to obtain reliable (if any) statistics of those events from a finite observational record. Therefore, the general task is to somehow extrapolate from the well-sampled center of a PDF to the scarcely or unsampled tails. The extrapolation into the more or less uncharted tails of a distribution can be roughly divided into four major, by no means mutually exclusive categories. In fact, the study of extreme events in weather and climate is most often done by combining the strategies of the following methods[3,4]

The *statistical approach (extreme value theory)* is solely based on mathematical arguments[3,5–8] It provides methods to extrapolate from the well-sampled center to the scarcely or unsampled tails of a PDF using mathematical tools. The key point of the statistical approach is that, in place of an empirical or physical basis, asymptotic arguments are used to justify the extreme value model. In particular, the generalized extreme value (GEV) distribution is a family of PDFs for the block maxima (or minima) of a large sample of independent random variables drawn from the same arbitrary distribution. Although the statistical approach is based on sound mathematical arguments, it does not provide much insight into the physics of extreme events. Extreme value theory is, however, widely used to explore climate extremes[9–11] In fact, the foundation of extreme value theory is very closely related to the study of extreme values in meteorological data.[7,8] Nowadays this is very often done in conjunction with the numerical modelling approach discussed below. That is, model output is analyzed using extreme value theory to see if statistics are altered in a changing climate.

The *empirical–physical approach* uses empirical knowledge and/or physical reasoning to provide a basis for an extreme value model. The key point here is that, in contrast to the purely statistical method that primarily uses asymptotic mathematical arguments to model only the tail of a PDF, empirical and/or physical reasoning is employed to model the entire PDF. The empirical–physical method can itself be further split into either empirical or physical strategies (or both), focusing on the empirical or physical aspects of the problem, respectively. For example, a purely empirical approach would simply fit a suitable PDF to given data, whereas a more physical ansatz would also determine the physical plausibility of a specific PDF. However, often a clear distinction between the physical and empirical aspects is impossible. The empirical–physical method lacks the mathematical rigor of the statistical method, but it provides valuable physical insight into relevant real world problems. An example for an empirical–physical application is the Gamma distribution, which is often used to describe atmospheric variables that are markedly asymmetric and skewed to the right. Often skewness occurs when there is a physical limit that is near the range of data.[6] Well-known applications are precipitation and wind speed, which are physically constrained to be non-negative. It should be noted that the empirical–physical approach can be, in principle, put on a more rigorous foundation using the principle of maximum entropy.[12–14] That is, given some physical information (i.e., constraints) of a process, the PDF that maximizes the information entropy under the imposed constraints is the one most likely found in nature (i.e., the least biased given the constraints). However, not all PDFs commonly used to explain weather and climate data can be justified this way. For example, the Gamma and Weibull distributions can be obtained by the principle of maximum entropy, but the constraints are not necessarily physically meaningful.[15]

The *numerical modelling* approach aims to estimate the statistics of extreme events by integrating a general circulation model (GCM) for a very long period.[16–18] That is, this approach tries to effectively lengthen the limited observational record with proxy data from a GCM, filling the unsampled tails of the observed PDF with probabilities from model data. Numerical modelling allows for a detailed analysis of the physics (at least model-physics) of extreme events. In addition, the statistical and empirical–physical methods can also be applied to model data, validating (or invalidating) the quality of the model. It is obvious that the efforts by the IPCC to understand and forecast the statistics of extreme weather and climate events in a changing climate fall into this category. Because very long model runs are needed to sample the tails of a PDF, global GCMs used for that purpose are currently run at a relatively coarse spatial resolution and are, therefore, unable to resolve important sub-grid scale features such as clouds, tornadoes, and local topography. Because of that, GCMs cannot be used for very localized studies of extreme events. To overcome this problem downscaling methods are commonly used, for which the prediction of extremes is not based on direct GCM output but on subsequent statistical or dynamical models to link the coarse GCM output to local events.[19]

The *non-Gaussian stochastic approach* makes use of stochastic theory to evaluate extreme events and the physics that governs these events.[4] Assuming that weather and climate dynamics are split into a slow (i.e., slowly decorrelating) and a fast (i.e., rapidly decorrelating) contribution, weather and climate variability can be approximated by a stochastic system with a predictable deterministic component and an unpredictable noise component. In general, the deterministic part is non-linear and the stochastic part is state dependent. The stochastic approach takes advantage of the non-Gaussian structure of the PDF by linking a stochastic model to the observed non-Gaussianity. This can be done in two conceptually different ways. On the one hand, if the deterministic component is non-linear and the stochastic component is state independent, the non-Gaussianity is due to the non-linear deterministic part. On the other hand, if the deterministic component is linear and the stochastic component is state dependent, the non-Gaussianity is due to the state dependent noise. Of course, any combination of the two mechanisms is also possible. Although the nonlinear approach with state-independent noise captures some types of non-Gaussian climate variability well,[20,21] it recently became clear that state-dependent (or multiplicative) noise plays a major role in describing weather and climate extremes.[4] The physical significance of multiplicative noise is that it has the potential to produce non-Gaussian statistics in linear systems. In particular, Sura and Sardeshmukh,[22] Sardeshmukh and Sura,[23] and Sura[4] attribute extreme anomalies to stochastically forced linear dynamics, where the strength of the stochastic forcing depends linearly on the flow itself (i.e., linear multiplicative noise). Most important, because the theory makes clear and testable predictions about non-Gaussian variability, it can be verified by analyzing the detailed non-Gaussian statistics of oceanic and atmospheric variability.

What do these approaches have in common? Every approach effectively extrapolates from the known to the scarcely known (or unknown) using certain assumptions and, therefore, requires a leap of faith. For the statistical approach the assumptions are purely mathematical. For example, the assumption of classical extreme value theory, that the extreme events are independent and drawn from the same distribution, and that sufficient data are available for convergence to a limiting distribution (the GEV distribution) may not be met.[5,6] The potential drawback of the empirical–physical approach is its lack of mathematical rigor (with the exception of the principle of maximum entropy); it primarily depends on empirical knowledge and physical arguments. The weakness of numerical modelling (including downscaling) lies in the largely unknown ability of a model to reproduce the correct statistics of extreme events. Currently, GCMs are calibrated to reproduce the observed first and second moments (mean and variance) of the general circulation of the ocean and atmosphere. Very little is known about the credibility of GCMs to reproduce the statistics of extreme events. Likewise, the non-Gaussian stochastic approach relies on the assumption that weather and climate variability can be modelled by a stochastic process. It has to be concluded that the common methods to study extreme events have some limitations and that the study of extreme weather and climate events is largely empirical. In particular, there exists no closed quantitative theory on how the statistics of weather and climate extremes might change in a warming climate.

Climate Change

What can be said about likely changes in frequency or intensity of climate extremes in a warming climate? Overall, not very much. In fact, there are only few processes we understand well enough to have some confidence in projected changes.[24] At the top of the list is an increase in the number of extremely warm days and heat waves. In fact, many land areas are already experiencing significant increases in maximum temperatures. This is also consistent with projected temperature changes obtained from GCM projections. Because warm air can hold more water, there is also an agreement that the intensity (mean and variability) of the hydrological cycle increases with increasing temperatures.[25] Given a more intense hydrological cycle, larger amounts of rainfall will come from heavy showers and more intense thunderstorms. Of course, more intense precipitation increases the likelihood of severe floods. Somewhat counter intuitively, the likelihood of severe drought will also increase in a warmer climate because, with a more intense hydrological cycle, a larger proportion of rain will fall in the more extreme events. In addition, higher temperatures will result in increased evaporation reducing the amount of moisture available at the surface. Note that the global spatial distribution of temperature and precipitation extremes (and the probability of droughts) is highly variable.[24,26] Of course, many people are mostly interested in the projected change of severe winds. Unfortunately, we have very little definite knowledge about how the strength and frequency of hurricanes and severe midlatitude storms might change under global warming. The reason for that is, that the genesis of tropical and midaltitude storms is controlled by many physical processes, including the large-scale atmospheric flow, whose interactions we currently do not fully understand. Also, there is a large uncertainty with regard to how our global climate models are capable of simulating the plethora of small- scale processes. That is also the reason why we cannot make a definitive projection for small-scale events such as tornadoes, hail, and thunderstorms.

Conclusion

Knowing the tails of weather and climate PDFs is an important goal in the atmospheric and ocean sciences because weather and climate risk assessment depends on understanding extremes. Although the commonly used definitions of extremes, and their conceptual implementation in a meteorological framework, are straightforward, it is very hard to obtain statistically significant information of extreme weather and climate events from scarce data. In particular, we only have a very limited physical and statistical understanding of how extremes are going to be altered in a changing climate. Many climate projections just look into the change of the mean and the variance, that is, assuming Gaussian statistics. However, the non-Gaussian statistics (the shape of the distribution) will most likely also be altered in a changing climate. More research (more observations, better theoretical and numerical models) is needed to improve our understanding of how a PDF might change in the future.

References

1. Wallace, J.M.; Hobbs, P.V. *Atmospheric Science: An Introductory Survey (Second Edition)*; Academic Press: 2006; 504 pp.
2. Hartmann, D.L. *Global Physical Climatology*; Academic Press: 1994; 411 pp.

3. Garrett, C.; Müller, P. Extreme events. Bull. Amer. Meteor. Soc. **2008**, *89*, ES45–ES56.

4. Sura, P. A general perspective of extreme events in weather and climate. Atmos. Res. **2011**, *101*, 1–21.

5. Coles, S. *An Introduction to Statistical Modeling of Extreme Values*; Springer-Verlag: 2001; 208 pp.

6. Wilks, D.S. *Statistical Methods in the Atmospheric Sciences*; Second Edition. Academic Press: 2006; 627 pp.

7. Gumbel, E.J. On the frequency distribution of extreme values in meteorological data. Bull. Amer. Meteor. Soc. **1942**, *23*, 95–105.

8. Gumbel, E.J. *Statistics of Extremes*; Columbia University Press: 1958; 375 pp.

9. Katz, R.W.; Parlange, M.B.; Naveau, P. Statistics of extremes in hydrology. Adv. Water Resour. **2002**, *25*, 1287–1304.

10. Katz, R.W.; Naveau, P. Editorial: Special issue on statistics of extremes in weather and climate. Extremes 2010, *13*, DOI 10.1007/s10 687-010-0111-9.

11. Smith, R.L. Extreme value statistics in meteorology and the environment. Environ. Stat. **2001**, 8, 300–357.

12. Jaynes, E.T. Information theory and statistical mechanics. Phys. Rev. **1957**, *106*, 620–630.

13. Jaynes, E. T. Information theory and statistical mechanics. II. Phys. Rev. **1957**, *108*, 171–190.

14. Jaynes, E.T. *Probability Theory: The Logic of Science*; Cambridge University Press: 2003; 758 pp.

15. Lisman, J.H.C.; van Zuylen, M.C.A. Note on the generation of most probable frequency distributions. Stat. Neerlandica **1972**, *26*, 19–23.

16. Easterling, D.R., Meehl, G.A.; Parmesan, C.; Changnon, S.A.; Karl, T.R.; Mearns, L.O. Climate extremes: Observations, modeling, and impacts. Science **2000**, *289*, 2068–2074.

17. Kharin, V. V.; Zwiers, F.W. Estimating extremes in transient climate change simulations. J. Clim. **2005**, *18*, 1156–1173.

18. Kharin, V.V.; Zwiers, F.W.; Zhang, X.; Hegerl, G.C. Changes in temperature and precipitation extremes in the IPCC ensemble of global coupled model simulations. J. Clim. **2007**, *20*, 1419–1444.

19. Wilby, R.L.; Wigley, T.M.L. Downscaling general circulation model output: a review of methods and limitations. Prog. Phys. Geogr. **1997**, *21*, 530–548.

20. Kravtsov, S.; Kondrashov, D.; Ghil, M. Multi-level regression modeling of nonlinear processes: derivation and applications to climate variability. J. Clim. **2005**, *18*, 4404–4424.

21. Kravtsov, S.; Kondrashov, D.; Ghil, M. Empirical model reduction and the modelling hierarchy in climate dynamics and the geosciences. In *Stochastic Physics and Climate Modelling*; Palmer, T.; Williams, P. Eds.; Cambridge University Press: 2010; 35–72.

22. Sura, P.; Sardeshmukh, P.D. A global view of non-Gaussian SST variability. J. Phys. Oceanogr. **2008**, *38*, 639–647.

23. Sardeshmukh, P.D.; Sura, P. Reconciling non-Gaussian climate statistics with linear dynamics. J. Clim. **2009**, *22*, 1193–1207.

24. Houghton, J. *Global Warming - The Complete Briefing. Fourth Edition*; Cambridge University Press: 2009; 438 pp.

25. Allen, M.R., Ingram, W.J. Constraints on future changes in climate and the hydrological cycle. Nature **2002**, *419*, 224–232.

26. Christensen, J.H.; Hewitson, B.; Busuioc, A.; Chen, A.; Gao, X.; Held, I.; Jones, R.; Kolli, R.K.; Kwon, W.-T.; Laprise, R.; Magana Rueda, V.T.; Mearns, L.; Menéndez, C.G.; Räisänen, J.; Rinke, A.; Sarr, A.; Whetton, P. Regional Climate Projections. In *Climate Change 2007: The Physical Science Basis. Contribution of Working Group I to the Fourth Assessment Report of the Intergovernmental Panel on Climate Change* Solomon, S.; Qin, D.; Manning, M.; Chen, Z.; Marquis, M.; Averyt, K.B.; Tignor, M.; Miller, H.L. Eds.; Cambridge University Press: Cambridge, United Kingdom and New York, NY, USA, 2007.

20

Climate and Climatology

Jill S. M. Coleman
Ball State University

Introduction

Atmospheric science examines the physical processes of the atmosphere, the influence of the atmosphere on other systems, and the impacts these other systems have on the atmosphere at a variety of spatial and temporal scales. Traditionally, the study of the atmosphere is subdivided into two main areas: meteorology and climatology. Meteorology emphasizes weather and weather forecasting, the short-term variation of the atmosphere on the order of hours, days, or a few weeks. Climatology or the study of climate focuses on the long-term or average weather conditions over an extended period of time, usually over multiple decades, and the processes that create those conditions. Specialization areas within climatology include paleoclimatology, the study of past climates using techniques (e.g., ice cores) prior to instrumental data; historical climatology, the study of earlier climates within the timeframe of the written record (i.e., the past few thousand years); hydroclimatology, the study of the interaction between the climate system and the hydrological cycle; bioclimatology, the study of the interaction between the climate system and living organisms; and climatology specializations based on a region (e.g., tropical climatology), atmospheric system size (e.g., synoptic climatology), unique environment (e.g., urban climatology), or application (e.g., tourism).

Climate describes the general state of the atmosphere or expected weather conditions. Whereas weather describes the immediate or near-future state of the atmosphere for a given time and place through atmospheric variables such as temperature, pressure, humidity, wind speed and direction, and cloud cover, climate relays the statistical properties and persistent behavior of the atmosphere. Climate statistics are often expressed as normals, extremes, and frequencies.[1] Climatic normals refer to the mean weather conditions of a location, usually averaged over a 30-year period; however, shorter or longer averaging periods are also used depending on data availability and timescale of comparison. Extremes show the range, variability, and anomalies in atmospheric conditions, such as the recorded low and high temperature. Frequencies give information on the probability or likelihood of a particular weather event (e.g., tornado) occurring. Climate is more than a descriptive science detailing statistics on typical or abnormal atmospheric conditions. Moreover, climate represents the coupling between the atmosphere and the surfaces of the Earth.

Climate System Controls

The dynamic linkages of the Earth–atmosphere system dictate the general atmospheric conditions or climate over a region. Climate is a direct response to the complex interactions of energy, mass, and momentum between the atmosphere and the other major Earth subsystems, the hydrosphere (water), biosphere (living organisms), and the lithosphere (land).[2] Several physical and geographical components of these subsystems combine to control the climate system. Global climate types and patterns are governed by features such as latitude, semipermanent pressure systems, water proximity, oceanic circulation, and local topography.

Latitude

Seasonal changes in the orientation of the Earth with respect to the Sun determine the amount and intensity of solar radiation different locations receive. Solar radiation or insolation receipt not only directly impacts diurnal temperature changes but also impacts atmospheric circulation, including pressure distribution, wind flow, precipitation, cloud formation, and other features of weather and climate. Sun angle and day length regulate the insolation amount and are largely a function of latitude, the geographic coordinates that specify the north–south position of a place relative to the equator.

Latitude is the angle between the equatorial plane and a location on the surface of the Earth with values ranging from a minimum of 0° at the equator to a maximum of 90° at the poles. As the Earth revolves around the Sun, the axial tilt of the Earth results in the Northern (Southern) Hemisphere directed toward the Sun for half of the year and away from the Sun for the other half of the year. Equatorial locations receive on average around 12 hours of daylight per day throughout the year, whereas polar locations fluctuate between periods of 24 hours of daylight or darkness.

While the amount of daylight hours is an important factor in insolation amount, equally as important is the intensity of the solar radiation. The higher the Sun remains above the horizon, the more intense the radiation. When the Sun is high in the sky, solar radiation beams are highly concentrated over a small area and are more effective at warming a surface than when the sun is low on the horizon and the energy from the beams is distributed over a greater area. In addition, low sun angles require insolation beams to pass through a thicker atmosphere, resulting in greater beam depletion from the scattering, reflection, and absorption by the atmosphere and less receipt than by the surface. On average, low latitudes of the tropics and subtropics receive the most insolation and experience higher temperatures over the course of a year, and high latitudes of the Arctic and Antarctic receive the least; hence, an inverse relationship generally exists between latitude and annual solar radiation receipt.

Global Atmospheric Circulation: Semipermanent Pressure Systems

Latitude and surface-type heating differences combined with the rotational speed of the Earth produce a well-defined pattern of atmospheric pressure systems and wind flow, often explained using the three-cell model. An idealized depiction of mean global atmospheric circulation, the three-cell model refers to the number of distinct circulation cells present in each hemisphere: the Hadley cell in the low latitudes, the Ferrell cell in the midlatitudes, and the polar cell in the high latitudes. These cells redistribute energy between the warmer regions that are energy rich and the colder regions that are energy poor, thereby preventing the tropics from becoming increasingly hot or the poles from becoming increasingly cold. The rising and sinking motions associated with each cell produce latitudinal zones dominated by either low (cyclonic) or high (anticyclonic) atmospheric pressure, varying in intensity and position with the seasons. Locations dominated by anticyclonic flow, such as the subtropics, typically have minimal cloud cover, lower precipitation frequency, and higher surface insolation receipt, whereas low-pressure regions of the tropics and high midlatitudes are often areas of rising motion and extensive cloud cover.

Water: Oceanic Circulation and Land Proximity

Similar to global wind and pressure patterns, oceanic circulation cells redistribute surplus energy from lower latitudes to high latitudes with an annual energy deficit. Each oceanic basin contains a system of circulation cells or gyres with warm and cold currents whose large-scale movements are dictated by the strength and position of the semipermanent subtropical anticyclones.[1,3] The eastern oceanic basins are dominated by cold ocean currents, such as the California Current off the western North American coast, that promote cooler temperatures and stable (nonprecipitation-forming) atmospheric conditions. In contrast, the western oceanic basins have warm ocean currents that produce warm, humid air masses prone to cloud formation and precipitation. Oceanic circulation patterns (and the prevailing climate) may also shift from interannual variations in salinity, surface water movement, and energy storage, which occur during El Niño or La Niña events of the equatorial Pacific.

Water also regulates other aspects of climate, particularly temperature. In comparison with most land surfaces, water has a high specific heat capacity or the amount of energy required to raise the temperature of a substance by 1°C. The specific heat capacity of water is approximately five times greater than that of land, meaning much more energy is required to heat and cool water than for land.[3] Consequently, coastal locations usually have narrower temperature ranges and fewer temperature extremes than locations further inland, an effect described as continentality. The degree of continentality is also moderated by other factors such as the prevailing wind and local topography. For instance, in the midlatitudes where the prevailing wind is from the west, east coast locations will have a diminished oceanic influence and enhanced continentality than the similar west coast locations.

Local Features

Regional topography and other site-specific factors can influence local climate characteristics, producing unique temperature and precipitation patterns within prevailing climate zones. The height above mean sea level or altitude of a location directly impacts temperature, with temperature decreasing by an average rate of 6.5°C per kilometer (or 3.6°F per 1,000 feet) in the lower troposphere.[3] For instance, despite being on the equator, Quito Ecuador has a relatively low mean annual maximum temperature (22°C or 72°F) primarily due to its high-altitude location at 2,879 m (9,446 ft).[4] High-altitude environments also have rapid evaporation rates and weaker radiation retention, which produces large diurnal temperature ranges.

Topography, moreover, influences insolation, precipitation patterns, and local winds. The orientation and steepness of mountain slopes determine direct daylight energy receipt, with south-facing slopes in the Northern Hemisphere receiving longer daylight hours and more direct sun angles than any other direction. Steep slopes can also generate katabatic or mountain winds, high-velocity winds driven by localized pressure differences between high- and low-elevation areas and intensified by nighttime cooling; Greenland and Antarctica are well known for this type of extreme wind. Locations oriented toward the prevailing wind flow generally receive much higher cloud cover, precipitation, and cooler temperatures than areas on the leeward side of a topographic barrier. Even in relatively flat regions, the prevailing wind flow may influence local moisture transport. For example, winds across the North American Great Lakes are predominantly from the west or northwest, thereby producing lake effect-enhanced snowfall for areas on the eastern sides of the water bodies.

Climate Classification

The Köppen classification system is perhaps the most widely known technique for grouping climates with similar thermal, moisture, and natural vegetation characteristics. Devised by Vladimir Köppen in early 20th century and subsequently modified, the multitiered method utilizes mean monthly temperature and precipitation to delineate climate regions into one of five major types (designated by a capital letter): tropical (A), dry (B), midlatitude-mild (C), midlatitude-cold (D), and polar (E).[3] A sixth major type, highland (H), is also commonly used for areas with large climate variation over short distances primarily due to altitude. Climate zones A, C, D, and E are differentiated according to latitude-based temperature characteristics, whereas climate zone B contains arid and semiarid areas with annual precipitation deficits. These broad categories are further subdivided (designated by lower-case letters) using criteria such as dry season duration and timing and seasonal temperature extremes. For example, the Mediterranean climate type common to southern Europe and the southwestern coasts of South America, Australia, and Africa is *Csb* in the Köppen system, indicating a midlatitude location with mild winters and warm, dry summers. Glen Trewartha and others have subsequently modified the Köppen system to allow for greater within-type variability and more consideration of natural vegetation types.

Based on a water balance approach, the Thornthwaite climate classification scheme examines the seasonal changes in the relationship between moisture availability and energy. Categories are based on several thermal and moisture indices that utilize the concept of potential evapotranspiration (PE) or the maximum amount of moisture that could be evaporated to the atmosphere from the surface (including vegetation) if moisture was not a limiting factor.[1,2,5] Humid conditions occur when precipitation (and soil water storage) exceeds the amount necessary for PE, whereas arid conditions arise when water sources are insufficient to meet PE requirements. Other indices generate divisions according to annual and seasonal thermal efficiency, with higher (lower) values indicating a greater (diminished) capacity for evapotranspiration and generally warmer (colder) temperatures. Analogous to the Köppen classification system, individual climate types are designated by a multilevel letter combination generated from the indices. The Thornthwaite classification system contains nine major divisions based on moisture availability and an additional nine based on thermal efficiency, of which both groups are further subdivided according to seasonal variation.

While the Köppen and Thornthwaite systems are the most extensively used global climate classification systems, other methods of climate delineation have gained popularity for their emphasis on local and regional climate system factors, dominant forcing mechanisms, and/or application. Hubert Lamb developed a manual climate classification scheme for the British Isles based on the midlatitude cyclone model whereby the relative position and the relative frequency of the synoptic-scale atmospheric circulation pattern determine climate parameters such as temperature, precipitation, and wind direction.[6,7] The Lamb catalogue consists of 26 unique weather types with regional atmospheric situations dominated by either anticyclonic, cyclonic, or directional flow. Similar regional weather typing classification schemes have been developed for the United States,[8–10] including the newer spatial synoptic classification (SSC) scheme.[11] The SSC identifies North American (and more recently Western Europe) locations with similar weather and air mass characteristics into one of seven major types: dry polar (DP), dry moderate (DM), dry tropical (DT), moist polar (MP), moist moderate (MM), moist tropical (MT), and transitional (TR). While regional climates can be classified according to annual and seasonal weather-type frequencies, the SSC also has the ability for operational (real-time) meteorology usage.[12]

Conclusion

Climate represents not only the long-term average weather conditions of a location or the climatic normal but also the extremes and frequencies of those conditions. As a discipline, climatology has evolved from a predominantly descriptive science detailing the properties and geographic distribution of climate types to that determining the physical processes that govern and modify the climate system. Consequently, several climatology subdisciplines have arisen to address the different interactions between the atmosphere and other Earth systems (i.e., hydrosphere, biosphere, and lithosphere) that create climate. While climate classification systems, such as the Köppen scheme, retain widespread implementation, other classification methods have gained popularity for their ability to incorporate region-specific attributes or provide explanatory power. Regardless of the system, a common theme is the integration of the major climate system controls such as latitude, water proximity, oceanic circulation, semipermanent pressure systems, and topography.

Investigation into the origins and modifications of past, current, and future climates will continue to be a major focus of climatology, particularly the impact of human activities on the climate system. Although the climate system has varied considerably throughout the geologic past from natural forcing mechanisms (e.g., variations in solar output), modern climate shifts have become readily apparent since the late 18th century when the industrialization era began to impact atmospheric composition and the climate system balance. As data resolution and technology capabilities continue to progress, scientists are increasingly more adept at modeling the impacts of increased greenhouse gas emissions (e.g., carbon dioxide and methane) and atmospheric pollutants on the Earth–atmosphere system. Understanding these potential impacts will be an important factor in environmental sustainability policy and practices.

References

1. Rohli, R.V.; Vega, A.J. *Climatology*; 4th Edition; Jones and Bartlett: Sudbury, MA, 2017.
2. Shelton, M.L. *Hydroclimatology: Perspectives and Applications*; Cambridge University Press: Cambridge, UK, 2009.
3. Aguado, E.; Burt, J.E. *Understanding Weather and Climate*; 7th Edition; Prentice Hall: Upper Saddle River, NJ, 2014.
4. Pearce, E.A.; Smith, G. *World Weather Guide*; Random House: New York, 1990.
5. Thornthwaite, C.W. An approach toward a rational classification of climate. *Geog. Rev.* **1948**, *38* (1), 55–94.
6. Lamb, H.H. *British Isles Weather Types and a Register of Daily Sequence of Circulation Patterns, 1861–1971*; Geophysical Memoir, Vol. 116, HMSO: London, UK, 1972.
7. Briffa, K.R. The simulation of weather types in GCMs: A regional approach to control-run validation. In *Analysis of Climate Variability: Applications of Statistical Techniques*; 2nd Edition; von Storch, H.; Navarra, A., Eds.; Springer Verlag: Heidelberg, New York, 1999; 119–138.
8. Muller, R.A. A synoptic climatology for environmental baseline analysis: New Orleans. *J Appl. Meteorol.* **1977**, *16* (1), 20–33.
9. Coleman, J.S.M.; Rogers, J.C. A synoptic climatology of the central United States and associations with Pacific teleconnection pattern frequency. *J. Clim.* **2007**, *20* (14), 3485–3497.
10. Keim, B.D.; Meeker, L.D.; Slater, J.F. Manual synoptic classification for the East Coast of New England (USA) with an application to PM$_{25}$ concentration. *Clim. Res.* **2005**, *28* (2), 143–153.
11. Sheridan, S.C. The redevelopment of a weather-type classification scheme for North America. *Int. J. Climatol.* **2002**, *22* (1), 51–68.
12. Sheridan, S.C. Spatial synoptic classification, http://sheridan.geog.kent.edu/ssc.html (accessed June 2019).

Bibliography

Bonan, G.B. Ecological Climatology: Concepts and Applications; Cambridge University Press: Cambridge, UK, 2008.
Bridgeman, H.A.; Oliver, J.E. *The Global Climate System: Patterns, Processes and Teleconnections*; Cambridge University Press: Cambridge, UK, 2006.
Hartmann, D.L. *Global Physical Climatology*; 2nd Edition; Elsevier Science: Waltham: MA, 2016.
Hidore, J.J.; Oliver, J.E.; Snow, M.; Snow, R. Climatology: An Atmospheric Science; 3rd Edition; Prentice Hall: Upper Saddle River, NJ, 2009.

21

Climatology: Moist Enthalpy and Long-Term Anomaly Trends

Souleymane Fall
Tuskegee University

Roger A. Pielke
University of Colorado at Boulder

Dev Niyogi
Purdue University

Gilbert L. Rochon
Tuskegee University

Introduction

Presently, air temperature is the key metric for assessing climate change and more specifically global warming over land. A huge body of studies dealing with surface air temperature trends suggest that at global scale, an increase took place over the last century.[1-5] This widely scrutinized warming observed at the surface and in the troposphere is associated with anthropogenic greenhouse forcing of the climate system,[6-8] although natural effects have also been suggested as being important.[9]

However, warming is related to atmospheric energy content and temperature is only one of its components, as emphasized in some studies.[10-13] Another component which plays an important role in the warming process is atmospheric moisture content, which has been reported to have increased during the past decades,[14-20] although recent studies suggest that this increase has stopped[21] and even reversed.[22] A broader assessment of warming could, therefore, take into consideration moist enthalpy, which includes both temperature and moisture and denotes the heat content of air.

Although this study is limited to the atmospheric heat content, it is worth mentioning that at global scale, ocean heat content 1) is the major contributor of the increase of heat content of the whole Earth system; 2) is found to be well correlated with the global net radiation flux; and 3) is the main driver of the variability of the Earth's climate system.[23-28]

So far, few studies have simultaneously quantified temperature and moisture by combining them into a single variable. Steadman[29,30] utilized a scale of apparent temperature (A), which expresses levels of human comfort. His method has been employed by the National Oceanic and Atmospheric Administration (NOAA) as a heat stress index. To investigate summertime heat stress over the United States, Gaffen and Ross[31] used a simplified version of Steadman's A and derived thresholds defined by the 85th percentile values of July and August daily temperature and A averages. They found that the annual frequency of days exceeding the thresholds as well as the number of high heat-stress nights did increase at most stations. Using observed temperature and humidity datasets from 188 first-order weather stations for the period 1961–1990, Gaffen and Ross[32] found upward trends in *A* over the United States, in accordance with upward temperature and humidity trends. More recent studies focus on moist enthalpy, which combines both air temperature and humidity in a single variable, to assess surface heating trends. At a global scale, Ribera et al.[33] used the NCEP/NCAR re-analysis temperature to study the relationships between equivalent temperature (*TE*) and modes of climate variability. Although an increase of the globally averaged *TE* was found, significant differences were observed between oceanic and continental areas. Pielke et al.[10] compared values of year 2002 daily temperature and moist enthalpy for the city of Fort Collins (Colorado) and the Central Plains Experimental Range of the U.S. Department of Agriculture's Agricultural Research Service located 60 km northeast of the city. Their results show that temperature and moist enthalpy are nearly equal when absolute humidity is low, but as humidity increases during the growing season, moist enthalpy values become much larger. Davey et al.[11] examined 1982–1997 temperature and moist enthalpy trend differences for surface sites in the eastern United States. They found that moist enthalpy trends are warmer than temperature trends during the winter season, but relatively cooler in the fall. Rogers et al.[34] analyzed 124-year records of summer moist enthalpy for Columbus (Ohio) and found that the highest values of moist enthalpy occurred during the summer of 1995 when both temperature and moisture were very high, in contrast with the hot summers of 1930–1936 which, despite high temperatures, experienced lower moist enthalpy because of relatively low or negative anomalies of absolute humidity. More recently, Fall et al.[12] used National Centers for Environmental Prediction North American Regional Reanalysis data to investigate temperature and moist enthalpy at near-surface and various upper-air standard levels. They noted that the moisture component induces larger trends and variability of moist enthalpy relative to temperature. Their results indicated that, while moist enthalpy values and trend were much larger than temperature ones at near-surface level, there was almost no difference at 300–200 mb. Peterson et al.[13] examined the energy content of the surface atmosphere, which is composed of temperature (enthalpy), kinetic energy, and latent heat. They found that the global surface atmospheric energy has increased since the 1970s, mainly because of increases in both enthalpy and latent heat, which equally contribute to the global increases in heat content. However, at regional scale, the two components were in some cases found to be of opposite signs: for example, Peterson et al.[13] observed that in Australia, despite the increase in the surface temperature, heat content was found to be decreasing.

Definition of Moist Enthalpy

Moist enthalpy, also referred to as moist static energy[10] or equivalent temperature (*TE*),[11,12,33,34] expresses the surface air heat content (*H*) by taking into account air temperature and moisture as single variable. A previous variant of moist enthalpy, namely A, was developed by Steadman[29,30] by combining four variables: summer temperature, humidity (vapor pressure), wind speed, and extra-radiation. Gaffen and Ross[31,32] simplified Steadman's *A* by ignoring the effects of wind and radiation and computing it from ambient temperature (*T*, in °C) and water vapor pressure (*e*, in kPa)

$$A = -1.3 + 0.92\,T + 2.2e \tag{21.1}$$

Recent studies have mainly focused on moist enthalpy,[10] which is written as:

$$H = CpT + Lvq \tag{21.2}$$

where *Cp* is the specific heat of air at constant pressure, *T* is the air temperature, Lv **is the latent heat of vaporization, and *q* is the specific humidity. Thus, it can be seen that, as described by Peterson et al.,[13] moist enthalpy is the sum of two terms: enthalpy (calculated from *T*) and latent heat.

The approximate value of Lv (30°C) has been used in most of the studies. Fall et al.,[12] following the Priestley–Taylor method, estimated Lv (J/kg) with the temperature function:

$$\text{Lv} = 2.5 - 0.0022 \times T \qquad (21.3)$$

Such an estimate accounts for the variation of Lv with temperature.

The specific humidity q can be computed from the dew-point temperature and surface pressure using Bolton's empirical relationship:[35]

$$q = \frac{0.62197e}{p - 0.37803e}, \quad \text{where } e = 6.112 \exp\left(\frac{17.67Td}{Td + 243.5}\right) \qquad (21.4)$$

where e is the saturated vapor pressure in hPa, p is the surface pressure in hPa, and Td is the dewpoint temperature in °C.[34,36]

H is the heat in units of Joule and must be scaled into degree units in order to obtain TE for easy comparison to air temperature.

$$TE = H / Cp \qquad (21.5)$$

Equation 21.5 can be also written as

$$TE = T + \left(\text{Lv}q / Cp\right) \qquad (21.6)$$

where Lv is in units of Joules per kilogram and Cp is in units of Joules per kilogram per degree K. As q is dimensionless (i.e., kg per kg), the ratio has units of degree K.

The above equations show that both sensible and latent heat contribute to the magnitude of TE. Pielke et al.[37] have shown that, for example, at 1000 mb, an increase of 2.5°C in air temperature will produce the same change in TE as a 1°C increase in dew point temperature. TE becomes larger as both T and q increase. Conversely, when T is abnormally high but out of phase with a low q, TE exhibits modest-to-low values.[34] However, in the long term, the combination of the two terms seems to result in larger trend and variability of TE, regardless of the magnitude of q. For example, in their comparison between T and TE over the United States (1979–2005), Fall et al.[12] found larger trends and variability of TE (relative to T), even though in terms of contribution to the magnitude of TE, the moisture component (Lvq) was much smaller than the enthalpy (or sensible heat: CpT).

Climatology

In general, moist enthalpy increases progressively from winter to summer and then decreases, thus exhibiting patterns that are similar to surface temperature annual cycle. A comparison between monthly mean T and TE over the United States shows that in winter and early spring there is almost no difference between the two variables. However, with increasing humidity from late spring to early fall, TE increases much more than T, in particular during summer (Figure 21.1). Therefore, large differences are observed during the growing season (up to 22.74°C in July according to results from Fall et al.).[12] The same patterns are noted at daily time scale and for both maximum and minimum T and TE.[10]

Moist Enthalpy Variability and Anomaly Trends

In general, at global scale, there is scientific consensus on an increase of the heat content of the ocean, especially during the latter half of the past century.[25] Findings from Levitus et al.[26] show that from 1955 to 1998, approximately 85% of this warming occurred in the world's oceans. With the ocean, only

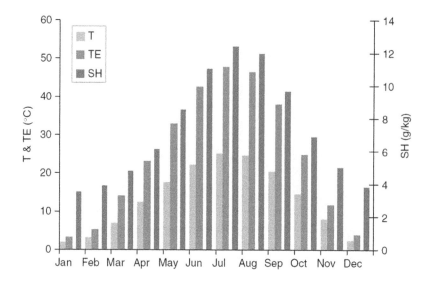

FIGURE 21.1 Monthly climatology of temperature (*T*), equivalent temperature (*TE*), and specific humidity (SH) at 2 m (average 1979–2005) over the United States. The ordinate scale on the right pertains to values of specific humidity.
Source: Reprinted with permission from Fall et al.[12]

the temperature is required to diagnose moist enthalpy. However, in the atmosphere, the humidity also must be included.

As shown in recent studies,[13,33] over the past decades moist enthalpy anomalies at global scale have generally exhibited warmer trends. Ribera et al.[33] found *TE* increments between +0.05 and +0.29°K/decade during the 1958–1998 period, with most of this increase occurring during the 1958–1978 period over densely populated coastal areas (e.g., Eastern Asia, Northern Europe, southeastern North America, and South Africa) and oceanic zones, in contrast with dry continental areas, which generally are characterized by negative trends (e.g., Sahara).

Positive *TE* trends are also found at regional and local scales, although Peterson et al.[13] indicate that the two terms can be of opposite signs in some regions (e.g., Southern Hemisphere sub-tropics, as shown in Figure 21.2). Results from Gaffen and Ross studies[31,32] indicate that the occurrence of hot and humid days increased in the United States during the past decades (1961–1995). As a result, extreme heat-stress events became more frequent, and because of a pronounced increase in atmospheric moisture content, summertime *TE* trends were positive and higher than summertime *T* ones. In more recent studies, comparisons between near-surface *T* and *TE* have confirmed that *TE* trends are generally warmer and significantly different from T.[11,12] However, differences between *T* and *TE* trends decrease gradually with altitude and almost disappear at 300–200 mb [about 9,000–12,000 m). For example, results from Fall et al.[12] who computed trend differences (*TE* minus *T*) at various levels for the United States during the period 1979–2005 (Figure 21.3) show not only the decreasing trend differences with altitude (e.g., 0.211°C/decade at near-surface and 0.005°C/decade at 300 mb), but also a shift to an opposite trend sign (−0.003°C/decade for both *T* and *TE* at 200 mb).

Seasonally, at global scale during the 1958–1998 period, Ribera et al.[33] found that *TE* trends remained positive for all seasons and most of the increase generally took place during the summer season; the largest trends were found over the continental southern hemisphere and oceanic areas. Over the United States, surface *TE* trends are found to be warmer in winter and cooler in fall and summer.[11,12] These seasonal patterns persist up to 700 mb and above that level, most of *TE* increase occurs during the fall season; at 200 mb, *TE* trends are negative, with the most substantial cooling taking place in winter.[12]

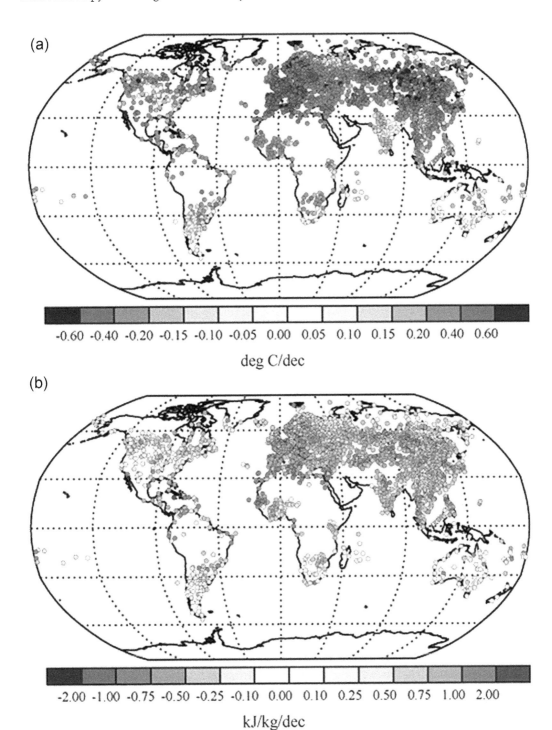

FIGURE 21.2 Decadal trends (1973–2003) calculated for HadCRUH stations using pentad anomaly-specific humidity and temperature and pentad climatologies (1974–2003) from Lott et al. (2008); (**a**) surface temperature trends (°C/decade); (**b**) heat content trends (kJ/kg/ decade).
Source: Reprinted with permission from Peterson et al.[13]

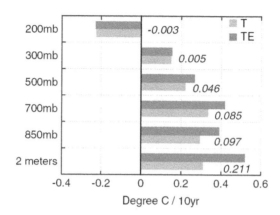

FIGURE 21.3 Decadal anomaly trends computed from monthly *T* and *TE* at different pressure levels (1979–2005; units: °C/10 yr). Italicized values denote the differences *TE – T*. All trends are significant at the 5% level (P value < 0.05).
Source: Reprinted with permission from Fall et al.[12]

Conclusion: On the Importance of Heat Content

The significance of considering heat content for a broader understanding of the Earth system climate variability has been demonstrated by various studies. The assessment of warming rates at global and regional scales requires considering the variability of ocean, atmosphere, and cryosphere heat content.[25–27] In particular, ocean heat content varies in conjunction with the annual cycle of the Earth net radiation balance and has been found to be a major source of variability of the global heat balance.[23,25,27,28,38] The observed ocean heat content variations also appear to be correlated with simulated variations in greenhouse gases, sulfate aerosols, solar irradiance, and volcanic aerosols.[27]

Several studies have highlighted the importance of atmospheric heat content (alternatively moist enthalpy or *TE*) in climate variability and change assessments. The relevance of *TE* includes, but is not limited to the following:

- Moist enthalpy, which expresses the combined effects of temperature and humidity, represents an important forcing of the general circulation of the atmosphere through different transport mechanisms and provides a better understanding of the large-scale dynamics in the atmospheric heat balance.[39,40] Ribera et al.[33] found a close relationship between *TE* and large scale modes of climate variability (the North Atlantic Oscillation and the Arctic Oscillation).

- Apparent temperature (A), a variant of *TE*, is closely related to human comfort. The *A* climatology developed by Gaffen and Ross[31,32] has been extended and adopted by NOAA as a useful index for heat stress and is periodically updated.

- Pielke et al.[10] stated that …*"The difference in temporal trends in surface and tropospheric temperatures [National Research Council,2000], which has not yet been explained, could be due to the incomplete analysis of the surface and troposphere for temperature, and not the more appropriate metric of heat content."* In a subsequent study, Fall et al.[12] compared *TE* and *T* at different atmospheric levels (up to 200 mb) and investigated the vertical structure of the combined effects of temperature and moisture. They analyzed the climatology, time series, and decadal trends of the two variables and found a contrast between 1) pronounced temporal and spatial differences at the near-surface level, and 2) almost no difference at 300–200 mb. More specifically, the thermal discrepancies between surface and upper air are much larger when *TE* is used instead of *T* alone. Fall et al.[12] concluded that the use of *TE* to assess tropospheric heating trends may "*help obtain an improved estimate of the impacts of surface properties on heating trends.*"

- Observation-and reanalysis-based studies have shown that moist enthalpy (*TE*) is more sensitive to surface vegetation properties than is air temperature (*T*). Davey et al.[11] found that over the eastern United States from 1982 to 1997, *TE* trends were similar or slightly cooler than *T* trends at predominantly forested and agricultural sites, and significantly warmer at predominantly grassland and shrubland sites. Results from Fall et al.[12] indicate that *TE* 1) is larger than *T* in areas with higher physical evaporation and transpiration rates (e.g., deciduous broadleaf forests and croplands) and 2) shows a stronger relationship than *T* to vegetation cover, especially during the growing season (biomass increase). These moist enthalpy-related studies confirm previous results showing that changes in vegetation cover, surface moisture, and energy fluxes generally lead to significant climatic changes[41–43] and responses, which can be of a similar magnitude to that projected for future greenhouse gas concentrations.[44,45] Therefore, it is not surprising that *TE*, which includes both sensible and latent heat, more accurately depicts surface and near-surface heating trends than *T* does.

In general, studies dealing with heat content have suggested that despite its undeniable relevance and popularity, temperature needs to be supplemented with additional metrics for assessing global warming. For this purpose, moist enthalpy, which includes both temperature and atmospheric moisture content, is a useful variable.[10–12] Using the Bowen ratio (ratio of sensible heat to latent heat) is another efficient way for analyzing the combined effects of temperature and moisture and their co-variation.[13,46]

Overall, a large majority of studies still use temperature as the key metric for assessing global warming. Global temperature is a convenient variable because of its higher availability and greater spatial/temporal coverage. However, several recent studies have addressed moist enthalpy (which is expressed in Joules and actually represents a robust measure of heat) and recommended both temperature and atmospheric moisture be considered in climate change assessments.

Acknowledgments

We thank Dallas Staley for her outstanding contribution in editing and finalizing the entry. The study benefited from the DOE ARM Program (08ER64674; Dr. Rick Petty and Dr. Kiran Alapaty), and in parts from NASA Terrestrial Hydrology Program (Dr. Jared Entin), and NSF CAREER-0847472 (Liming Zhou and Jay Fein). Roger Pielke, Sr. acknowledges support from NSF Grant 0831331 and received support from the University of Colorado at Boulder (CIRES/ATOC).

References

1. Crowley, T.J.; Lowery, T. How warm was the medieval warm period? Ambio **2000**, *29*, 51–54.
2. Mann, M.E.; Jones, P.D. Global surface temperatures over the past two millennia. Geophysical Res. Lett. **2003**, *30* (15), 1–4.
3. Soon, W-H.; Legates, D.R.; Baliunas, S. Estimation and representation of long-term (>40 year) trends of Northern Hemisphere-gridded surface temperature: a note of caution. Geophysical Res. Lett. **2004**, *31*, L03209, doi: 1029/ 2003GRL019141.
4. Moberg, A.; Sonechkin, D.M.; Holmgren, K.; Datsenko, N.M.; Karlen, W. Highly variable Northern Hemisphere temperatures reconstructed from low-and high-resolution proxy data. Nature **2005**, *433*, 613–618.
5. Intergovernmental Panel on Climate Change (IPCC). Summary for policymakers. In *Climate Change: The Physical Science Basis. Contribution of Working Group I to the Fourth Assessment Report of the Intergovernmental Panel on Climate Change;* Solomon, S., Qin, D., Manning, M., Chen, Z., Marquis, M., Averyt, K.B., Tignor, M., Miller, H.L., Eds.; Cambridge University Press: Cambridge, New York, 2007.
6. Mears, C.A.; Wentz, F.J. The effect of diurnal correction on satellite-derived lower tropospheric temperature. Science **2005**, *309*, 1548–1551.

7. Sherwood, S.C.; Lanzante, J.R.; Meyer, C.L. Radiosonde daytime biases and late 20th century warming. Science **2005**, *309*, 1556–1559.
8. Santer, B.D.; Thorne, P.W.; Haimberger, L.; Taylor, L.K.E.; Wigley, T.M.L.; Lanzante, J.R.; Solomon, S.; Free, M.; Gleckler, PJ.; Jones, P.D.; Karl, T.R.; Klein, S.A.; Mears, C.; Nychka, D.; Schmidt, G.A.; Sherwood, S.C.; Wentz, F.J. Consistency of modelled and observed temperature trends in the tropical troposphere. Int. J. Climatol. **2008**, *28*, 1703–1722.
9. Spencer, R.W.; Braswell, W.D. On the misdiagnosis of surface temperature feedbacks from variations in earth's radiant energy balance. *Remote Sensing* **2011**, *3*, 1603–1613.
10. Pielke, R.A., Sr.; Davey, C.; Morgan, J. Assessing "global warming" with surface heat content. Eos Trans. **2004**, *85* (21), 210–211.
11. Davey, C.A.; Pielke R.A., Sr.; Gallo, K.P. Differences between near surface equivalent temperature and temperature trends for the eastern United States: equivalent temperature as an alternative measure of heat content. Global Planetary Change **2006**, *54*, 19–32.
12. Fall, S.; Diffenbaugh, N.; Niyogi, D.; Pielke, R.A., Sr.; Rochon, G. Temperature and equivalent temperature over the United States (1979–2005). Int. J. Climatol. **2010**, *30*, 2045–2054, doi: 10.1002/joc.2094.
13. Peterson, T.C.; Willett, K.M.; Thorne, P.W. Observed changes in surface atmospheric energy over land. Geophysical Res. Lett. **2011**, *38*, L16707, doi: 10.1029/2011 GL048442.
14. Wentz, F.J.; Schabel, M.C. Precise climate monitoring using complementary satellite data sets. Nature **2000**, *403*, 414–416.
15. Trenberth, K.E.; Fasullo, J.; Smith, L. Trends and variability in column-integrated atmospheric water vapor. Climate Dynamics **2005**, *24*, 741–758.
16. Held, I.M.; Soden, B.J. Robust responses of the hydrological cycle to global warming. J. Climate **2006**, *19* (21), 5686–5699.
17. Santer, B.D.; Mears, C.; Wentz, F.J.; Taylor, K.E.; Gleckler, P.J.; Wigley, T.M.L.; Barnett, T.P.; Boyle, J.S.; Bruegge-mann, W.; Gillett, N.P.; Klein, S.A.; Meehl, G.A.; Nozawa, T.; Pierce, D.W.; Stott, P.A.; Washington, W.M.; Wehner, M.F. Identification of human-induced changes in atmospheric moisture content. Proc. Natl. Acad. Sci. **2007**, *104*, 15248–15253, doi: 10.1073/pnas.0702872104.
18. Wentz, F.J.; Ricciardulli, L.; Hilburn, K.; Mears C. How much more rain will global warming bring? Science **2007**, *317*, 233–235.
19. Willett, K.M.; Jones, P.D.; Gillett, N.P.; Thorne, P.W. Recent changes in surface humidity: development of the HadCRUH dataset. J. Climate **2008**, *21*, 5364–5383.
20. Willett, K.M.; Jones, P.D.; Thorne, P.W.; Gillett, N.P. A comparison of large scale changes in surface humidity over land in observations and CMIP3 GCMs. Environ. Res. Lett. **2010**, *5*, 025210 doi:10.1088/1748–9326/5/2/025210.
21. Wang, J.-W., Wang, K., Pielke, R.A., Sr; Lin, J.C., Matsui, T. Towards a robust test on North America warming trend and precipitable water content increase. Geophysical Res. Lett. **2008**, *35*, L18804, doi: 10.1029/2008GL034564.
22. Solomon, S.; Rosenlof, K.; Portmann, R.; Daniel, J.; Davis, S.; Sanford, T; Plattne, G-K. Contributions of stratospheric water vapor to decadal changes in the rate of global warming. Science **2010**, *327* (5970), 1219–1223, doi: 10.1126/ science.1182488.
23. Ellis, J.S.; Vonder Haar, T.H.; Levitus, S.; Oort, A.H. The annual variation in the global heat balance of the Earth. J. Geophysical Res. **1978**, *83*, 1958–1962.
24. Piexoto, J.P.; Oort, A.H. Physics of climate. Am. Inst. Phys. 1992, 520 pp.
25. Levitus, S.; Antonov, J.I.; Boyer, T.P.; Stephens, C. Warming of the World Ocean. Science **2000**, *287*, 2225–2229.
26. Levitus, S.; Antonov, J.; Boyer, T. Warming of the world ocean, 1955–2003. Geophysical Res. Lett. **2005**, *32*, L02604, doi: 10.1029/2004GL021592.
27. Levitus, S.; Antonov, J.; Wang, J.; Delworth, T.L.; Dixon, K.W.; Broccoli, A.J. Anthropogenic warming of Earth's climate system. Sciences **2001**, *292*, 267–270.

28. Pielke, R.A., Sr. Heat storage within the Earth system. Bull. Am. Meteorol. Soc. **2003**, *84*, 331–335.

29. Steadman, R.G. The assessment of sultriness, part 2: effects of wind, extra radiation and barometric pressure on apparent temperature. J. Appl. Meteorol. **1979**, *18*, 874–885.

30. Steadman, R.G. A universal scale of apparent temperature. J. Appl. Meteorol. Climatol. **1984**, *23*, 1674–1687.

31. Gaffen, D.J.; Ross, R.J. Increased summertime heat stress in the U.S. Nature **1998**, *396*, 529–530.

32. Gaffen, D.J.; Ross, R.J. Climatology and trends in U.S. surface humidity and temperature. J. Climate **1999**, *12*, 811–828.

33. Ribera, P.; Gallego, D.; Gimeno, L.; Perez, J.F.; Garcia, R.; Hernandez, E.; de la Torre, L.; Nieto, R.; Calvo, N. The use of equivalent temperature to analyze climate variability. Studia Geophysica et Geodaetica **2004**, *48*, 459–468.

34. Rogers, J.C.; Wang, S.H.; Coleman, J.S.M. Evaluation of a long-term (1882–2005) equivalent temperature time series. J. Climate **2007**, *20*, 4476–4485.

35. Bolton, D. The computation of potential equivalent temperature. Monthly Weather Rev. **1980**, *108*, 1046–1053.

36. Pielke, R.A., Sr.; Wolter; K.; Bliss, O.; Doesken, N.; McNoldy, B. The July 2005 Denver heat wave: How unusual was it? Natl. Weather Digest. **2006**, *31*, 24–35.

37. Pielke, R.A., Sr. Influence of the spatial distribution of vegetation and soils on the prediction of cumulus convective rainfall. Rev. Geophys. **2001**, *39*, 151–177.

38. Rossby, C. Current problems in meteorology. In *The Atmosphere and Sea in Motion*. Rockefeller Inst. Press: New York 1959; 9–50.

39. Riehl, H.; Malkus, J. On the heat balance in the equatorial trough zone. Geophysica **1958**, *6*, 3–4.

40. Tian, B.; Zhang, G.; *Ramanathan*, V. Heat balance in the Pacific warm pool atmosphere during TOGA COARE and CEPEX. J. Climate **2001**, *14*, 1881–1893.

41. Bonan, G.B. Effects of land use on the climate of the United States. Climatic Change **1997**, *37*, 449–486.

42. Bounoua, L.; Collatz, G.J.; Los, S.; Sellers, P.J.; Dazlich, D.A.; Tucker, C.J.; Randall, D.A. Sensitivity of climate to changes in NDVI. J. Climate **2000**, *13*, 2277–2292.

43. Niyogi, D; Kishtawal, C.M.; Tripathi, S.; Govindaraju, R.S. Observational evidence that agricultural intensification and land use change is reducing the Indian summer monsoon rainfall. Water Resources Res. **2010**, *46*, W03533, 17 pp., doi: 10.1029/2008WR007082.

44. Feddema, J.J.; Oleson, K.W.; Bonan, G.B.; Mearns, L.O.; Buja, L.E.; Meehl, G.A.; Washington, W.M. The importance of land-cover change in simulating future climates. Science **2005**, *310*, 1674–1678.

45. Diffenbaugh, N.S. Atmosphere-land cover feedbacks alter the response of surface temperature to CO2 forcing in the western United States. Climate Dynamics **2005**, *24*, 237–251, doi: 10.1007/s00382-004-0503-0.

46. Pielke, R.A., Sr; Davey, C.; Niyogi, D.; Fall, S.; Hubbard, K.; Lin, X.; Cai, M.; Lim, Y-K.; Li, H.; Nielsen-Gammon, J.; Gallo, K.; Hale, R.; Angel, J.; Mahmood, R.; Foster, S.; Steinweg-Woods, J.; Boyles, R.; McNider, R.T.; Blanken, P. Unresolved issues with the assessment of multi-decadal global land surface temperature trends. J. Geophysical Res. **2007**, *112*, D24S08, doi: 10.1029/2006JD008229.

22

Crops and the Atmosphere: Trace Gas Exchanges

Jürgen Kreuzwieser
and Heinz
Rennenberg
University of Freiburg

Introduction

The atmosphere mainly consists of nitrogen (78% by volume), oxygen (21% by volume) and the noble gas argon (0.93% by volume) together making up >99.9% of the atmosphere's composition; the remainder is known as trace gases. Trace gases are typically present in the range of parts per trillion by volume (pptv) to parts per million by volume (ppmv); they include the greenhouse gases carbon dioxide (CO_2), methane (CH_4), nitrous oxide (N_2O), ethane, water vapor, and ozone (O_3); the air pollutants sulfur dioxide (SO_2), ammonia (NH_3), nitric oxide (NO), nitrogen dioxide (NO_2), peroxyacyl nitrates (PAN), nitric acid (HNO_3), and carbon monoxide (CO); and a number of volatile organic compounds (VOCs) (Table 22.1). Among VOCs are isoprenoids (mainly isoprene and monoterpenes) and many oxygenated species such as alcohols, aldehydes, and organic acids.[2] Because of their reactivity, VOCs strongly affect the oxidation capacity of the troposphere and influence the concentration and distribution of several other trace gases, including CH_4 or CO.[3] On a regional scale, VOCs significantly contribute to the formation of tropospheric O_3.[4] The main source of VOCs (~90%) is natural emission by vegetation.[2]

Impacts of Trace Gases on Crops

The effects of atmospheric trace gases on crops are quite diverse and depend on the type of gas, its concentration, the duration of exposure, and the amount taken up, as well as a range of plant internal factors. Direct phytotoxic effects due to exposure to high concentrations of pollutants such as O_3 and SO_2 on crop plants include, among others, visible leaf injury; changes in chloroplast structure and cell membranes; disturbances of stomatal regulation, respiration, and photosynthesis; and reductions in growth and yield.[5] However, because sulfur (S) is an essential nutrient for plants, SO_2 absorbed by foliage may also be used as an additional source of sulfur in polluted areas, in addition to sulfate from the soil.[6] The same principle applies to nitrogen (N). Thus, effects of trace gases can be divided into phytotoxic effects caused by protons, organic compounds, SO_2, NO_2, NH_3, and O_3; and nutritional effects caused by S-and N-containing gases and CO_2.[7]

TABLE 22.1   Range of Ambient Concentrations of Different Trace
Gases in the Atmosphere

Trace Gas	Ambient Concentrations
Ar	9,340 ppm
CO_2	379 ppm (2005)
CH_4	1774 ppb (2005)
N_2O	319 ppb (2005)
Isoprene	ppt–several ppb
Monoterpenes	ppt–several ppb
Alcohols	1–30 ppb
Carbonyls (aldehydes, ketones)	1–30 ppb
Alkanes	1–3 ppb
Alkenes	1–3 ppb
Esters	ppt–several ppb
Carbonic acids	0.1–16 ppb
SO_2	0.5–20 ppb
H_2S	0–0.2 ppb
NO_2	0.5–15 ppb
O_3	20–80 ppb

Source: Data from Forster, et al.,[1] Guenther, et al.,[2] Kesselmeier &
Staudt,[3] Bender & Weigel,[5] and Herschbach, et al.[6]

Factors Controlling Trace Gas Exchange

Trace gases can be exchanged between the atmosphere and aboveground plant parts by 1) dry deposition
as gases or aerosols; 2) wet deposition as dissolved compounds in rainwater or snow; or 3) interception of
compounds dissolved in mist or cloud water[7] (Figure 22.1). The direction of the exchange (i.e., emission
vs. deposition) and its velocity (i.e., the exchange rates) is controlled by the physicochemical conditions
and by internal plant factors.

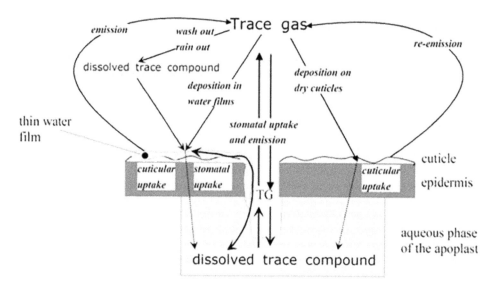

FIGURE 22.1   Main access routes for trace gases to enter a leaf: Uptake via the stomata and cuticular uptake.
Source: Redrawn from Rennenberg and Geßler.[8]

TABLE 22.2 Typical Range of Rates of Exchange for Different Trace Gases between Crops and the Atmosphere

Trace Gas	Range of Exchange [μg g^{-1} (leaf d.wt.) h^{-1}]
SO_2	−300 to 0
H_2S	−2 to 2
$O3$	−300 to 0
NO_2	−3 to 1
NH_3	−3 to 3
Isoprene	0 to 0.1
Monoterpenes	0 to 20
Oxygenated VOCs	−50 to 50

Negative values indicate deposition and positive values indicate emission of respective trace gas.

The gradient in the gas concentration between substomatal cavities and the atmosphere is the driving force for gas exchange. The gas flux is a diffusive (passive) process and can be described by Fick's law of diffusion. Accordingly, the net flux of a trace gas is zero when the substomatal concentration is equal to the concentrations in the surrounding ambient air. This concentration is referred to as the compensation point for the particular gas. When trace gas concentrations outside the leaves are higher than those in the substomatal cavities, a net flux into the leaves will take place (deposition), and vice versa. Therefore, crops may act both as a source (if ambient concentrations are lower than substomatal concentrations) or as a sink of a specific gas (Table 22.2). This dual behavior has been observed for a variety of gases (SO_2, H_2S, NO_2, NH_3, organic acids, and aldehydes). Compensation points for pollutants such as NH_3, e.g., range between 1 and 6 parts per billion by volume (ppbv) (see Husted et al.[9] and references therein). They depend mainly on the plant species or cultivar, development stage, temperature, and status of N nutrition of the plants. Generally, compensation points increase with increasing availability of nutrients in the soil, which suggests that this is one mechanism by which plants cope with excess nutrient supply.[8]

Plant Physiological Controls

The existence of a compensation point depends on the capacity of a plant to produce the trace gas to be exchanged, or to consume it. Therefore, compensation points do not exist for compounds that cannot be produced (e.g., O_3) or consumed (e.g., isoprene). It is evident that for each of the many gases exchanged between crop plants and the atmosphere, specific metabolic pathways exist, not all of which are understood yet. As an example, some details on the exchange of nitrogen compounds are presented (Figure 22.2).

Both NO_2 and NH_3 can be taken up by aboveground parts of plants, mainly via the stomata of leaves.[7] In the aqueous phase of the apoplast, NO_2 is either disproportionated yielding equal amounts of NO_2^- and NO_3^-, or it reacts with apoplastic ascorbate.[10] Because disproportionation of NO_2 in water is slow at atmospheric NO_2 concentrations, the reaction with ascorbate may be of more importance. Upon conversion to either NO_3^- or NO_2^-, these anions are transported to the cytoplasm, where they are reduced by the assimilatory nitrate reduction pathway yielding NO_4^+ and the amino acid glutamine. Atmospheric NH_3 dissolves in the aqueous phase of the apoplastic space to yield NO_4^+, which is then taken up into the cytoplasm.

Both NH_3 and NO_2 can also be emitted by plants. NH_3 may be released from cellular NO_4^+ pools when plants are exposed to excess nitrogen in the soil. In addition, it can be released from drying water films at the leaf surface. During this process, the remaining NH_4^+ concentrations on the surface will increase. By contrast, the chemical source of NO_2 emitted by the leaves is largely unknown; it has been proposed that nitrate reductase may be involved in the reduction of NO_2^- to NO_2.[8]

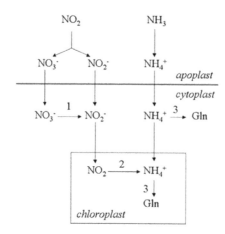

FIGURE 22.2 Main processes involved in the assimilation of atmospheric NH_3 and NO_2 taken up by the leaves. 1: nitrate reductase, 2: nitrite reductase, 3: glutamine synthetase (cytoplasmatic and chloroplastic isoforms). **Source:** Redrawn from Rennenberg, and Geßler.[8]

The rate of trace gas emission does not necessarily depend on the actual rate of production. Some volatile compounds are produced and then stored in particular pools. For example, some plant reservoirs contain high amounts of monoterpenes, which can be emitted throughout the day and night independent of light-dependent biosynthesis.[3]

Through their effect on biochemical pathways, biotic and abiotic factors (e.g., stress factors) and the developmental stage of plants influence the rate of trace gas emission. For example, stress caused by wounding, chilling, iron deficiency, O_2 deficiency, or induction of oxidative stress caused by exposure to O_3 or SO_2 can lead to the production of the VOCs hexenal, hexanal, formaldehyde, formate, ethene, ethane, ethanol, and acetaldehyde.[3]

Internal Transfer Resistances

The transfer of gases in and out of plants is often described by a resistance analogy. Along the path from the sites of production (or consumption) to the atmosphere, a series exists of mainly internal plant resistances. Because biosynthesis and consumption of volatile compounds usually take place in the cytoplasm or in other compartments of the cell, the gas must pass across the bordering membranes. The diffusive flux through these membranes is determined by the molecular size and the lipophilic character of the particular compound. Polar molecules such as organic and inorganic acids are not likely to be dissolved in the lipophilic membranes; therefore, the diffusive flux should be slow. Carrier proteins can facilitate the transport of these polar compounds across membranes from the cytoplasm into the apoplastic space, similar to the transport of organic acids.[11]

The compounds are transferred from the liquid phase in the apoplastic space to the gaseous phase in the substomatal cavity, or vice versa. The volatilization into the internal leaf air space depends on 1) chemical properties of both the aqueous phase and the individual compound to be emitted (e.g., its solubility in the apoplastic solution); and 2) physical factors such as the ambient concentration of the gas, its vapor pressure, and temperature.[11] A reduction of the apoplastic pH reduces the resistance for inorganic and organic acids, because these compounds become protonated and thereby more volatile. When present in the gaseous phase of the apoplastic space, trace gases can escape from the leaves through either the cuticle or the stomata. However, because of their polarity, the lipophilic cuticle constitutes a strong barrier, and therefore diffusion through the stomata is the main pathway. However, nonstomatal emission and deposition have also been observed in air pollutants such as NO,

NO_2, and SO_2, but at much lower rates than stomatal exchange. For this reason, factors that influence the stomatal aperture exert a strong influence on the rate of gas exchange between plants and the atmosphere. The control by the stomata of emission and deposition of NO_2, SO_2, O_3, PAN, and other trace gases was observed in many studies, including in investigations of crop plants. Thus, concentration and time of exposure are not the only factors determining the effect of an air pollutant on vegetation. Plants usually close their stomata during hot, dry conditions when, for instance, O_3 levels are high. This may provide some protection for the plants from O_3 injury. Alternatively, in northern Europe where O_3 concentrations are lower than in southern and central Europe, the potential O_3 uptake at a given O_3 concentration may be higher because of higher air humidity, leading to high rates of stomatal O_3 uptake.

Conclusion

Trace gases influence not only natural ecosystems but also agricultural crops. Future studies should focus on the impact of different combinations of air pollutants (e.g., increased nitrogen input combined with elevated O_3 concentrations) on plants and should include aspects of global climate change (e.g., higher temperatures, droughts, and increased frequency of heavy rainfalls and droughts in combination with elevated CO_2 concentrations). Future management practices should allow for optimum plant growth, while simultaneously reducing the loss of pollutants from the plant–soil system.[12]

References

1. Forster, P.; Ramaswamy, V.; Artaxo, P.; Berntsen, T.; Betts, R.; Fahey, D.W.; Haywood, J.; Lean, J.; Lowe, D.C.; Myhre, G.; Nganga, J.; Prinn, R.; Raga, G.; Schulz, M.; Van Dorland, R. Changes in Atmospheric Constituents and in Radiative Forcing. In: *Climate Change 2007: The Physical Science Basis. Contribution of Working Group I to the Fourth Assessment Report of the Intergovernmental Panel on Climate Change.* Solomon, S.; Qin, D.; Manning, M.; Chen, Z.; Marquis, M.; Averyt, K.B.; Tignor, M.; Miller, H.L. Eds.; Cambridge University Press: Cambridge and New York, 2007; 137–153 pp.
2. Guenther, A.B.; Hewitt, C.N.; Erickson, D.; Fall, R.; Geron, C.; Graedel, T.; Harley, P.; Klinger, L.; Lerdau, M.; McKay, W.A.; Pierce, T.; Scholes, B.; Steinbrecher, R.; Tallamraju, R.; Taylor, J.; Zimmerman, P. A. Global model of natural volatile organic compound emissions. J. Geophys. Res. **1995**, *100*, 8873–8892.
3. Kesselmeier, J.; Staudt, M. Biogenic volatile organic compounds (VOC): An overview on emission, physiology and ecology. J. Atmos. Chem. **1999**, *33*, 23–88.
4. Ashworth, K.; Wild, O.; Hewitt, C.N. Impacts of biofuel cultivation on mortality and crop yields. Nat. Clim. Change **2013**, *3*, 492–496.
5. Bender, J.; Weigel, H.-J. Changes in atmospheric chemistry and crop health: A review. Agron. Sustain. Dev. **2011**, *31*, 81–89.
6. Herschbach, C.; De Kok, L.J.; Rennenberg, H. Net uptake of sulfate and its transport to the shoot in tobacco plants fumigated with H_2S or SO_2. Plant Soil. **1995**, *175*, 75–84.
7. Wellburn, A.R. Why are atmospheric oxides of nitrogen usually phytotoxic and not alternative fertilizers? New Phytol. **1990**, *115*, 395–429.
8. Rennenberg, H.; Geßler, A. Consequences of N Deposition to Forest Ecosystems—Recent Results and Future Research Needs. In *Forest Growth Responses to the Pollution Climate of the 21st Century;* Sheppard, L.J.; Neil Cape, J., Eds.; Kluwer Academic Publication: Dordrecht, the Netherlands, 1999; 47–64.
9. Husted, S.; Schjoerring, J.K.; Nielsen, K.H.; Nemitz, E.; Sutton, M.A. Stomatal compensation points for ammonia in oilseed rape plants under field conditions. Agr. Forest Meteorol. **2000**, *105*, 371–383.

10. Ramge, P.; Badeck, F.W.; Ploechl, M.; Kohlmaier, G.H. Apoplastic antioxidants as decisive elimination factors within the uptake of nitrogen dioxide into leaf tissues. New Phytol. **1993**, *125*, 771–785.
11. Gabriel, R.; Schafer, L.; Gerlach, C.; Rausch, T.; Kesselmeier, J. Factors controlling the emissions of volatile organic acids from leaves of *Quercus ilex* L. (Holm oak). Atmos. Environ. **1999**, *33*, 1347–1355.
12. Rogasik, J.; Schroetter, S.; Schnug, E. Impact of air pollutants on agriculture. Phyton **2002**, *42*, 171–182.

23

Drought: Management

Donald A. Wilhite
and Michael J. Hayes
*National Drought
Mitigation Center*

Introduction

Some would argue that drought cannot be "managed." Yes, it is true droughts are a normal part of climate for virtually all areas of the world (e.g., Figure 23.1), and that droughts affect more people worldwide than any other natural hazard.[1] It is also true that officials from both developing and developed nations struggle to deal with the wide range of economic, environmental, and social impacts related to droughts. However, these officials are not powerless to reduce the impacts of drought. Rather, there are important management actions that officials at local, regional, and national levels can take to reduce the impacts from droughts. The approach taken to address drought impacts and reduce their effects is called drought management, or perhaps more appropriately, drought risk management. The long-term goal is to reduce the impacts of drought through the adoption of drought preparedness plans.

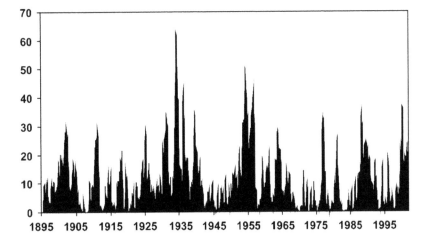

FIGURE 23.1   The percent area of the United States in severe to extreme drought by month from 1895 through March 2002. Similar periodic patterns appear on graphs depicting regional hydrological basins in the United States, and would likely appear for most regions in the world.
Source: Adapted from National Climatic Data Center, Asheville, North Carolina, U.S.A.

Shifting the Emphasis from Crisis to Drought Risk Management

Traditionally, droughts have been viewed as unusual occurrences that creep up on officials who are typically unprepared to deal with the impacts droughts create. This is why drought has been called the "creeping phenomenon."[1] In reality, drought is a normal feature for virtually all climates. Officials often react to the occurrence of drought through "crisis management." After a drought is over, officials turn their attention to the next crisis, and any lessons learned about responding to the drought are most likely lost and forgotten. This crisis management approach is illustrated in the "Hydro-Illogical Cycle" (Figure 23.2). Crisis management approaches to dealing with droughts are reactive, poorly coordinated and targeted, untimely, and generally too late. As a result, they are largely ineffective.

In order to break the Hydro-Illogical Cycle, officials around the world at local, regional, and national scales need to adopt a drought risk management approach. Drought risk management involves taking actions before droughts occur in order to reduce the drought impacts. It has three main components: 1) a comprehensive drought monitoring and early warning system; 2) planning and building the institutional capacity to respond to droughts; and 3) identification and implementation of mitigation actions and policies that can be taken before the next drought. These components will be discussed in greater detail.

A comprehensive drought monitoring and early warning system is a critical component of drought risk management because effective, timely decisions related to droughts can only be made if officials have an accurate assessment of the potential or developing drought event. This early warning system must incorporate all of the critical components of the hydrologic system (e.g., precipitation, streamflow, groundwater, snowpack, soil moisture, and reservoir and lake levels) because drought severity cannot be defined by precipitation deficiencies alone. A comprehensive system will assist officials by providing appropriate "triggers" for actions that the officials need to take, or by identifying when particular impacts are going to occur. An effective drought monitoring and early warning system requires synthesis and analysis of timely data and an efficient dissemination system to communicate this information (e.g., the media, extension services, or the World Wide Web).

Drought planning is a very important component of drought risk management because it establishes and preserves the institutional capacity with which officials can respond to droughts and reduce drought impacts. There are many benefits of a drought plan. A drought plan serves as the organizational

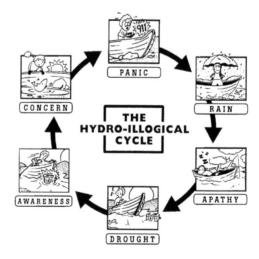

FIGURE 23.2   The Hydro-Illogical Cycle.
Source: National Drought Mitigation Center, University of Nebraska, Lincoln, Nebraska, U.S.A.

framework for dealing with droughts and improving the coordination between and within levels of government. In addition, drought plans enable proactive mitigation and response to droughts; enhance early warning through integrated monitoring efforts; involve stakeholders, which are necessary for successful programs; identify areas, groups, and sectors particularly at risk; improve information dissemination by outlining the information delivery systems and strategies; and build public awareness of the need for improved drought and water management.

Several methodologies exist for assisting officials with the development of drought plans. One of these methodologies, described by Wilhite et al.,[2] is a 10-step drought planning process that targets drought planners in the United States and elsewhere (http://drought.unl.edu/center/pdfpubs/10step.pdf) (accessed April 2002). The process was designed to be generic and adaptable because it is important for planners to develop a plan appropriate for their regional and governmental structures. These plans must be dynamic, reflecting changing government policies, technologies, personnel, and natural resources management practices.

The third important component of drought risk management is mitigation. Mitigation is defined as the policies and actions taken before a drought that will reduce drought impacts. Sometimes, if officials are alert enough and can see the development of drought in its early stages, mitigation can take place during the drought's early stages and may be very effective in reducing impacts as the drought becomes more severe. Otherwise, actions taken during a drought are generally responses directly related to the drought's severity and impacts. These responses are important, of course, and need to be well documented ahead of time within a drought plan. But it is important to keep in mind that mitigation is most effective if it takes place during times when drought is not occurring and officials are not responding to drought during a crisis. Mitigation actions should address vulnerabilities associated with drought with the goal of reducing impacts in future events.

What are some examples of drought mitigation? Certainly the development of a comprehensive drought monitoring and early warning system and the development of a drought plan, as described above, are two examples of mitigation. Both of these actions should be taken before a region is experiencing drought. Other broad categories for potential drought mitigation actions include revising or developing legislation or public policies related to drought and water supplies; water supply augmentation and the development of new supplies; demand reduction and the development of water conservation programs; public education and awareness programs; specific priorities for water allocations; and water use conflict resolution.[3]

As with drought planning, there are several methodologies for identifying the appropriate mitigation actions to take in a region. In 1998, as part of the activities of the Western Drought Coordination Council (WDCC), a methodology was developed to look at drought risk. An important part of this methodology was the identification of mitigation actions and how these actions would be implemented.[4] The methodology also involves identifying and understanding the people and sectors that are vulnerable to droughts and why, allowing officials to target their mitigation efforts more effectively.

Drought Risk Challenges and Opportunities

Serious challenges still remain in drought risk management. One of these challenges is the acceptance of drought as a natural hazard, and a hazard that needs to be prepared for. Fragmented resource management and numerous federal programs present challenges, as do the declining financial and human resources. Confusion over the difference between mitigation and response is a challenge, and many times officials have a difficult time identifying innovative mitigation actions and implementing new policies and programs. Stakeholder involvement and acceptance still needs improvement. Perhaps one of the biggest challenges is to maintain the momentum for risk management in a changing political climate.

Some progress toward drought risk management is being made around the world. Australia and New Zealand, for example, have had national drought policies and strategies to reduce drought impacts.[5,6] Other nations are looking at establishing national drought policies. A global drought preparedness network is in the development stages; this network would assist nations by promoting drought risk management and sharing lessons learned about drought monitoring, planning, and mitigation. The network, based at the National Drought Mitigation Center/International Drought Information Center at the University of Nebraska, would be made up of regional networks coordinated by institutions around the world. Collectively this network of regional networks may enhance the drought management capability of many nations.

In the United States, three states had drought plans in 1982. By 2002, 33 states had drought plans, and 6 of those states incorporate mitigation actions into their plans. In 1998, New Mexico became the first state to develop drought plan that emphasizes mitigation. Five states are currently in the process of developing drought plans, and it is hoped that mitigation will be a major component of each of these new plans. A number of Native American nations in the southwestern United States have developed drought mitigation plans as well. In addition, improved coordination has occurred within federal agencies and between federal and state governments.

New drought monitoring efforts and products have been developed. One of the best examples of progress in this area is the Drought Monitor product, developed to assess current drought conditions in the United States. The first Drought Monitor map was issued in August 1999, and a weekly update is posted every Thursday morning (http://drought.unl.edu/dm/) (accessed April 2002). The unique feature of this product is that four agencies rotate creating the map: the National Drought Mitigation Center, the United States Department of Agriculture, the Climate Prediction Center, and National Climatic Data Center of the National Oceanic and Atmospheric Administration. In addition, a feedback network of more than 160 local experts provides input about the map's portrayal of drought conditions before the map is released each week.

Conclusion

Clearly, there is reason for optimism about drought risk management and reducing drought impacts in the future. But it is also clear that officials around the world need to take proactive steps to develop comprehensive and integrated drought monitoring and early warning systems, determine who and what is at risk to droughts and why, and create drought mitigation plans with specific actions that address these risks with the goal of reducing the impacts of future drought events. There is a growing recognition that drought risk management is a critical ingredient of sustainable development planning and must be addressed systematically through risk-based policies and plans.

References

1. Wilhite, D.A. Drought as a natural hazard: concepts and definitions. In *Drought: A Global Assessment*; Wilhite, D.A., Ed.; Routledge: New York, 2000; Vol. 1, 1–18.
2. Wilhite, D.A.; Hayes, M.J.; Knutson, C.; Smith, K.H. Planning for drought: moving from crisis to risk management. J. Am. Water Res. Assoc. **2000**, *36* (4), 697–710.
3. Wilhite, D.A. State actions to mitigate drought: lessons learned. In *Drought: A Global Assessment*; Wilhite, D.A., Ed.; Routledge: New York, 2000; Vol. 2, 149–157.
4. Knutson, C.; Hayes, M.J.; Phillips, T. How to Reduce Drought Risk, Preparedness and Mitigation Working Group of the Western Drought Coordination Council: Lincoln, NE; 1998; 1–43, http://drought.unl.edu/handbook/risk.pdf (accessed April 2002).
5. O'Meagher, B.; Smith, M.S.; White, D.H. Approaches to integrated drought risk management: Australia's national drought policy. In *Drought: A Global Assessment;* Wilhite, D.A., Ed.; Routledge: New York, 2000; Vol. 2, 115–128.
6. Haylock, H.J.K.; Ericksen, N.J. From state dependency to self-reliance: agricultural drought policies and practices in New Zealand. In *Drought: A Global Assessment;* Wilhite, D.A., Ed.; Routledge: New York, 2000; Vol. 2, 105–114.

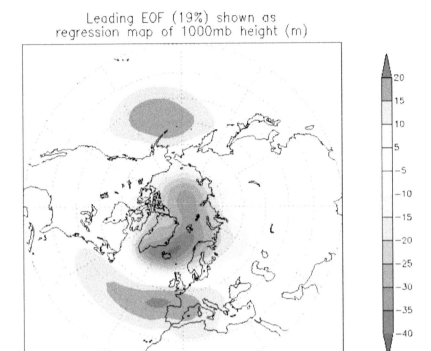

FIGURE 5.2 The seesaw-like pattern of the Arctic Oscillation.
Source: Data adapted from http://www.cpc.ncep.noaa.gov/prod-ucts/precip/CWlink/daily_ao_index/ao.shtml.[4]

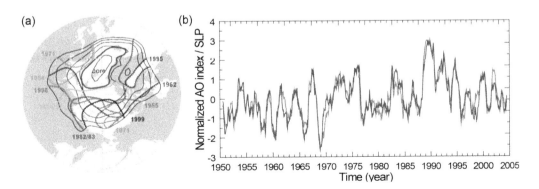

FIGURE 5.7 (a) AOCR and (b) the normalized AO and negative average SLP of AOCR.
Source: Data adapted from Zhao et al.[33]

(a)

Hydrogen ion concentration as pH from measurements made at the Central Analytical Laboratory, 1994

Sites not pictured:
Alaska 01	5.2
Alaska 03	5.1
Puerto Rico 20	5.2

Lab pH
- ≤ 4.1
- 4.5
- 4.9
- 5.3
- ≥ 5.7

National Atmospheric Deposition Program/National Trends Network
http://nadp.slh.wisc.edu

(b)

Hydrogen ion concentration as pH from measurements made at the Central Analytical Laboratory, 2017

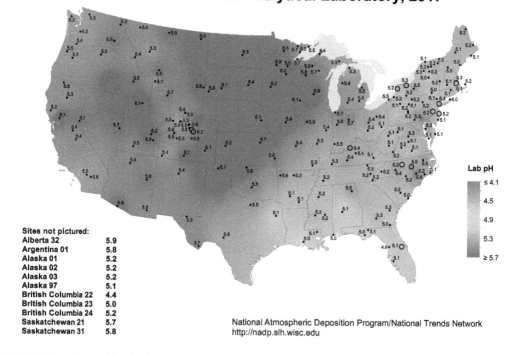

Sites not pictured:
Alberta 32	5.9
Argentina 01	5.8
Alaska 01	5.2
Alaska 02	5.2
Alaska 03	5.2
Alaska 97	5.1
British Columbia 22	4.4
British Columbia 23	5.0
British Columbia 24	5.2
Saskatchewan 21	5.7
Saskatchewan 31	5.8

Lab pH
- ≤ 4.1
- 4.5
- 4.9
- 5.3
- ≥ 5.7

National Atmospheric Deposition Program/National Trends Network
http://nadp.slh.wisc.edu

FIGURE 7.5 National levels of pH in precipitation for **(a)** 1994 and **(b)** 2017. (Adapted from NADP[15].)

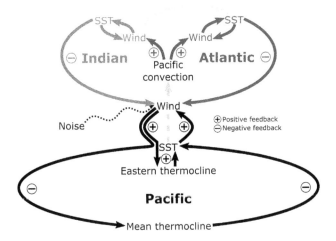

FIGURE 9.1 Interconnectivity among interannually varying tropical circulation patterns is illustrated through this schematic. The black loop [Bottom set of loops] represents internal feedbacks in the Pacific: Fast positive feedbacks are represented by short arrows; delayed negative feedbacks, by the long arrows. Interbasin feedbacks (positive) from the Pacific onto both the Atlantic and the IOs are shown by the short arrows in center of loop (Pacific convection). In contrast, negative feedbacks onto the Pacific from both adjacent oceans are represented by the upper loop, left side arrows (Indian Ocean) and upper loop, right side arrows (Atlantic Ocean). Note the positive feedback from the Atlantic onto the IO is shown in center-to-center right arrows at very top of upper loop. The overall result of the tropical interconnectivity is a warming tropical Pacific that warms adjacent oceans; in turn, through both direct and indirect means, mostly through the atmosphere, both the warmed Atlantic and the warmed IO negatively feed back onto the Pacific, leading to a termination in a warm ENSO phase, sometimes leading to a La Nina. (Reprinted with permission from AAAS: Cai et al. (2019).[1])

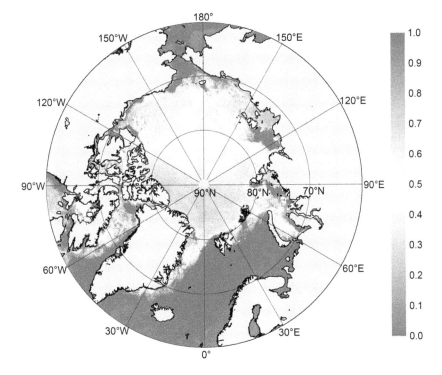

FIGURE 11.2 Shortwave albedo map of the Arctic sea-ice zone. (Qu et al., 2016.)

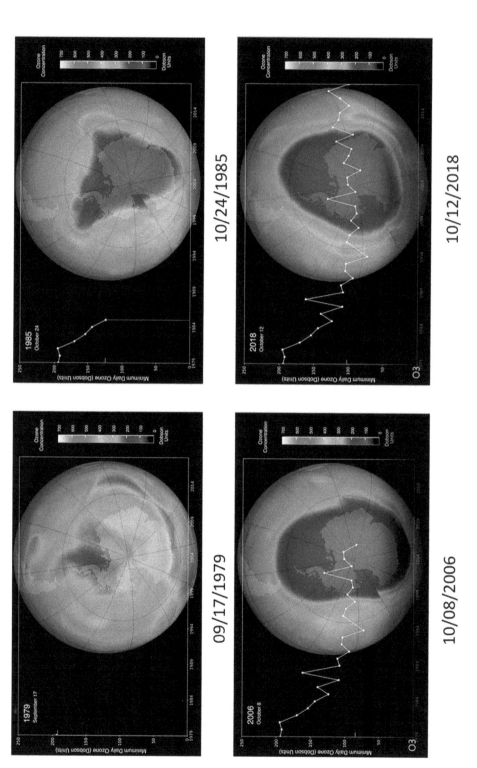

FIGURE 14.3 The trend of ozone hole over Antarctic between 1979 and 2018 observed by NASA satellites with images in 1979, 1985, 2006, and 2018. (NASA's Goddard Space Flight Center, https://svs.gsfc.nasa.gov/12816.)

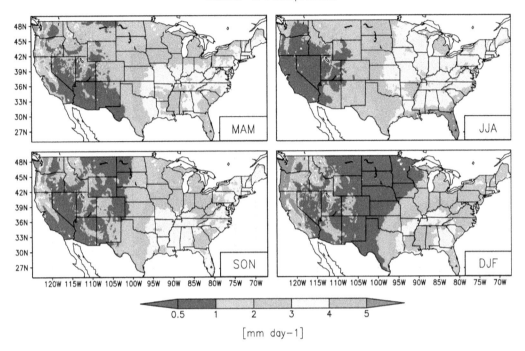

FIGURE 15.1 Thirty-year seasonal climatology (Mar 1979–Feb 2009) of surface precipitation in NLDAS Phase 2.

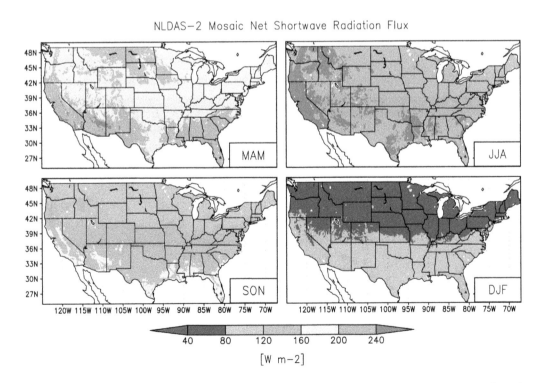

FIGURE 15.2 Thirty-year seasonal climatology (Mar 1979–Feb 2009) of surface net solar radiation from the Mosaic LSM in NLDAS Phase 2.

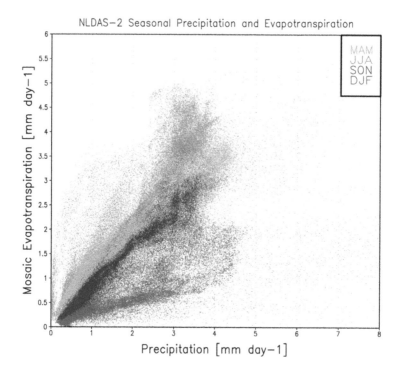

FIGURE 15.5 Scatter plot between surface precipitation and Mosaic *Et* for the MAM, JJA, SON, and DJF periods in NLDAS Phase 2. Each scattered point represents a grid point over CONUS, and all values are based on a 30-year seasonal climatology (Mar 1979–Feb 2009).

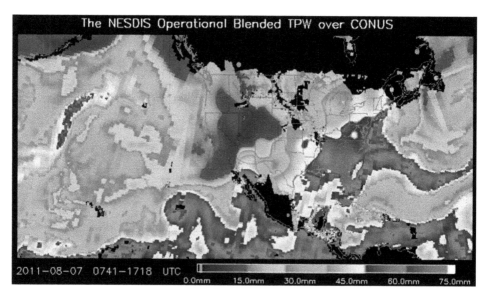

FIGURE 16.1 Enhanced total water vapor over the continental United States on a summer day (7 August 2011). NESDIS Operational Blended TPW Products.

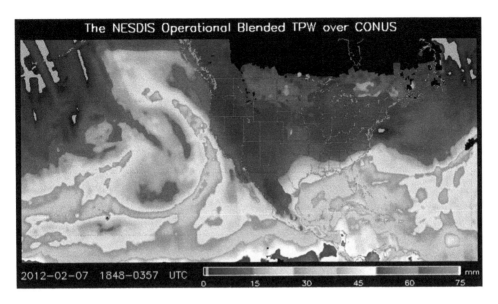

FIGURE 16.2 Reduced total water vapor over the continental United States on a winter day (7 February 2012). NESDIS Operational Blended TPW Products.

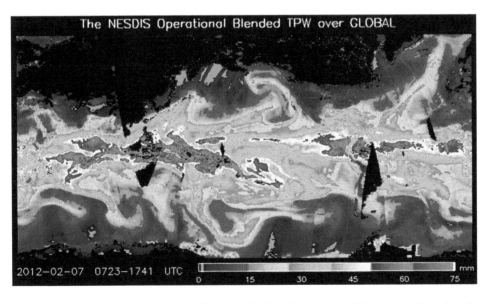

FIGURE 16.3 Expanded view of Figure 16.2 showing global total water vapor. Note the significantly enhanced moisture over the tropics in the Intertropical Convergence Zone. Dark areas, mainly over land, indicate an absence of data (7 February 2012). NESDIS Operational Blended TPW Products.

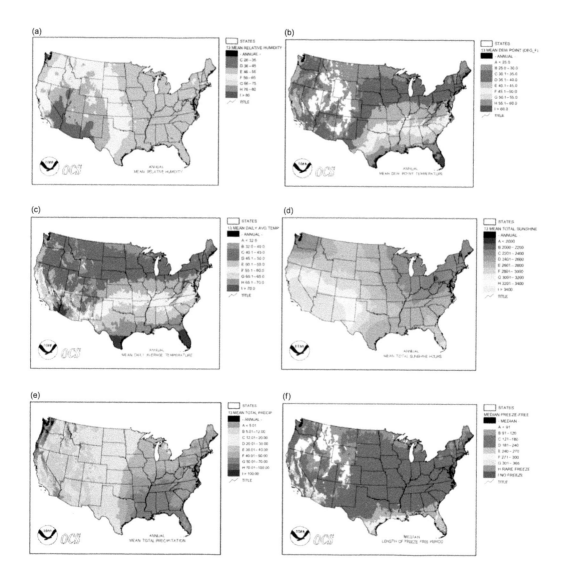

FIGURE 17.4 Examples of climatology maps collected from the National Climatic Data Center's Climate Maps of the United States database. Maps are developed from the 1961 to 1990 period of record from official weather and climate station sites unless otherwise noted. (a) Annual mean relative humidity in percent, (b) annual mean dew point temperature in degrees Fahrenheit, (c) annual mean daily average temperature in degrees Fahrenheit, (d) annual mean total sunshine hours, (e) annual mean total precipitation in inches, and (f) median length of freeze-free period in days.

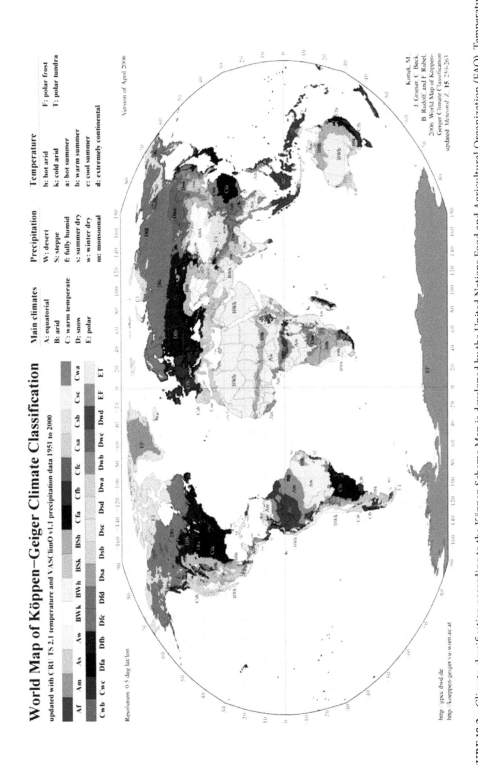

FIGURE 18.2 Climate classification according to the Köppen Scheme Map is developed by the United Nations Food and Agricultural Organization (FAO). Temperature and rainfall data are based on the Climate Research Unit (CRU) and the Global Precipitation Climatology Centre (GPCC) data, respectively.

Source: Grieser et al.,[11] Mitchell et al.,[12] and Beck et al.[13]

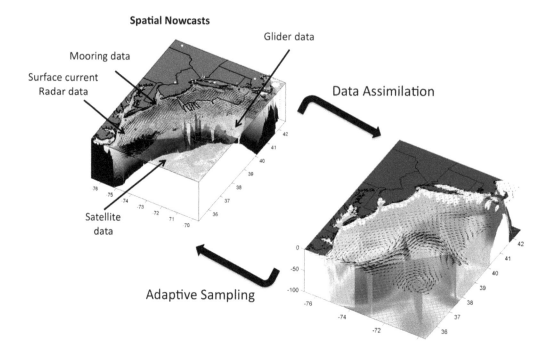

Spatial Nowcasts

Glider data

Mooring data

Surface current
Radar data

Data Assimilation

Satellite
data

Adaptive Sampling

FIGURE 29.1 An example of a coupled ocean observation and modelling network along the East coast of the United States. The spatial nowcast consists of observational data delivered in near real-time to shore. The nowcast consists of satellite data (sea surface temperature is shown), surface current radar (black arrows on the ocean), mooring data (black dots), subsurface data collected from gliders (water temperature shown here), and weather data collected by shore based stations (small white dots). The nowcast provides data to simulation models via data assimilation. The simulation model provides a 48-hour forecast, which is used to redistribute observational assets to provide an improved data set before the next forecast cycle.

FIGURE 29.2 A Webb underwater glider being deployed in the Ross Sea Antarctica in February 2011.
Source: Photo credit Chris Linder.

FIGURE 37.1 Appalachian Trail HUC-10 shell, FIA plot distribution, and the plots with tree of heaven observed.

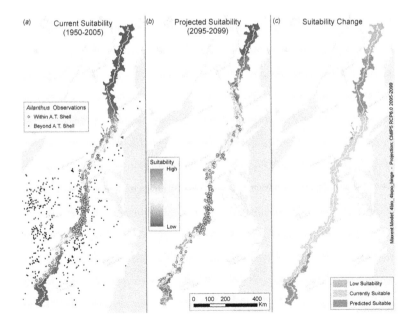

FIGURE 37.3 Maxent modeling-projected suitability of tree of heaven habitats for the current (1950–2005) (a), the future (2095–2099) (b), and the change (c).

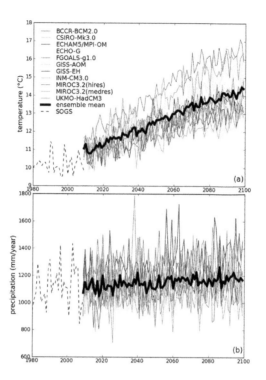

FIGURE 38.2 Projected mean temperature (a) and precipitation (b) through the end of the 21st century downscaled from CMIP3 multimodel dataset of SRES A1B scenario for 11 global climate models (GCMs) (Hashimoto et al., 2011).

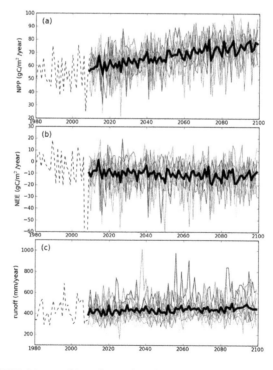

FIGURE 38.3 Projected NPP (a), NEE (b), and runoff (c) through the end of the 21st century downscaled from CMIP3 multimodel dataset of SRES A1B scenario for 11 GCMs (Hashimoto et al., 2011).

Global Land Cover Types

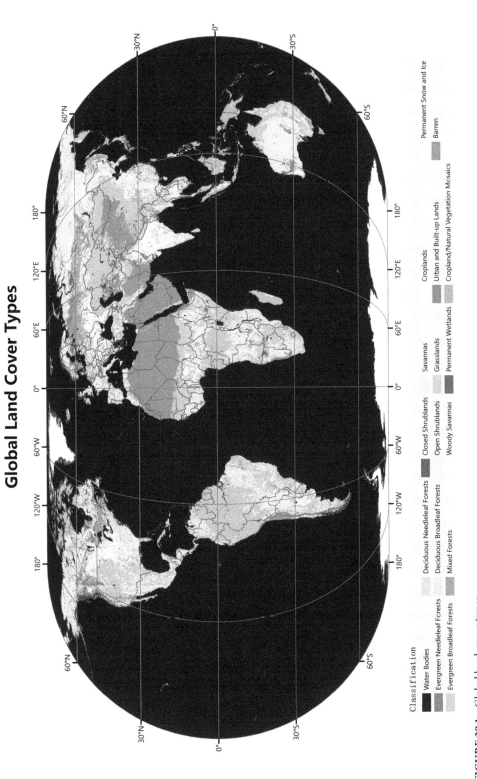

Classification

Water Bodies

Evergreen Needleleaf Forests

Evergreen Broadleaf Forests

Deciduous Needleleaf Forests

Deciduous Broadleaf Forests

Mixed Forests

Closed Shrublands

Open Shrublands

Woody Savannas

Savannas

Grasslands

Permanent Wetlands

Croplands

Urban and Built-up Lands

Cropland/Natural Vegetation Mosaics

Permanent Snow and Ice

Barren

FIGURE 39.1 Global land-cover types.

(a) The Start of the Growing Seasons (SOS) Distribution from 1982 to 2015

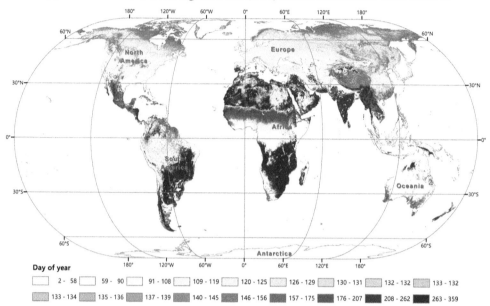

Day of year

2 - 58	59 - 90	91 - 108	109 - 119	120 - 125	126 - 129	130 - 131	132 - 132	133 - 132
133 - 134	135 - 136	137 - 139	140 - 145	146 - 156	157 - 175	176 - 207	208 - 262	263 - 359

(b) The End of the Growing Seasons (EOS) Distribution from 1982 to 2015

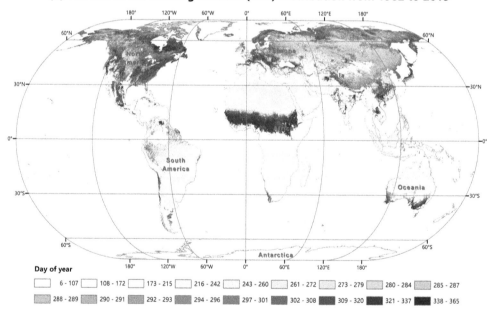

Day of year

6 - 107	108 - 172	173 - 215	216 - 242	243 - 260	261 - 272	273 - 279	280 - 284	285 - 287
288 - 289	290 - 291	292 - 293	294 - 296	297 - 301	302 - 308	309 - 320	321 - 337	338 - 365

FIGURE 39.2 Averages of the SOS (a) and the EOS (b), around the globe between 1982 and 2015.

(a) The Start of the Growing Seasons (SOS) Trend from 1982 to 2015

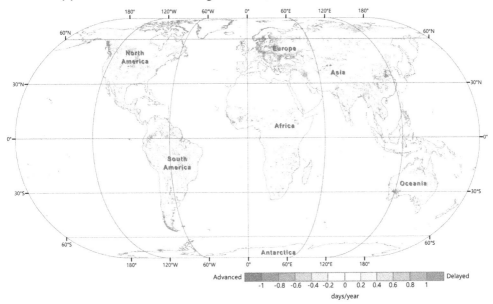

(b) The End of the Growing Seasons (EOS) Trend from 1982 to 2015

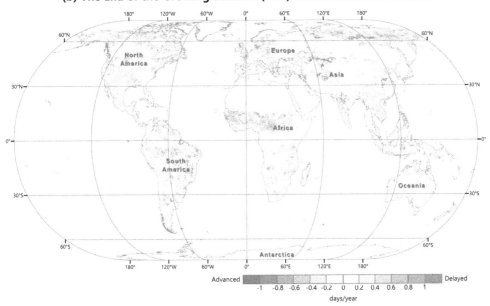

FIGURE 39.5 Trends of the SOS (a) and the EOS (b), from 1982 to 2015.

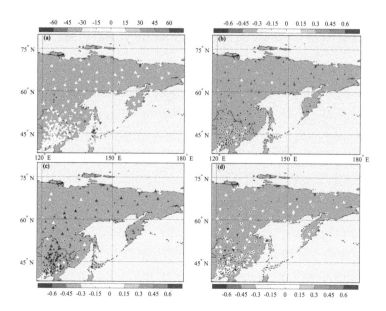

FIGURE 40.2 Spatial distribution of the changes in annual PRCP, T_{max}, T_{min}, and DTR in northeastern Eurasia (a, b, c, and d, respectively) from 1961 to 2010. The upward-pointing and downward-pointing triangles represent increasing and decreasing trends, respectively. The change magnitudes are shown in different colors. The units are mm per decade for precipitation and °C per decade for temperature. The overlying dots on triangles indicate significant trends ($p < 0.05$) estimated by a Mann–Kendall test.

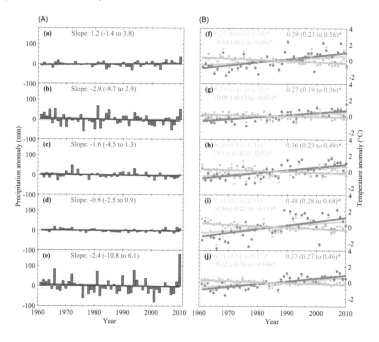

FIGURE 40.3 Regionally averaged seasonal and annual precipitation (A) and temperature (B) time series for northeastern Eurasia during the period from 1961 to 2010. Subfigures from up to down represent spring, summer, autumn, winter, and annual time series, respectively. Anomalies were relative to the average during the period from 1961 to 2010. Slope indicates the linear trend of time series, and values in the parentheses represent the 5%–95% uncertainty range of the linear regression slope. The units of slope were mm per decade for precipitation and °C per decade for temperature. No trends were significant at a 95% confidence level for precipitation. For temperature, T_{max}, T_{min}, and DTR are shown in orange, carmine, and green, respectively. The symbol "*" indicates that the linear trend was significant at the 0.05 level.

24

Drought: Precipitation, Evapotranspiration, and Soil Moisture

Joshua B. Fisher
and Konstantinos
M. Andreadis
*California Institute
of Technology*

Introduction

Drought ranks as one of the most expensive natural disasters in terms of human welfare and food security. For example, in the United States the annual cost of drought relief measures has been estimated between US$6 and US$8 billion. Droughts can cover very large areas and last for several years with so-called megadroughts documented from medieval times.[1] In general terms, drought is caused by extremes within the natural variability of climate, but can be exacerbated by human activity (e.g., deforestation). The literature on drought is extensive, with definitions categorically ranging from meteorological (or, climatological, atmospheric), agricultural, hydrologic, and socio-economic (e.g., management based),[2–5] but we focus here on the vegetation transpiration and evaporation (or, actual evapotranspiration, ET) component of drought.[6] From a vegetation perspective in general, physical drought is the drying of soil such that the overlying vegetation experiences physiological water stress manifested in a reduction of productivity, loss of leaves/needles, and, ultimately, mortality. As such, a given soil moisture content (SMC) would correspond to different classes of drought depending on the ability of vegetation to adapt to decreased soil moisture. For instance, some species or stages of succession may have plants with deep roots that can tap deep sources of soil moisture, even the groundwater table, so the ability of these plants to withstand what would otherwise be considered a drought may, in fact, not be considered a drought for some time.[7] Other species, however, may be poorly adapted to low levels of soil moisture through sparse root distribution, low water use efficiency, or high temperature sensitivity, and thus may enter into a drought much more quickly than other, better adapted, species.

SMC is inherently coupled directly to precipitation (PPT) and ET, with PPT as the moisture input, and ET as the moisture withdrawal; soil water holding capacity acts as the intermediate "bucket" size if considering a bucket model of SMC change. Among those three variables (i.e., SMC, PPT, ET), the end members of SMC can be outlined: 1) for two areas with equivalent PPT and ET, SMC may be different because of different SMC retention properties (e.g., sandy soils may hold less water than soil with more clay); 2) for two areas with equivalent soils and PPT, SMC may be different because of different ET; and 3) for two areas with equivalent soils and ET, SMC may be different because of different PPT.

The same exercise can be applied for areas to have similar SMC (e.g., two areas with equivalent soils, one with high PPT and high ET, and the other with low PPT and low ET). By this definition, and with reference to the title of this entry, ET can vary under drought or non-drought situations, although ET will go to zero under persistent and intense drought. Between the ET components, transpiration will go to zero as drought persists (assuming no deep water sources), because stomata close to avoid water loss (assuming no leaky stomata),[8] and plants will maintain respiration through carbon stores; evaporation from the soil surface will also go to zero as the soil dries out (assuming no hydraulic lift/redistribution from deep water sources).[9] The role of ET in drought is particularly pertinent in already water-limited environments where increasing temperatures over time accelerate ET, which leads to greater drought severities.[10]

Plants are typically able to withstand relatively short periods of SMC decline, so a given day with no PPT and high ET may not be considered a drought. Although there are different metrics of drought that take ET into account such as the widely used Palmer Drought Severity Index,[2] which uses the potential ET, it is the *cumulative water deficit* (CWD) that plants respond to – that is, the summation of days in which the soil water deficit (SWD) is below a critical water stress threshold. SWD may be calculated as follows:

$$\text{IF PPT} - \text{ET} > 0, \text{THEN SWD}_1 = 0,$$

$$\text{ELSE SWD}_1 = P - ET + SWD_0$$

where the subscripts indicate adjacent time steps. The maximum CWD (MCWD) reached during the time period of interest relative to the time-averaged climatological MCWD may be considered "drought." [11] Both the length of the CWD and the MCWD in a given period must surpass the long-term means of the two for that period to be considered a drought, although a corresponding vegetation water stress response is also necessary.

An example of MCWD drought and vegetation response is shown in the *Science* paper by Phillips et al.,[11] "Drought sensitivity of the Amazon rainforest." Here, we describe how measurements of anomalously low PPT indicated the possibility of an intense drought over Amazonia – the lowest PPT at the time, in fact, in the past 100 years. With few soil moisture measurements available, SWD was constructed from measured PPT, estimated ET using meteorological measurements,[12,13] and measurements of the soil

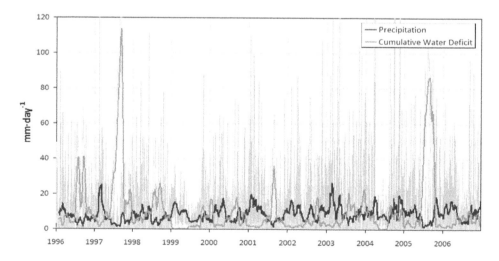

FIGURE 24.1 Precipitation (black) and cumulative water deficit (gray) over 10 years for a site in Amazonia, described in Phillips et al. [11] The light colors are the daily values, and the dark colors are the 30-day moving averages.

water holding capacity. We calculated MCWD at 136 sites where we also observed vegetation response, and determined that sites experiencing the greatest hydrologic drought as defined by MCWD also had the greatest vegetation response, specifically mortality and biomass loss. It can be seen at one of the sites, for example (Figure 24.1), that PPT varies seasonally, as does CWD, but in 2005 (also 1997 and 2001) CWD spikes well beyond the mean CWD for the 10-year record. In this analysis, the length and peak (MCWD) of the CWD spike vary by site and are proportional to the vegetation response (e.g., mortality).

Conclusion

Evaporation and transpiration are critical components to drought, although many traditional definitions of drought ignore ET. Vegetative drought inherently implies a response from vegetation, and this response must be calculated from a cumulative water deficit, as ET > P adds up over time, drying out the soil. In the absence of soil moisture measurements, soil moisture may be calculated from precipitation and ET. Even with soil moisture measurements, the understanding of the bioclimatic variables controlling ET helps elucidate and predict how drought will change given changes in the controlling factors.

Acknowledgments

This entry was written by the Jet Propulsion Laboratory, California Institute of Technology, under a contract with the National Aeronautics and Space Administration.

References

1. Cook, E.R.; Woodhouse, C.A.; Eakin, C.M.; Meko, D.M.; Stahle, D.W. Long-term aridity changes in the western United States, Science **2004**, *306* (5698), 1015–1018.
2. Palmer, W.C. Meteorological drought Rep., U.S. Department of Commerce, 1965; 58.
3. Dracup, J.A.; Lee, K.S.; Paulson, Jr., E.G. On the definition of droughts. Water Resour. Res. **1980**, *16* (2), 297–302.
4. Wilhite, D.A.; Glantz, M.H. Understanding the drought phenomenon: the role of definitions. Water Int. **1985**, *10*, 111–120.
5. McKee, T.B.; Doesken, N.J.; Kleist, J. The relationship of drought frequency and duration to time scales, in *Eight Conference on Applied Climatology;* American Meteorological Society: Anaheim, CA,1993; 179–184 pp.
6. Fisher, J.B.; Whittaker, R.H.; Malhi, Y. ET Come Home: A critical evaluation of the use of evapotranspiration in geographical ecology, Global Ecol. Biogeogr. **2012**, *20*, 1–18.
7. Fisher, R.A.; Williams, M.; Da Costa, A.L.; Malhi, Y.; Da Costa, R.F.; Almeida, S.; Meir, P. The response of an Eastern Amazonian rain forest to drought stress: results and modelling analyses from a throughfall exclusion experiment, Global Change Biol. **2007b**, *13* (11), 2361–2378.
8. Fisher, J.B.; Baldocchi, D.D.; Misson, L.; Dawson, T.E.; Goldstein, A.H. What the towers don't see at night: Nocturnal sap flow in trees and shrubs at two AmeriFlux sites in California., Tree Physiol. **2007a**, *27* (4), 597–610.
9. Dawson, T.E. Hydraulic lift and water use by plants: implications for water balance, performance and plant-plant interactions, Oecologia **1993**, *95* (4), 565–574.
10. Andreadis, K.M.; Lettenmaier, D.P. Trends in 20th century drought over the continental United States., Geophys. Res. Lett. **2006**, *33* (L10403), doi:10.1029/2006GL025711.
11. Phillips, O.L.; Aragão, L.E.O.C.; Lewis, S.L.; Fisher, J.B.; Lloyd, J.; López-González, G.; Malhi, Y.; Monteagudo, A.; Peacock, J.; Quesada, C.; van der Heijden, G.; Almeida, S.; Amaral, I.; Arroyo, L.; Aymard, G.; Baker, T.R.; Bánki, O.; Blanc, L.; Bonal, D.; Brando, P.; Chave, J.; Oliveira, A.C.A.; Cardozo, N.D.; Czimczik, C.I.; Espejo, J.; Feldpausch, T.; Freitas, M.A.; Higuchi, N.; Jiménez, E.; Lloyd, G.; Meir, P.; Mendoza, C.; Morel, A.; Neill, D.; Nepstad, D.; Patiño, S.; Peñuela,

M.C.; Prieto, A.; Ramírez, F.; Schwarz, M.; Silveira, M.; Thomas, A.S.; ter Steege, H.; Stropp, J.; Vásquez, R.; Zelazowski, P.; Dáavila, E.A.; Andelman, S.; Andrade, A.; Chao, K.-J.; Erwin, T.; Di Fiore, A.; Honorio, E.; Keeling, H.; Killeen, T.; Laurance, W.; Cruz, A.P.; Pitman, N.; Vargas, P.N.; Ramírez-Angulo, H.; Rudas, A.; Salamão, R.; Silva, N.; Terborgh, J.; Torres-Lezama, A. Drought sensitivity of the Amazon rainforest., Science **2009**, *323* (5919), 1344–1347.

12. Fisher, J.B.; Tu, K.; Baldocchi, D.D. Global estimates of the land-atmosphere water flux based on monthly AVHRR and ISLSCP-II data, validated at 16 FLUXNET sites., Remote Sensing Env. **2008**, *112* (3), 901–919.

13. Fisher, J.B.; Malhi, Y.; de Araújo, A.C.; Bonal, D.; Gamo, M.; Goulden, M.L.; Hirano, T.; Huete, A.; Kondo, H.; Kumagai, T.; Loescher, H.W.; Miller, S.; Nobre, A.D.; Nouvellon, Y.; Oberbauer, S.F.; Panuthai, S.; von Randow, C.; da Rocha, H.R.; Roupsard, O.; Saleska, S.; Tanaka, K.; Tanaka, N.; Tu, K.P. The land-atmosphere water flux in the tropics, Global Change Biol. **2009**, *15,* 2694–2714.

25

Drought: Resistance

Graeme C. Wright
and Nageswararao
C. Rachaputi
*Queensland Department
of Primary Industries*

Introduction

Drought can be considered as a set of climate pressures that can result from a combination of heat, aerial, or soil water deficits, as well as salinity. The diversity of drought created from these phenomena has led to the selection of numerous types of resistance mechanisms that operate at different levels of life organization (molecule, cell, organ, plant, and crop). Decades of research have been dedicated to the understanding of these mechanisms, with a premise that the improved understanding would contribute to the long-term improvement of plant and crop production under drought conditions.[1]

Discussion

This entry will concentrate on crop production, and in that context drought is a term used to define circumstances in which growth or yield of the crop is reduced because of insufficient water supply to meet the crop's water demand. During the 1960s to 1980s most of the drought research was dedicated to understanding the mechanisms of survival and growth under drought conditions. It is only in fairly recent times that attention has been given to recognizing the complex nature of drought and to separating the *productivity* of crop plants under drought from *survival* mechanisms. Drought resistance in modern agriculture requires sustainable and economically viable crop production, despite stress. However, plant survival can be a critical factor in subsistence agriculture, where the ability of a crop to survive drought and produce some yield is of critical importance.

Hence, in the context of agricultural production, drought resistance in a crop can be best defined in terms of the optimization of crop yield in relation to a limiting water supply.[2] Multitudes of options exist for farmers to alleviate the effects of drought on crop yield, depending on the probability of drought. These can be categorized into management and genetic options, although they can be integrated into a package to manage drought in the target environment. The basis of most management technologies adopted by farmers revolves around optimizing water conservation and its subsequent utilization by the crop. Examples include the use of deep tillage to increase rainfall infiltration, stubble retention to minimize soil evaporation,[3] and intercropping.[4] Genetic options include the use of the best locally adapted varieties or landraces, as well as relaying and intercropping varieties with varying phenology to exploit differences in timing and severity of drought patterns.[5]

Future advances in crop drought resistance and associated improvements in productivity under drought are most likely to come from genetic improvement programs that can apply the wealth of knowledge created over the past century. As Richards[6] states, however, it will never be possible to *overcome* the effects of drought, any progress is likely to be slow, and the gains will only be small. The following sections will therefore concentrate on opportunities and emerging technologies for the improvement of drought resistance in crop plants, using genetic enhancement.

Drought Resistance Traits

Levitt[7] has proposed a terminology for drought resistance and its subdivision into different categories based on different mechanisms. These three categories of drought resistance have been widely accepted, and they continue today in a slightly modified form to provide a framework for evaluating potential traits for use in crop breeding programs.[8–10] They are drought escape, dehydration postponement, and dehydration tolerance.

Drought Escape

Matching the phenology to the expected water supply in a given target environment has been an important strategy for improving productivity in water-limited environments.[11] In most crop species there is large genetic variability in phenological traits, and these traits are highly heritable and amenable to selection in large-scale breeding programs.[12] Matching phenology has proven to be a highly successful approach in environments that have a high probability of end-of-season drought stress pattern.[13]

Dehydration Postponement

Crops use a variety of mechanisms to maintain turgor in leaves and reproductive structures despite declining water availability. They can effectively regulate water loss from leaves via stomatal control, with large varietal differences in stomatal conductance in response to leaf water potential recorded in cereals[14] and grain legumes.[15]

Production of abscisic acid (ABA) has been implicated as a mechanism behind stomatal control.[16] Other benefits of ABA, including maintenance of turgor in wheat spikelets and subsequent grain set, have also been reported.[17] Subsequent stimulation of research activity into the use of the ABA trait in crop breeding programs has followed. However, Blum[18] recently concluded that while ABA is undoubtedly involved in plant response to drought stress and even perhaps in desiccation tolerance, its value in the context of drought resistance breeding is still questionable.

Osmotic adjustment (OA) has been reported as an important drought-adaptive mechanism in crop plants where solutes accumulate in response to increasing water deficits, thereby maintaining tissue turgor despite decreases in plant water potential.[19] OA has been shown to maintain stomatal conductance, photosynthesis, and leaf expansion at low water potential,[20] as well as reducing flower abortion[21] and improving soil-water extraction despite declining water availability.[22,23] Research has confirmed that OA is directly associated with grain yield in a number of crops, including wheat,[24,25] sorghum,[26] and chickpea.[27]

Dehydration Tolerance

The ability of cells to continue metabolism at low-leaf water potential is known as dehydration tolerance. Membrane stability and the associated leakage of solutes from the cell[28] provide one measure of the ability of crop/genotypes to withstand dehydration. Sinclair and Ludlow[29] have suggested that the lethal water potential is a key measure of dehydration tolerance, with significant variation among crops and cultivars observed. Some stages of the plant's life cycle are less susceptible to large reductions in water content. For instance, prior to establishment of a large root system, seedlings may often survive large reductions in water content.[30] Protection against lethal damage in seedlings and seeds is correlated with the accumulation of sugars and proteins.[31] Proline has also been implicated in cellular survival of water deficits, and has also been involved in osmotic adjustment. Its role as a selection trait for enhanced drought resistance has been questioned.[32] Although the work on drought resistance mechanisms has produced a few promising leads, their application in practical breeding programs has been limited.

Hierarchy of Drought Resistance Traits

Richards[6] suggests there are two major principles to consider when identifying critical traits to use in breeding programs aimed at improving productivity under drought, namely, the influence of the trait in relation to time scale and to level of organization.

Time Scale

Traits that influence drought resistance in crops can span a wide range of time scales. Short-term responses to water deficit, for example, include many of the processes covered earlier (heat-shock proteins, stomatal closure, OA, ABA), which Passioura[2] suggests are often primarily concerned with "metabolic housekeeping." Although these processes are important, they tend to be associated with crop survival rather than with events that influence crop productivity. At the other end of the time scale are longer-acting processes such as control of leaf area development, which can be modulated by the crop to adjust water supply to prevailing demand.[33] It is not always clear which processes are operating to control these balances, but presumably hormonal signals are involved. Passioura[2] argues that researchers need to distinguish between traits linked to short-term responses that might be important for overall drought resistance and those that are unimportant when integrated over longer time scales.

Level of Organization

The capacity of the trait to influence yield is related to the level of organization (molecule–cell–organ–plant–crop) in which the trait is likely to be expressed.[34] Richards[6] cites an example that despite the doubling of crop yields since 1900, the rate of leaf photosynthesis, which is expressed at the cellular-organ level, has remained the same or decreased. Increases in leaf area during this period have been largely responsible for the yield increases. It is concluded that the closer the trait is to the level of organization of the crop the more influence it will have on productivity (Figure 25.1).

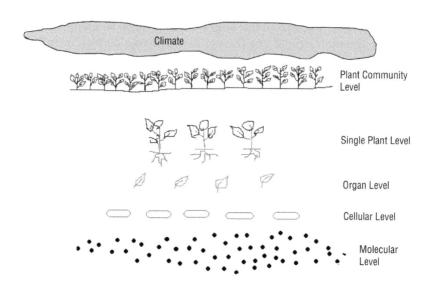

FIGURE 25.1 The hierarchy of processes leading to crop yield.

Drought Resistance Traits in Terms of the Passioura Water Model

Passioura[2,34,35] argues that there are no traits that confer global drought resistance. Also, given the earlier discussion which clearly suggests that short term responses to drought stress operating at the cellular level may have no bearing on the yield of water-limited crops, crop productivity is best analyzed from the top down. Here the thinking has shifted from understanding defense mechanisms for survival to applying this knowledge to optimize economic productivity under a given water-limited condition.

Passioura proposed that when water is the major limit, grain yield (GY) is a function of the amount of water transpired by the crop (T), the efficiency of use of this water in biomass production (WUE), and the proportion of biomass that is partitioned into grain, or harvest index (HI), thus:

$$GY = T \times WUE \times HI \tag{25.1}$$

It was argued that individual traits could be assessed in terms of their contribution to each of these functional yield components and thereby increase yield under drought. This identification has provided a framework to more critically identify and evaluate important drought resistance traits. The following examples demonstrate the utility of this approach.

In a number of crops, improvement in the WUE trait has been achieved via selection for carbon isotope discrimination,[36] or correlated surrogate measures.[37–39] Readers are referred to reviews for more details on this subject.[6,40,41]

In maize, increased partitioning to the grain (or higher HI) has been brought about by using an ideotype selection index that focused on a reduction in anthesis to silking interval.[42] Grain yield of the final selections increased by 108 kg/ha/cycle and came about by an increase in HI, with no change in final biomass relative to the parents.

Wright et al.[43] proposed that estimates of each of the water model components for a peanut crop could be derived from simple and low-cost measurements of total and pod biomass at harvest, from estimates of WUE from correlated measures of specific leaf area, and by reverse engineering the TDM component of the water model, such that T=TDM/TE. A selection study is being conducted on four peanut populations, in which a selection index approach utilizing T, WUE, and HI is assessing the value of these traits in a large-scale breeding program in India and Australia.[44]

Conclusion

Although drought is commonly referred to as "prolonged water deficit" periods during a crop's life, it is indeed a complex syndrome with various climate pressures operating together in infinite combinations, resulting in significant reductions in crop performance. Historically, options for managing agricultural drought have revolved around management techniques such as deep tillage, mulching, intercropping, etc. However, future options for drought management will increasingly be based on the genetic improvement of crops for targeted environments. With genetic options, matching phenology to water availability in target environments has proven to be highly a successful approach. Although our understanding of drought resistance mechanisms has improved significantly, it is essential to distinguish the traits linked to short-term responses from those which are important when integrated over longer time scales.

References

1. Schulze, E.D. Adaptation Mechanisms of Non-Cultivated Arid-Zone Plants: Useful Lessons for Agriculture. In *Drought Research Priorities for the Dryland Tropics*; Bidinger, F.R., Johansen, C., Eds.; ICIRSAT: Patancheru, A.P. 502324, India, 1988; 159–177.
2. Passioura, J.B. Drought and Drought Tolerance. In *Drought Tolerance in Higher Plants: Genetical, Physiological, and Molecular Biological Analysis*; Belhassen, E., Ed.; Kluwer Academic Publishers: the Netherlands, 1996; 1–6.
3. Unger, W.P.; Jones, O.R.; Steiner, J.L. Principles of Crop and Soil Management Procedures for Maximising Production per Unit Rainfall. In *Drought Research Priorities for the Dry Land Tropics*; Bidinger, F.R., Johansen, C., Eds.; ICIRSAT: Patancheru, A.P. 502324, India, 1988; 97–112.
4. Natarajan, M.; Willey, R.W. The effects of water stress on yield advantages of intercropping systems. Field Crops Res. **1986**, *13*, 117–131.
5. Nageswara Rao, R.C.; Wadia, K.D.R.; Williams, J.H. Intercropping of short and long duration groundnut genotypes to increase productivity in environments prone to end-of-season drought. Exp. Agric. **1990**, *26*, 63–72.
6. Richards, R.A. Defining Selection Criteria to Improve Yield Under Drought. In *Drought Tolerance in Higher Plants: Genetical, Physiological, and Molecular Biological Analysis*; Belhassen, E., Ed.; Kluwer Academic Publishers: the Netherlands, 1996; 71–78.
7. Levitt, J. *Response of Plants to Environmental Stresses*; Academic Press: New York, 1972.
8. Turner, N.C. Crop water deficits: A decade of progress. Adv. Agron. **1986**, *39*, 1–51.
9. Turner, N.C.; Wright, G.C.; Siddique, K.H.M. Adaptation of grain legumes (pulses) to water-limited environments. Adv. Agron. **2001**, *71*, 193–231.
10. Turner, N.C. Drought Resistance: A Comparison of Two Frameworks. In *Management of Agricultural Drought: Agronomic and Genetic Options*; Saxena, N.P., Johansen, C., Chauhan, Y.S., Rao, R.C.N. Eds.; Oxford and IBH: New Delhi, 2000.
11. Subbarao, G.V.; Johansen, C.; Slinkard, A.E.; Rao, R.C.N.; Saxena, N.P.; Chauhan, Y.S. Strategies for improving drought resistance in grain legumes. Crit. Rev. Plant Sci. **1995**, *14*, 469–523.
12. Jackson, P.; Robertson, M.; Cooper, M.; Hammer, G. The role of physiological understanding in plant breeding; From a breeding perspective. Field Crops Res. **1996**, *49*, 11–37.
13. Siddique, K.H.M.; Loss, S.P.; Regan, S.P.; Jettner, R. Adaptation of cool season grain legumes in Mediterranean type environments of south-western Australia. Aust. J. Agric. Resour. **1999**, *50*, 375–387.

14. Wright, G.C.; Smith, R.C.G.; Morgan, J.M. Differences between two grain sorghum genotypes in adaptation to drought stress. III. Physiological responses. Aust. J. Agric. Resour. **1983**, *34,* 637–651.

15. Flower, D.J.; Ludlow, M.M. Contribution of osmotic adjustment to the dehydration tolerance of water-stressed pigeon pea (*Cajanus cajan* (L.) millsp.) leaves. Plant Cell Environ. **1986**, *9,* 33–40.

16. Davies, W.J.; Tardieu, F.; Trejo, C.L. How do chemical signals work in plants that grow in drying soil? Plant Physiol. **1994**, *104,* 309–314.

17. Morgan, J.M. Possible role of abscisic acid in reducing seed set in water-stressed wheat plants. Nature (Lond.) **1980**, *285,* 655–657.

18. Blum, A. Crop Responses to Drought and Interpretation of Adaptation. In *Drought Tolerance in Higher Plants: Genetical, Physiological, and Molecular Biological Analysis;* Belhassen, E., Ed.; Kluwer Academic Publishers: the Netherlands, 1996; 57–70.

19. Morgan, J.M. Osmoregulation and water stress in higher plants. Annu. Rev. Plant Physiol. **1984**, *35,* 299–319.

20. Jones, M.M.; Rawson, H.M. Influence of the rate of development of leaf water deficits upon photosynthesis, leaf conductance, water use efficiency, and osmotic potential in sorghum. Physiol. Plant. **1979**, *45,* 103–111.

21. Morgan, J.M.; King, R.W. Association between loss of leaf turgor, abscisic acid levels and seed set in two wheat cultivars. Aust. J. Plant Physiol. **1984**, *11,* 143–150.

22. Morgan, J.M.; Condon, A.G. Water use, grain yield and osmoregulation in wheat. Aust. J. Plant Physiol. **1986**, *13,* 523–532.

23. Wright, G.C.; Smith, R.C.G. Differences between two grain sorghum genotypes in adaptation to drought stress. II. Root water uptake and water use. Aust J. Agric. Resour. **1983**, *34,* 627–636.

24. Morgan, J.M. Osmoregulation as a selection criterion for drought tolerance in wheat. Aust. J. Agric. Resour. **1983**, *34,* 607–614.

25. Blum, A.; Mayer, J.; Gozlan, G. Associations between plant production and some physiological components of drought resistance in wheat. Plant Cell Environ. **1983**, *6,* 219–225.

26. Santamaria, J.M.; Ludlow, M.M.; Fukai, S. Contributions of osmotic adjustment to grain yield in *Sorghum bicolor* (L.) Moench under water-limited conditions. I. Water stress before anthesis. Aust. J. Agric. Resour. **1990**, *41,* 51–65.

27. Morgan, J.M.; Rodriguez-Maribona, B.; Knights, E.J. Adaptation to water-deficit in chickpea breeding lines by osmoregulation: Relationship to grain-yields in the field. Field Crops Res. **1991**, *27,* 61–70.

28. Blum, A. *Plant Breeding for Stress Environments;* CRC Press: Boca Raton, FL, 1988.

29. Sinclair, T.R.; Ludlow, M.M. Influence of soil water supply on the plant water balance of four tropical grain legumes. Aust. J. Plant Physiol. **1986**, *13,* 329–341.

30. Chandler, P.M.; Munns, R.; Robertson, M. Regulation of Dehydrin Expression. In *Plant Responses to Cellular Dehydration During Environmental Stress;* Close, T.J., Bray, E.A., Eds.; American Society of Plant Physiologists: Rockville, MD, 1993; 159–166.

31. Close, T.J.; Fenton, R.D.; Yang, A.; Asghar, R.; DeMason, D.A.; Crone, D.E.; Meyer, N.C.; Moonan, F. Dehydrin: The Protein. In *Plant Responses to Cellular Dehydration During Environmental Stress;* Close, T.J., Bray, E.A., Eds.; Amer. Society of Plant Physiology, 1993; 104–118.

32. Hanson, A.D.; Hitz, W.D. Metabolic responses of mesophytes to plant water deficits. Annu. Rev. Plant Physiol. **1982**, *33,* 163–203.

33. Mathews, R.B.; Harris, D.; Williams, J.H.; Nageswara Rao, R.C. The physiological basis for yield differences between four groundnut genotypes in response to drought. II. Solar radiation interception and leaf movement. Exp. Agric. **1988**, *24,* 203–213.

34. Passioura, J.B. The Interaction between Physiology and the Breeding of Wheat. In *Wheat Science— Today and Tomorrow;* Evans, L.T., Peacock, W.J., Eds.; Cambridge University Press: Cambridge, 1981; 191–201.

35. Passioura, J.B. Grain yield, harvest index and water use of wheat. J. Aust. Inst. Agric. Sci. **1977**, *43*, 117–120.
36. Farquhar, G.D.; Richards, R.A. Isotopic composition of plant carbon correlates with water-use efficiency in wheat genotypes. Aust. J. Plant Physiol. **1984**, *11*, 539–552.
37. Nageswara Rao, R.C.; Wright, G.C. Stability of the relationship between specific leaf area and carbon isotope discrimination across environments in peanuts. Crop Sci. **1994**, *34*, 98–103.
38. Masle, J.; Farquhar, G.D.; Wong, S.C. Transpiration ratio and plant mineral content are related among genotypes of a range of species. Aust. J. Plant Physiol. **1992**, *19*, 709–721.
39. Clark, D.H.; Johnson, D.A.; Kephart, K.D.; Jackson, N.A. Near infrared reflectance spectroscopy estimation of 13C discrimination in forages. J. Range Manag. **1995**, *48*, 132–136.
40. Subbarao, G.V.; Johansen, C.; Rao, R.C.N.; Wright, G.C. Transpiration Efficiency: Avenues for Genetic Improvement. In *Handbook of Plant and Crop Physiology*; Pessarakli, M., Ed.; Marcel Dekker: New York, NY, 1994; 785–806.
41. Wright, G.C.; Rachaputi, N.C. Transpiration Efficiency. In *Water and Plants (Biology) for the Encyclopedia of Water Science*; Marcel Dekker, Inc., 2003; 982–988.
42. Fischer, K.S.; Johnson, E.C.; Edmeades, G.O. Breeding and Selection for Drought Resistance in Tropical Maize. In *Drought Resistance in Crops with Emphasis on Rice*; International Rice Research Institute: Los Banos, 1983; 377–399.
43. Wright, G.C.; Rao, R.C.N.; Basu, M.S. A Physiological Approach to the Understanding of Genotype by Environment Interactions—A Case Study on Improvement of Drought Adaptation in Groundnut. In *Plant Adaptation and Crop Improvement*; Cooper, M., Hammer, G.L., Eds.; CAB International: Wallingford, 1996; 365–381.
44. Nigam, S.N.; Nageswara Rao, N.C.; Wright, G.C. In *Breeding for Increased Water-Use Efficiency in Groundnut*, New Millennium International Groundnut Workshop, Shandong Peanut Research Institute, Qingdao, China, Sept. 4–7, 2001; 1–2.

26

El Niño, La Niña, and the Southern Oscillation

Felicity S. Graham
CSIRO Marine and
Atmospheric Research
University of Tasmania

Jaclyn N. Brown
CSIRO Marine and
Atmospheric Research

Introduction

The equatorial Pacific Ocean and atmosphere are the stage for one of the most important phenomena affecting the world's climate: El Niño-Southern Oscillation (ENSO). The term El Niño refers to a period of prolonged warmer-than-usual sea-surface temperatures in the central-to-east-ern equatorial Pacific Ocean, typically peaking around December. El Niño is strongly linked to an atmospheric phenomenon known as the Southern Oscillation, part of which is the Walker circulation, a zonal atmospheric cell characterized by ascending motion in the western Pacific and descending motion in the eastern Pacific, connected at the surface by the trade winds. The opposite phase of El Niño, denoted La Niña, is characterized by cooler-than-usual sea-surface temperatures in the central and eastern equatorial Pacific. The whole phenomenon—El Niño, La Niña, and the Southern Oscillation—is known as ENSO and affects the Pacific ocean–atmosphere system on interannual timescales. However, ENSO events have far broader-reaching effects on climate and weather patterns than just the Pacific Ocean. For instance, extreme weather associated with ENSO events range from droughts to increased rainfall and severe flooding.[1–4] Furthermore, ENSO is associated with changes in the incidence of tropical cyclones[5–7] and affects weather and climate in Pacific Island countries,[8] the United States,[9] South America,[10] Australia,[11] and even Europe.[12] The biogeochemical structure of the tropical ocean is altered by ENSO, which in turn affects the habitat availability for fish such as tropical tunas.[13] This entry summarizes the current understanding of ENSO in the context of the physical conditions of the tropical Pacific and its global significance.

What Is ENSO?

The term "El Niño" was originally used to describe warming of a local current along the west coast of South America in December (El Niño in Spanish is "the Child Jesus"). However, it was not until the 1960s that scientists realized El Niño had a much broader impact on sea-surface temperatures than along

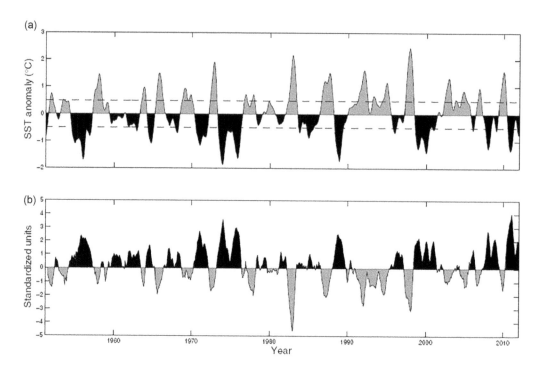

FIGURE 26.1 (a) Time series of sea-surface temperature (SST) anomalies from 1951 to 2012 relative to the mean seasonal climatology over the whole period. Data is averaged over the Niño-3.4 region (120°W–170°W and 5°S–5°N). Events with sea-surface temperature anomalies equal to or greater (less) than 0.5°C (–0.5°C) for a period of 6 consecutive months are defined as El Niño (La Niña). (b) Time series of the Southern Oscillation Index, based on sea-surface pressure observations at Tahiti and Darwin from 1951 to 2012. The zero corresponds to the mean over the same period. El Niño events have negative units; La Niña events have positive units.
Source: Data provided by the NOAA/OAR/ESRL PSD, Boulder, Colorado, U.S.A, from their website at http://www. esrl.noaa.gov/psd/.

the Peruvian coast.[10] During El Niño years, there tends to be broad warming of sea-surface temperatures from the International Date Line to the western coast of South America.[10] These warming events occur on an interannual timescale, approximately every 2–7 years, with the event lasting about 1 year. Cool ENSO events called La Niña (Spanish for "the girl child") events, are characterized by basin-scale cooling of sea-surface temperatures. There are many different definitions for what constitutes an ENSO event;[14] however, one common criterion used to define an El Niño is that the 5-month running mean of the sea-surface temperature anomaly in the eastern Pacific in the region bound by 120°W–170°W and 5°S–5°N (denoted Niño-3.4) is at least +0.5°C higher in that region for a period of 6 consecutive months. [15] The converse is true for La Niña. Using this definition, ENSO events can be identified in a time series of the sea-surface temperature averaged in the Niño-3.4 region over the last 60 years (Figure 26.1a). Corresponding to the changes in the sea-surface temperature during ENSO events is a shift in atmospheric conditions. The Southern Oscillation is a see-saw oscillation of the surface air pressure between the western and eastern equatorial Pacific with two centers of action over Indonesia in the west and Tahiti in the east.[10] El Niño events and the Southern Oscillation are closely related,[16] with anomalously high (low) sea-level pressures in the west (east) during El Niño years. Increased sea-level pressures over Australia and southeast Asia during El Niño years result in drier-than-normal conditions, whereas in the low-pressure region over western parts of South America, prevailing conditions are wet (Figure 26.2a). The reverse is true during La Niña years (Figure 26.2b). The Southern Oscillation Index (SOI) is defined as the pressure anomaly at Tahiti minus that at Darwin, Australia (Figure 26.1b), and is strongly anticorrelated with Niño-3.4. Understanding the underlying characteristics of the tropical Pacific ocean–atmosphere system is key to understanding why ENSO has such broad-reaching effects on weather and climate around the globe.

Mean Tropical Pacific Ocean–Atmosphere Interactions

Sea-surface temperatures of approximately 28°C typically occur in the western tropical Pacific, in a body of warm water known as the "warm pool" (Figure 26.3a). In the eastern Pacific, a "tongue" of cool water extends west along the equator with a sea-surface temperature of approximately 22°C. Zonal variations in the depth of the thermocline (the depth of the sharpest change in temperature in the upper ocean) partly drive the differences between eastern and western Pacific sea-surface temperatures. In the western Pacific, the thermo-cline is relatively deep (on average, ~200 m below the surface), and prevents cooler subsurface waters from mixing with the warm waters above (Figure 26.3d). On the other hand, a shallow thermocline in the eastern Pacific allows cooler subsurface waters to be transported upward ("upwelled") into the surface layer where they are readily mixed to bring about the cold "tongue."

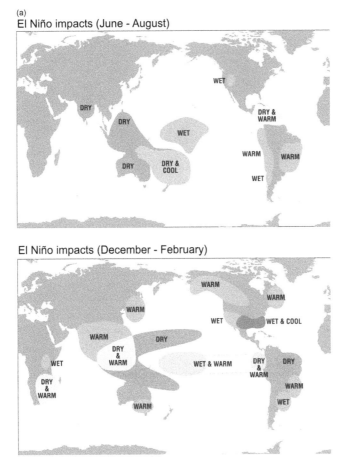

FIGURE 26.2 (a) El Niño regional impacts in June–August (upper panel) and December–February (lower panel). (b) La Niña regional impacts in June–August (upper panel) and December–February (lower panel).
Source: Image modified from NOAA/OAR/ESRL PSD, Boulder, Colorado, U.S.A., from their website at http://www.ncdc.noaa.gov/paleo/ctl/clisci10.html, based on research by Halpert and Ropelewski[54] and Ropelewski and Halpert.[55]

(Continued)

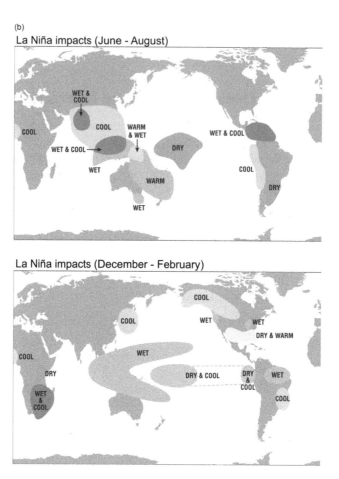

FIGURE 26.2 (CONTINUED)   (a) El Niño regional impacts in June–August (upper panel) and December–February (lower panel). (b) La Niña regional impacts in June–August (upper panel) and December–February (lower panel).
Source: Image modified from NOAA/OAR/ESRL PSD, Boulder, Colorado, U.S.A., from their website at http://www.ncdc.noaa.gov/paleo/ctl/clisci10.html, based on research by Halpert and Ropelewski[54] and Ropelewski and Halpert.[55]

The atmospheric response to the contrast in sea-surface temperatures along the equatorial Pacific is for regions of low sea-level pressure to be located in the western equatorial Pacific above warm sea-surface temperatures and for regions of high sea-level pressure to be located in the east. The prevailing winds in the equatorial region—the equatorial trade winds—are influenced by the atmospheric pressure contrast ("gradient") and blow toward low-pressure regions where air converges and is convected. As a result, the western equatorial Pacific has a higher incidence of rainfall, thunderstorms, and tropical cyclones than the eastern equatorial Pacific. The trade winds reinforce the east–west difference in sea-surface temperature by piling up warm water in the west and driving upwelling of cool waters in the east and along the equator. It is clear then that atmospheric responses to an anomalously warm oceanic state drive further sea-surface temperature increases in the ocean, through a positive feedback.

Much of our understanding of tropical Pacific ocean–atmosphere interactions has been through the success of the Tropical Ocean Global Atmosphere (TOGA) program (see the special volume 103 of the *Journal of Geophysical Research,* June 1998, for a comprehensive review). TOGA was initiated in 1985 to study ENSO. The backbone of the program is an array of buoys known as the Tropical

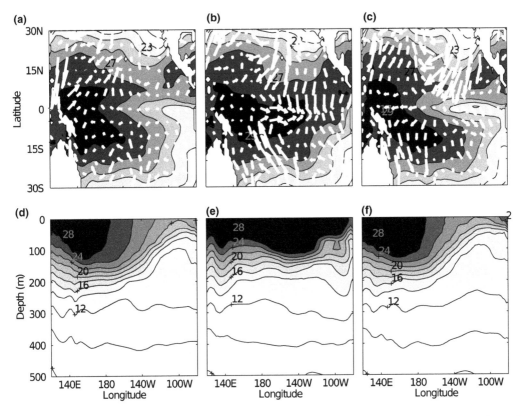

FIGURE 26.3   The contours in the upper panels are of sea-surface temperature anomaly (SST) (in °C) with zonal wind stress (in N m−2) overplotted in each. The contours in the lower panels represent the subsurface temperature profile (in °C) along the equator. The left-most panels (**a** and **d**) are for December 1996, which was before the peak of the 1997–1998 El Niño event. The middle panels (**b** and **e**) show conditions during the peak of the 1997–1998 El Niño in December, and the right-most panels (**c** and **f**) correspond to the peak of the December 1998–1999 La Niña event. **Source:** Data prepared by Neville Smith's group at the Australian Bureau of Meteorology Research Centre (BMRC).[56–60]

Atmosphere-Ocean (TAO) array. Each of the buoys measures temperature in the upper 500 m of the ocean, surface winds, sea-surface temperature, surface air temperature, and humidity, and relay hourly averaged data back to the United States National Oceanic and Atmospheric Administration's (NOAA) Pacific Marine Environmental Laboratory (PMEL). The ENSO observing system also includes a Voluntary Observing Ship (VOS) program, an island tide-gauge network, surface drifters, and satellite data. The coupled nature of the tropical Pacific ocean–atmosphere described in the preceding text is important for the onset and development of an ENSO event, as well as its center of action and magnitude.

ENSO Evolution

No two ENSO events are the same; rather, ENSO events vary considerably in how and where they evolve, their intensity, and the regions that they affect.[17] Furthermore, ENSO and its impacts vary from decade to decade and from generation to generation.[18–20] However, there are key ocean–atmosphere interactions involved in the onset, growth, and termination of an ENSO event. Here we describe these interactions for an El Niño event that evolves in the eastern equatorial Pacific.

Consider a slightly warmer sea-surface temperature in the central-to-eastern equatorial Pacific. This initial anomalous warming reduces the zonal gradient in sea-surface temperatures between the east and west. As a result, the air above the sea-surface also warms, reducing the overlying atmospheric pressure gradient and leading to a change in the atmospheric Walker circulation. The easterly trade winds reduce, moving the center of precipitation eastward and deepening the thermocline in the eastern Pacific. A deeper thermocline prevents cooler subsurface waters entering the mixed layer in the eastern Pacific. Consequently, the original warm sea-surface temperature anomaly is reinforced, and the associated responses in the atmosphere and ocean circulation are strengthened (Figure 26.3b and e).[16] A similar but reversed mechanism applies for La Niña (Figure 26.3c and f), where a slightly cooler sea-surface temperature in the central-to-eastern equatorial Pacific strengthens the trade winds and raises the thermocline in the east. Increased upwelling in the east further cools the sea-surface temperature anomaly, growing the event.

Recently, a modified type of El Niño has been identified, having its strongest signature in the central Pacific.[21] This type of El Niño is termed the central Pacific El Niño or El Niño Modoki (Japanese for "similar but different") and, as its name suggests, it has some dynamical similarities to the canonical, eastern Pacific El Niño.[22]

There are a number of mechanisms thought to be important in the termination of an ENSO event. [23–27] One important mechanism is the displacement of warm water (above the thermocline or 20°C isotherm).[28,29] During El Niño, the trade winds slacken, and the thermocline deepens in the east, causing the thermocline to flatten along the equator. A flattened thermocline changes the oceanic pressure gradient and causes the transport of warm water north and south from the equator. Such a discharge of warm water is sufficient to overcome the ocean-atmosphere interactions that grow an El Niño event, and so the El Niño event begins to decline[29,30] The discharge mechanism works in reverse during a La Niña event; at the termination of the event, the equatorial Pacific is left in a "recharged" state. Other mechanisms that have been proposed include wave reflection at the western boundary[23,24] and the zonal movement of mean and anomalous currents along with wave reflection at the eastern boundary.[31] Some attempt has been made to unify these mechanisms in a simple model.[32] The characteristics of the onset, development, and termination of El Niño and La Niña events, including their quasi-periodicity, asymmetry, and decadal and generational variability, have been well documented in the literature.[33–36] Leading coupled models are now at a stage where they effectively simulate ENSO dynamics; however, they still suffer from a range of issues including overly cool water along the equator known as the cold tongue bias.[37]

Predicting El Nino Events

El Niño events tend to be followed by a La Niña event (Figure 26.1a). The initiation of El Niño events, however, is still a subject of ongoing research. The ocean is capable of sitting in a weak La Niña or neutral ocean state for a number of years without an El Niño occurring.[38] In many situations, a westerly wind burst associated with a tropical cyclone or Madden–Julian Oscillation weather system can provide enough of a reduction in the easterly trades to trigger the positive feedback mechanism to begin generating an El Niño.[39–42]

Seasonal forecasting models are becoming more adept at forecasting El Nino events.[43] It still remains challenging to predict the generation of an El Niño through the March–April–May period known as the "Spring Predictability Barrier."[44] This period is key for making cropping decisions over the Austral winter period if an El Niño is to develop later in the year.[45]

The tropical Pacific Ocean is projected to warm under increasing greenhouse gas scenarios.[46] It is not clear how ENSO will change in response to this increased warming.[47–49] ENSO is expected to continue occurring; however, leading climate models disagree on whether it will be stronger or weaker in amplitude and more or less frequent.

Conclusion

ENSO events have broad-reaching impacts: they affect where rainfall occurs, determine the availability of food for fisheries in the eastern Pacific by suppressing or increasing the supply of nutrients, and set up conditions favorable for the development of tropical cyclones. ENSO can have devastating impacts on communities and the economy as well: the 1997–1998 extreme El Niño is estimated to have taken 32,000 lives, displaced 300 million people from their homes, and brought about total losses of US$ 89 billion.[50] Scientists' understanding of ENSO over the past few decades has progressed significantly through the establishment of the TOGA observing system in the tropical Pacific Ocean, advances in theory,[23,24,29,51,52] and our ability to simulate ENSO,[53] all of which have also enabled early detection of the onset and development of an ENSO event. However, there are still many aspects of ENSO that are poorly understood; for example, we have limited understanding of how the mechanisms of ENSO might be impacted by climate change. Given the potential impacts of ENSO on the environment, health, and the economy, better understanding and predicting ENSO should be a key driver for future research.

References

1. McBride, J.L.; Nicholls, N. Seasonal relationships between Australian rainfall and the Southern Oscillation. Monthly Weather Rev. **1983,** *111*, 1998–2004.
2. Ropelewski, C.F.; Halpert, M.S. Global and regional scale precipitation patterns associated with the El Nino/Southern Oscillation. Monthly Weather Rev. **1987,** *115* (8), 1606–1626.
3. Zhang, X.-G.; Casey, T.M. Long-term variations in the Southern Oscillation and relationships with Australian rainfall. Aust. Meteorol. Mag. **1992,** *40,* 211–225.
4. Andrews, E.D.; Antweiler, R.C.; Neiman, P.J.; Ralph, F.M. Influence of ENSO on flood frequency along the California coast. J. Climate **2004,** *17* (2), 337–348.
5. Nicholls, N. Predictability of interannual variations of Australian seasonal tropical cyclone activity. Monthly Weather Rev.**1985,** *113,* 1144–1149.
6. Chen, G.; Tam, C.-Y. Different impacts of two kinds of Pacific Ocean warming on tropical cyclone frequency over the Western North Pacific. Geophysical Res. Lett. **2010,** *37,* 1–6.
7. Callaghan, J.; Power, S.B. Variability and decline in the number of severe tropical cyclones making land-fall over Eastern Australia since the late nineteenth century. Climate Dynamics **2011,** *37,* 647–662.
8. Australian Bureau of Meteorology and CSIRO. Climate change in the Pacific: Scientific Assessment and New Research. Volume 1: Regional Overview. Volume 2: Country Reports. 2011; http://www.cawcr.gov.au/projects/PCCSP/publications.html#1 (accessed December 2012).
9. Horel, J.D.; Wallace, J.M. Planetary-scale atmospheric phenomena associated with the Southern Oscillation. Monthly Weather Rev. **1981,** *109* (4), 813–829.
10. Philander, S.G. *El Niño, La Niña, and the Southern Oscillation.* Academic Press: San Diego, **1990.**
11. Nicholls, N. Historical El Niño/Southern Oscillation Variability in the Australian Region. In *El Niño, Historical and Paleoclimatic Aspects of the Southern Oscillation;* Diaz, F, Margrav, V., Eds.; Cambridge University Press: Cambridge, UK, 1992; 151–174.
12. Fraedrich, K. An ENSO Impact on Europe? Tellus. Series A Dynamic Meteorol. Oceanography **1994,** *46* (4), 541–552.
13. Ganachaud, A.; Sen Gupta, A.; Brown, J.N.; Evans, K.; Maes, C.; Muir, L.C.; Graham. F.S.; Projected changes in the tropical Pacific Ocean of importance to tuna fisheries. Climatic Change **2012,** *115,* 1–17.
14. Meyers, G.; McIntosh, P.; Pigot, L.; Pook, M. The years of El Niño, La Niña, and interactions with the tropical Indian Ocean. J. Climate **2007,** *20,* 2872–2880.
15. Trenberth, K.E. The definition of El Niño. Bull. Am. Meteorol. Soc. **1997,** *78* (12), 2771–2777.

16. Bjerknes, J. Atmospheric teleconnections from the equatorial Pacific. Monthly Weather Rev. **1969,** *97* (3), 163–172.

17. Brown, J.N.; McIntosh, P.C.; Pook, M.J.; Risbey, J.S. An investigation of the links between ENSO flavors and rainfall processes in Southeastern Australia. Monthly Weather Rev. **2009,** *137* (11), 3786–3795.

18. Power, S.; Folland, C.; Colman, A.; Mehta, V. Inter-decadal modulation of the impact of ENSO on Australia. Climate Dynamics **1999,** *15,* 319–324.

19. Power, S.B.; Smith, I.N. Weakening of the Walker Circulation and apparent dominance of El Niño both reach record levels, but has ENSO really changed? Geophysical Res. Lett. **2007,** *34,* GL30854.

20. Wang, B. Interdecadal changes in El Niño onset in the last four decades. J. Climate **1995,** *8,* 267–285.

21. Ashok, K.; Behera, S.K.; Rao, S.A.; Weng, H.; Yamagata, T. El Niño Modoki and its possible tele-connection. J. Geophysical Res. **2007,** *112* (C11).

22. Kug, J.-S.; Jin, F.F.; An, S.-I. Two types of El Niño events: cold tongue El Niño and warm pool El Niño. J. Climate **2009,** *22* (6), 1499–1515.

23. Battisti, D.S.; Hirst, A.C. Interannual variability in a tropical atmosphere-ocean model: influence of the basic state, ocean geometry and nonlinearity. J. Atmospheric Sci. **1989,** *46* (12), 1687–1712.

24. Suarez, M.J.; Schopf, P.S. A delayed action oscillator for ENSO. J. Atmospheric Sci. **1988,** *45* (21), 3283–3287.

25. Boulanger, J.P.; Menkes, C.; Lengaigne, M. Role of high- and low-frequency winds and wave reflec-tion in the onset, growth and termination of the 1997–1998 El Niño. Climate Dynamics **2004,** *22* (2–3), 267–280.

26. McPhaden, M.J.; Yu, X. Equatorial waves and the 1997–1998 El Niño. Geophysical Res. Lett. **1999,** *26* (19), 2961–2964.

27. Harrison, D.E.; Vecchi, G.A. On the termination of El Niño. Geophysical Res. Lett. **1999,** *26* (11), 1593–1596.

28. Wyrtki, K. Water displacements in the Pacific and the genesis of El Niño cycles. J. Geophysical Res. Biogeosci. **1985,** *90* (C4), 7129–7132.

29. Jin, F.-F. An equatorial ocean recharge paradigm for ENSO. Part I: conceptual model. J. Atmospheric Sci. **1997,** *54,* 19.

30. Meinen, C.S.; McPhaden, M.J. Observations of warm water volume changes in the equatorial pacific and their relationship to El Niño and La Niña. J. Climate **2000,** *13,* 9.

31. Picaut, J. An advective-reflective conceptual model for the oscillatory nature of the ENSO. Science **1997,** *277,* 663–666.

32. Wang, C. A unified oscillator model for the El Niño – Southern Oscillation. J. Climate **2001,** *14,* 98–115.

33. Rasmusson, E.M.; Carpenter, T.H. Variations in tropical sea surface temperature and surface wind fields associated with the Southern Oscillation/El Niño. Monthly Weather Rev. **1982,** *110,* 354–384.

34. Harrison, D.E.; Larkin, N.K. El Niño-Southern Oscillation sea surface temperature and wind anomalies, 1946–1993. Rev. Geophysics (1985) **1998,** *36* (3), 353–399.

35. Larkin, N.K.; Harrison, D.E. ENSO warm (El Niño) and cold (La Niña) event life cycles: ocean surface anomaly patterns, their symmetries, asymmetries, and implications. J. Climate **2002,** *15* (10), 1118–1140.

36. Dommenget, D.; Bayr, T.; Frauen, C. Analysis of the nonlinearity in the pattern and time evolution of El Niño-Southern Oscillation. Climate Dynamics **2012,** *14,* 1457.

37. Brown, J.N.; Sen Gupta, A.; Brown, J.R.; Muir, L.C.; Risbet, J.S.; Whetton, P.; Zhang, X.; Ganachaud, A.; Murphy, B.; Wijffels. S.E. Implications of CMIP3 model biases and uncertainties for climate projections in the western tropical Pacific. Climatic Change **2012,** 1–15.

38. Kessler, W.S. Is ENSO a cycle or a series of events? Geophysical Res. Lett. **2002,** *29* (23).
39. Luther, D.S.; Harrison, D.E.; Knox, R.A. Zonal winds in the central equatorial Pacific and El Niño. Science **1983,** *222,* 327–330.
40. Lengaigne, M.; Guilyardi, E.; Boulanger, J.-P., Menkes, C.; Delecluse, P.; Innesse, P.; Cole, J.; Slingo, J. Triggering of El Niño by westerly wind events in a coupled general circulation model. Climate Dynamics **2004,** *23* (6), 601–620.
41. McPhaden, M.J.; Zhang, X.B.; Hendon, H.H., Wheeler, M.C. Large scale dynamics and MJO forcing of ENSO variability. Geophysical Res. Lett. **2006,** *33* (16).
42. Latif, M.; Biercamp, J.; Von Storch, H. The response of a coupled ocean-atmosphere general circulation model to wind bursts. J. Atmospheric Sci. **1988,** *45* (6), 964–979.
43. Wang, W. Simulation of ENSO in the new NCEP coupled forecast system model (CFS03). Monthly Weather Rev. **2005,** *133* (6), 1574–1593.
44. Webster, P.J.; Yang, S. Monsoon and ENSO: selectively interactive systems. Q. J. R. Meteorol. Soc. **1992,** *118* (507), 877–926.
45. McIntosh, P.C.; Pook, M.J.; Risbey, J.S.; Lisson, S.N.; Reb-beck, M. Seasonal climate forecasts for agriculture: towards better understanding and value. Field Crops Res. **2007,** *104,* 130–138.
46. IPCC. Climate Change 2007: the physical science basis. In *Series Climate Change 2007: the Physical Science Basis;* Solomon, S., Qin, D., Manning, M., et al., Eds.; Cambridge University Press: Cambridge, United Kingdom and New York, NY, USA, 2007; http://www.ipcc.ch/publications_and_data/publications_ipcc_fourth_assessment_report_wg1_report_the_physical_science_basis.htm (December 2012).
47. Vecchi, G.A.; Wittenberg, A.T. El Niño and our future climate: where do we stand? Wiley Interdiscip. Rev. **2010,** *1* (2), 260–270.
48. Guilyardi, E.; Bellenger, H.; Collins, M.; Ferrett, S.; Cai, W, Wittenberg, A. A first look at ENSO in CMIP5. CLIVAR Exchanges No. 58 **2012,** *17* (1), 29–32.
49. Collins, M.; An, S.-I.; Cai, W.; Ganachaud, A.; Guilyardi, E.; Jin, F.-F.; Jochum, M.; Lengaigne, M.; Power, S.; Tim-mermann, A.; Vecchi, G.; Wittenberg, A. The impact of global warming on the tropical Pacific ocean and El Niño. Nat. Geosci. **2010,** *3,* 391–397.
50. Trenberth, K.E. The extreme weather events of 1997 and 1998. Consequences **1999,** *5* (1), 3–15.
51. Neelin, J.D.; Battisti, D.S.; Hirst, A.C.; Jin, F.-F.;Wakata, Y.; Yamagata, T.; Zebiak, S.E. ENSO Theory. J. Geophysical Res. **1998,** *103* (C7), 14261–14314.
52. Power, S.B. Simple analytic solutions of the linear delayed-action oscillator equation relevant to ENSO theory. Theor. Appl. Climatol. **2011,** *104* (1), 251–259.
53. Zebiak, S.E.; Cane, M.A. A model EI Nino-Southern Oscillation. Monthly Weather Rev. **1987,** *115* (10), 2262–2278.
54. Halpert, M.S.; Ropelewski, C.F. Surface temperature patterns associated with the Southern Oscillation. J. Climate **1992,** *5* (6), 577–593.
55. Ropelewski, C.F.; Halpert, M.S. Precipitation patterns associated with the high index phase of the Southern Oscillation. J. Climate **1989,** *2* (3), 268–284.
56. Smith, N.R. An improved system for tropical ocean subsurface temperature analyses. J. Atmospheric Oceanic Technol. **1995,** *12* (4), 850–870.
57. Smith, N.R. Objective quality controls and performance diagnostics of an oceanic subsurface thermal analysis scheme. J. Geophysical Res. Biogeosciences **1991,** *96,* 3279–3287.
58. Smith, N.R. The BMRC ocean thermal analysis system. Aust. Meteorol. Mag. **1995,** *44,* 93–110.
59. Smith, N.R.; Blomley, J.E.; Meyers, G.A Univariate statistical interpolation scheme for subsurface thermal analyses in the tropical oceans. Prog. Oceanography **1991,** *28* (Pergamon), 219–256.
60. Meyers, G.; Phillips, H.; Smith, N.R.; Sprintall, J. Space and time scales for optimum interpolation of temperature - tropical Pacific ocean. Prog. Oceanography **1991,** *28* (Pergamon), 189–218.

Bibliography

Clarke, A.J. *An Introduction to the Dynamics of El Niño Southern Oscillation.* Academic Press. **2008.**

Fedorov, A.V.; Brown, J.N. Equatorial waves. In *Encyclopedia of Ocean Sciences, Second Edition*; Steele, J., Ed.; Academic Press, **2009,** 3679–3695.

McPhaden, M.J.; Busalacchi, A.J.; Cheney, R., Donguy, J.-R.; Gage, K.S.; Halpern, D.; Ji, P.; Meyers, G.; Mitchum, G.T.; Niiler, P.P.; Picaut, J.; Reynolds, R.W.; Smith, N.; Takeuchi, K. The tropical ocean-global atmosphere observing system: a decade of progress. J. Geophysical Res. **1998,** *103* (C7), 14169–14240.

Neelin, J.D.; Battisti, D.S.; Hirst, A.C.; Jin, F.-F.;Wakata, Y.; Yamagata, T.; Zebiak, S.E. ENSO theory. J. Geophysical Res. Biogeosci. **1998,** *103* (c7), 14261–14290.

Wang, C.; Picaut, J. Understanding ENSO physics—a review. *Earth's Climate: The Ocean-Atmosphere Interaction*; Wang, C., Xie, S.-P., Carton, J.A., Eds.; AGU: Washington, D.C., **2004;** 147, 21–48.

27

Meteorology: Tropical

Zhuo Wang
*University of Illinois at
Urbana-Champaign*

Introduction

Tropical meteorology is the study of weather and climate systems in the tropics. There are different definitions of the tropics. One may define the tropics as the region between 23.5°S and 23.5°N, or the region between 30°S and 30°N. In the former definition, the tropics are the region where the sun reaches the zenith twice a year, whereas the latter definition divides the globe into two equal halves, tropics and extratropics. A more meteorological definition of the "tropics" was provided by Riehl,[1] which defined the tropics as "that part of the world where atmospheric processes differ decidedly and sufficiently from those in higher latitudes."

Atmospheric processes in the tropics differ from those in higher latitudes in various ways. In the tropics, the Coriolis force is small (zero along the equator). The quasi-geostrophic theory, which provides a dynamic framework for understanding the large-scale phenomena in midlatitudes, is not readily applicable. Compared with the midlatitudes, the pressure gradient and temperature gradient are generally very weak in the tropics (except for tropical cyclones).[2] Also contrary to the midlatitudes, baroclinic instability does not play an important role in large-scale circulations in the tropics. On the other hand, the tropics are the source of angular momentum and energy for the global atmospheric system. Figure 27.1 shows the distributions of the zonally averaged absorbed solar flux and the emitted thermal infrared flux at the top of the atmosphere as a function of latitude. There is a surplus of radiative energy in the tropics and a net deficit in middle and high latitudes, which requires on average a poleward transport of energy. The tropics are also characterized by warm sea surface temperature (SST). Moist convection interacts with circulations of various spatial scales in the tropics and plays an important role in driving the global atmospheric circulation system.

Tropical meteorology studies atmospheric circulations of different spatial and temporal scales. Some of the major weather and climate systems in the tropics are described in the following sections.

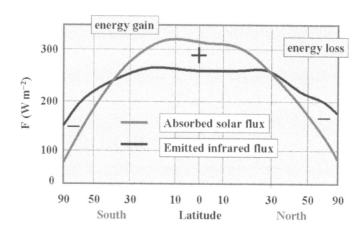

FIGURE 27.1 Zonally averaged components of the absorbed solar flux and emitted thermal infrared flux at the top of the atmosphere. The plus and minus signs denote energy gain and loss, respectively.
Source: Adapted from Vonder Haar and Suomi,[3] courtesy of Roger Smith. (© American Meteorological Society. Used with permission).

Major Climate and Weather Systems in the Tropics

The Hadley Circulation and the ITCZ

The Hadley circulation is the mean meridional circulation in the tropics. It has two thermally direct circulation cells: air rises near the equator, goes poleward near the tropopause (~15–16 km) in both hemispheres, descends in the subtropics (~25–30°N/S), and then moves equatorward near the surface. Due to the Coriolis force, the poleward flow near the tropopause turns eastward and leads to the subtropical westerly jet in the upper troposphere. Similarly, the equatorward flow turns westward and results in the prevailing trade winds in the lower troposphere. The upper-level poleward flow transports energy from the tropics to higher latitudes (meridional energy transport is also carried out by eddy motions and oceanic currents[4] whereas the trade winds act as accumulators of latent heat[5] and transport moisture equatorward, fueling convection in the rising branch of the Hadley circulation.

The convergence zone in the lower troposphere near the equator is the so-called intertropical convergence zone (ITCZ), otherwise known by sailors as the Doldrums. It is a band of heavy precipitation (Figure 27.2), cyclonic relative vorticity, and relatively low surface pressure in the monthly or seasonal mean map. In a snap shot of an infrared image, the ITCZ is manifested as a zone of transient cloud clusters. In contrast, the subtropics are characterized by relatively high surface pressure and arid climates, where evaporation exceeds precipitation and most of the world's large deserts reside.

The Walker Circulation

The Walker circulation is the east–west overturning circulation in the tropics. It was discovered by the British meteorologist Sir Gilbert Walker. The Walker circulation is driven by the zonal variations of SST and differential heating due to land distribution and topography. The ascending branches of the Walker circulation occur over the western North Pacific-Maritime continent, equatorial South America, and equatorial Africa, and the descending motions occur over the East Pacific, the East Atlantic, and the West Indian Ocean, where the SST is relatively cold. The most extensive Walker cell is over the Pacific basin, where SST is relatively warm in the west and cold in the east. The western North Pacific-Maritime

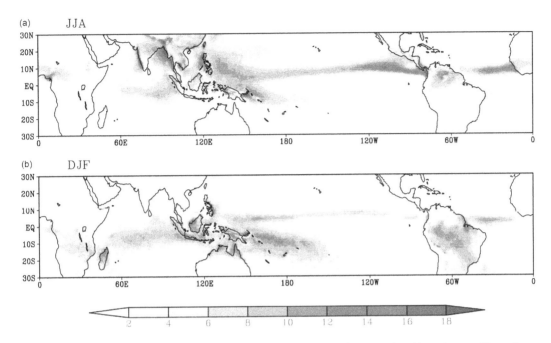

FIGURE 27.2 Seasonal mean precipitation in boreal summer (June–July–August) and boreal winter (December–January–February) derived from TRMM 3B43 (averaged over 1998–2009; units: mm/day).

continent region receives more precipitation than the East Pacific (Figure 27.2). The large amount of latent heat release warms the atmospheric column over the western North Pacific and results in the eastward pressure gradient force in the upper troposphere and westward pressure gradient force in the lower troposphere, which lead to eastward winds in the upper troposphere and westward winds in the lower troposphere. The interannual variability of the Walker cell over the Pacific basin is closely related to the SST anomalies over the equatorial central and eastern Pacific, or the El Niño-Southern Oscillation (see the El Niño-Southern Oscillation section).

Monsoons

Monsoon is traditionally defined as the seasonal reversal of the prevailing surface wind over southern Asia and the Indian Ocean. The seasonal variations of the surface wind are accompanied by variations of precipitation: a rainy summer season with onshore flow and a dry winter season with offshore flow. A more general definition of monsoon has been adopted in recent years, which denotes the seasonal variations of wind and precipitation. Besides the Asian–Australian monsoon, the global monsoon system includes the African monsoon, the North American monsoon, and the South American monsoon.[6,7] Compared with the Asian–Australian monsoon, the other monsoon systems are less extensive and may not have a winter counterpart.

The large-scale monsoon circulation is driven by seasonally varying, continental-scale distributions of sensible and latent heating and cooling. In particular, the different heat capacity between land and ocean enhances the temperature gradient due to differential solar heating. The land-sea thermal contrast leads to horizontal pressure gradient and drives a regional circulation under the influence of the Coriolis force. Moist convection and cloud radiative processes also modulate the differential heating and provide feedback to the monsoon circulation.[8] At the regional scales, the monsoon circulation and precipitation are affected by local topography and SST gradient.[9,10]

Monsoon is important because it is one of the most energetic components of the global climate system, and also because it has profound social and economical impacts in many tropical regions. On the interannual time scales, the Asian–Australian monsoon is subject to the impacts of ENSO, the Indian dipole pattern, and the tropical tropospheric biennial oscillation (TBO); on the intraseasonal time scales, the Asian–Australian monsoon has active and break cycles in terms of rainfall and is influenced by the Madden–Julian oscillation.[11] Precipitation anomalies of the Asian summer monsoon also have remote impacts over North America.[12]

El Niño-Southern Oscillation

El Niño describes the unusual SST warming over the equatorial eastern and central Pacific. This phenomenon typically occurs in boreal winter around Christmas, and is hence named El Niño (Spanish for "the boy child"). La Niña is the counterpart to El Niño and is characterized by SST cooling across much of the equatorial eastern and central Pacific. The Southern Oscillation (SO) refers to an oscillation in sea-level pressure (SLP) between the tropical eastern and western Pacific, which is closely related to the variability of the Walker circulation. Bjerknes[13] pointed out that the El Niño and the SO are strongly coupled to each other: SST cooling over the equatorial eastern Pacific (i.e., La Niña) would enhance the zonal SST and SLP gradients across the Pacific basin, lead to a stronger Walker circulation and enhance the trade wind easterlies along the equator. The enhanced easterly wind can further amplify the east–west SST gradient through the effects of oceanic advection, upwelling, and thermocline displacement, leading to a positive feedback loop. Conversely, the oceanic warming over the tropical eastern Pacific during an El Niño event is associated with a weaker Walker circulation (the negative phase of the Southern Oscillation) and reduced surface easterly winds, which would in turn further amplify the anomalous warming over the equatorial eastern Pacific. The term El Niño-Southern Oscillation, or ENSO, refers to these coupled atmospheric and oceanic processes.

ENSO is the most important mode of interannual variability in the tropics. Although the strong SST anomalies are confined to the equatorial eastern and central Pacific, ENSO affects the Asian–Australian monsoon and tropical cyclones over both the Pacific and the Atlantic; through teleconnection patterns or modulations of the subtropical jet stream, it also has remote impacts in many extratropical regions.[4-16] As a low-frequency mode (its warming or cooling cycle occurs every 2-7 years), it is an important source of climate predictability in the tropics.

Madden-Julian Oscillation

The Madden-Julian oscillation (MJO), which was first documented by Madden and Julian,[17,18] is the dominant mode of intraseasonal variability in the tropics. It is a quasi-periodic oscillation, with the period varying between 30 and 60 days, and is characterized by the eastward progression of large areas of suppressed and enhanced precipitation (Figure 27.3). The MJO originates over the western equatorial Indian Ocean. Over the Indian Ocean–western Pacific warm pool region, perturbations in the zonal wind and surface pressure fields are coupled with convection and propagate eastward with a phase speed 5–8 m/s.[19] East of the dateline, the convective signals of the MJO usually vanish, and the signals in the wind and surface pressure fields continue to propagate eastward at a much faster speed.

The MJO plays a significant role in the tropical weather and climate.[20,21] It has impacts on monsoons,[22,23] and modulates tropical convection and tropical cyclone activity over the East Pacific and the Atlantic.[24,25] Some studies suggest that it plays an active role in the onset and development of ENSO.[26] The MJO also has remote impacts on the midlatitude weather.[27]

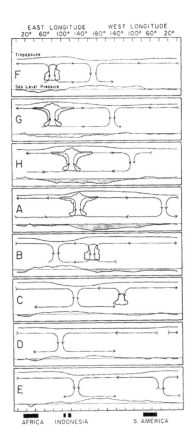

FIGURE 27.3   Longitude-height schematic diagram of the MJO life cycle (from top to bottom). Arrows indicate the zonal circulation, and clouds represent regions of enhanced convection, and the curves below and above the circulation indicate the pressure perturbations in the upper troposphere and at the sea level.
Source: Adapted from Madden and Julian.[18] © American Meteorological Society. Used with permission.

Equatorial Waves

Equatorial waves are large-scale waves trapped in the equatorial region, which include equatorial Rossby waves, Kelvin waves, westward and eastward propagating inertiogravity waves, and mixed Rossby–gravity waves. Away from the equator the amplitude of these waves decays rapidly. The seminal paper by Matsuno[28] laid the theoretical foundation for understanding the dynamics of equatorial waves. Using a shallow-water model on an equatorial betaplane, Matsuno[28] extracted different modes of waves, which were later confirmed by observational analysis. Gill[29] extended this work and interpreted the atmospheric response to steady heating in the tropics in terms of equatorial waves. Recent studies have shown that some types of equatorial waves are coupled with moist convection, which modifies the wave structure and tends to slow down the wave propagation.

Equatorial waves excited by convective heating can affect weather and climate in remote regions. Some low-frequency Tropical Cyclones equatorial waves contribute to the intraseasonal variability in the tropics. Equatorial waves also modulate tropical cyclone activities or serve as synoptic-scale precursors for tropical cyclone formation.[30,31] Oceanic Kelvin waves are believed to play a role in the ENSO dynamics.[32]

TABLE 27.1 Classification of Tropical Cyclones Based on the Maximum Sustained Surface Wind Speed

Category	Sustained Winds
Tropical depression	Less than 38 mph (34 kt or 62 km/hr)
Tropical storm	39–73 mph 35–63 kt 63–118 km/hr
Category 1 hurricane	74–95 mph 64–82 kt 119–153 km/hr
Category 2 hurricane	96–110 mph 83–95 kt 154–177 km/hr
Category 3 hurricane	111–129 mph 96–112 kt 178–208 km/hr
Category 4 hurricane	130–156 mph 113–136 kt 209–251 km/hr
Category 5 hurricane	157 mph or higher 137 kt or higher 252 km/hr or higher

The hurricane categories follow the Saffir–Simpson Hurricane scale.

Tropical Cyclones

A tropical cyclone is a non-frontal, cyclonic circulation (counterclockwise swirling winds in the Northern Hemisphere and clockwise swirling winds in the Southern Hemisphere) in the tropics with organized convection. Based on the maximum sustained surface wind speed, tropical cyclones can be ranked into different categories (Table 27.1). A tropical cyclone with maximum sustained surface wind speed exceeding 33 m/s is called a hurricane over the Atlantic and the East Pacific, and a typhoon over the western North Pacific.

An intense tropical cyclone usually has an eye, which is a cloud-free region in satellite imagery. The eyewall of a tropical cyclone is a ring of deep convection and is characterized by heavy precipitation and strong surface wind. Outside the eyewall, deep convection tends to organize into long, narrow rainbands that are oriented in the same direction of the horizontal wind. Tropical cyclones have a warm core vertical structure, and the cyclonic circulation weakens with height. Although the radius of maximum wind is of the order of 100 km, the low-level cyclonic wind extends to much larger radii (up to 1000 km).

Tropical cyclones form over warm oceans, which provide the energy source for the storms through the release of the latent heat of condensation. The formation of a tropical cyclone also requires a pre-existing synoptic-scale disturbance (such as tropical easterly waves), weak vertical wind shear and moist unstable air.[33] Tropical cyclones rarely form within 5° of the equator due to the weak Coriolis force (or the weak planetary vorticity) near the equator.[34] The track of a tropical cyclone is mainly determined by the steering flow and the beta-effect. Tropical cyclones usually move westward or northwestward (southwestward in the Southern Hemisphere) before being carried eastward by the extratropical westerly steering flow. Tropical cyclones can induce coastal flooding due to storm surges and torrential rainfall, and they may have significant downstream impacts if undergoing extratropical transition.

Deep Cumulonimbus Clouds: Hot Towers

Riehl and Malkus[5] hypothesized that buoyant updrafts in deep cumulonimbus clouds carry energy-rich air from the boundary layer to the upper troposphere with little or no dilution and are essential to the vertical energy transport and the general circulation in the tropics. The updrafts in these convective clouds are referred to as "hot towers." Riehl and Malkus[5] estimated that between 1500 and 5000 hot towers were needed to balance the global heat budget. Recent studies confirmed the role of hot towers in vertical energy transport, but suggested that undiluted convective cores rarely exist in the tropics.[35,36] Although air parcels get diluted by mixing with the drier and low-energy environmental air, they become more buoyant in the upper troposphere due to latent heat release by ice microphysical processes.

Hot towers are also believed to play an important role in the formation and intensification of tropical cyclones. In a vorticity-rich environment, cumulonimbus clouds can generate small-scale, intense, cyclonic vortex tubes, the so-called vortical hot towers.[37] The merger of these vortices leads to the formation of a tropical cyclone proto vortex. The latent heat release associated with these cumulonimbus clouds collectively drives a transverse circulation with low-level inflow, which concentrates the low-level vorticity and intensifies the cyclonic wind.

Conclusion

Tropical meteorology is the study of tropical atmospheric processes of various spatial and temporal scales. The tropics are the source of angular momentum and energy for the global atmospheric circulation. Weather and climate systems in the tropics may have remote impacts over extratropical regions. Advances in tropical meteorology in recent decades, aided by satellite observations and modern numerical models, have significantly increased our knowledge of the tropical atmosphere and contributed to improved weather and climate prediction in both the tropics and extratropics.

References

1. Riehl, H. *Climate and Weather in the Tropics.* Academic Press, 1979; 611 pp.
2. Sobel, A.H.; Nilsson, J.; Polvani, L.M. The weak temperature gradient approximation and balanced tropical moisture waves. J. Atmos. Sci. **2001**, *58*, 3650–3665.
3. Vonder H.; Thomas H.; Suomi, V.E. Measurements of the earth's radiation budget from satellites during a five-year period. Part I: extended time and space means. J. Atmos. Sci. **1971**, *28*, 305–314.
4. Fasullo, J.T.; Trenberth, K.E. The annual cycle of the energy budget: Meridional structures and poleward transports. J. Clim **2008**, *21*, 10, 2314–2326.
5. Riehl, H.; Malkus, J. On the heat balance in the equatorial trough zone. Geophysica **1958**, *6*, 503–538.
6. Chang, C.-P.; Wang, B.; Lau, N.-C. *The Global Monsoon System: Research and Forecast. Report of the International Committee of the Third International Workshop on Monsoons (IWM-III),* Vol. WMO/TD No. 1266 (TMRP Report No. 70); Secretariat of the World Meteorological Organization, 2005a; 542.
7. Chang, C.-P., Ding, Y.H., Johnson, R.H., Lau, G.N.-C., Wang, B., Yasunari, T., Eds.; *The Global Monsoon System: Research and Forecast,* 2nd Ed.; World Scientific, 2011; 608 pp.
8. Webster, P. The elementary monsoon. In *Monsoons;* Fein, J.S., Stephens, P.L., Eds.; John Wiley & Sons, 1987; 3-32 pp.
9. Chang, C.-P.; Wang, Z.; McBride, J.; Liu, C.H. Annual cycle of Southeast Asia-Maritime Continent rainfall and the asymmetric monsoon transition. J. Clim. **2005b**, *18,* 287–301.
10. Wang, Z.; Chang, C.P. A numerical study of the interaction between the large-scale monsoon circulation and orographic precipitation over South and Southeast Asia. J. Clim. **2012**, *25,* 2440–2455.
11. Pai, D.; Bhate, J.; Sreejith, O.; Hatwar, H. Impact of MJO on the intraseasonal variation of summer monsoon rainfall over India. Clim. Dyn. **2009**, doi:10.1007/s00382-009-0634-4.
12. Wang, Z. *Summertime Teleconnections Associated with U.S. Climate Anomalies and Their Maintenance.* Ph.D. Dissertation, Dept. of Meteorology, University of Hawaii: Honolulu, 2004; 195 pp.
13. Bjerknes, J. Atmospheric teleconnections from the equatorial Pacific. Mon. Wea. Rev. **1969**, *97,* 163–172.
14. Gray, W.M. Atlantic seasonal hurricane frequency. Part I: El Niño and the 30 mb quasi-biennial oscillation influences. Mon. Wea. Rev. **1984**, *112,* 1649–1668.
15. Lau, N.-C.; Wang, B. Interactions between Asian monsoon and the El Nino-Southern Oscillation. In *The Asian Monsoon;* Wang, B. Ed.; Springer/Praxis Publishing: New York, 2006; 478–512 pp.
16. Wang, Z.; Chang, C.P.; Wang, B. Impacts of El Niño and La Niña on the U.S. Climate during Northern Summer. J. Clim. **2007**, *20,* 2165–2177.
17. Madden, R.A.; Julian, P.R. Detection of a 40–50 day oscillation in the zonal wind in the Tropical Pacific. J. Atmos. Sci. **1971**, *28,* 702–708.
18. Madden, R.A.; Julian, P.R. Description of global-scale circulation cells in the tropics with a 40–50 day period. J. Atmos. Sci. **1972**, *29,* 1109–1123.

19. Zhang, C. Madden-Julian Oscillation. Rev. Geophys. **2005**, *43*, RG2003, doi:10.1029/2004RG000158.
20. Lau, K.-M., Waliser, D.E., Eds., *Intraseasonal Variability of the Atmosphere-Ocean Climate System;* Springer: Heidelberg, Germany, 2005; 474 pp.
21. Wang, B.; Ding, Y. An overview of the Madden-Julian oscillation and its relation to monsoon and mid-latitude circulation. Adv. Atmos. Sci. **1992**, *9*, 1, 93–111.
22. Yasunari, T. Cloudiness fluctuations associated with the Northern Hemisphere summer monsoon. J. Meteor. Soc. Jpn. **1979**, *57*, 227–242.
23. Lau, K.-M.; Chan, P.H. Aspects of the 40–50 day oscillation during the Northern summer as inferred from outgoing longwave radiation. Mon. Wea. Rev. **1986**, *114*, 1354–1367.
24. Chen, S.S.; Houze, R.A.; Mapes, B.E. Multiscale variability of deep convection in relation to large-scale circulation in TOGA COARE. J. Atmos. Sci. **1996**, *53*, 1380–1409.
25. Maloney, E.D.; Hartmann, D.L. Modulation of eastern north pacific hurricanes by the Madden-Julian oscillation. J. Clim. **2000**, *13*, 1451–1460.
26. Zhang, C.; Gottschalck, J. SST anomalies of ENSO and the Madden-Julian oscillation in the equatorial pacific. J. Clim. **2002**, *15*, 2429–2445.
27. Liebmann, B.; Hartmann, D.L. An observational study of tropical-midlatitude interaction on intraseasonal time scales during winter. J. Atmos. Sci. **1984**, *41*, 3333–3350.
28. Matsuno, T. Quasi-geostrophic motions in the equatorial area. J. Meteor. Soc. Jpn. **1966**, *44*, 25–43.
29. Gill, A.E. Some simple solutions for heat-induced tropical circulation. Quart. J. Roy. Meteor. Soc **1980**, *106*, 447–462.
30. Frank, W.M.; Roundy, P.E. The role of tropical waves in tropical cyclogenesis. Mon. Wea. Rev. **2006**, *134*, 2397–2417.
31. Dunkerton, T.J.; Montgomery, M.T.; Wang, Z. Tropical cyclogenesis in a tropical wave critical layer: easterly waves. Atmos. Chem. Phys. **2009**, *9*, 5587–5646.
32. Jin, F. An equatorial ocean recharge paradigm for ENSO. Part I: Conceptual model. J. Atmos. Sci. **1997**, *54*, 811–829.
33. Gray, W.M. Global view of the origin of tropical disturbances and storms. Mon. Wea. Rev. **1968**, *96*, 669–700.
34. Elsberry, R.L.; Foley, G.; Willoughby, H.; McBride, J.; Ginis, I.; Chen, L. *Global Perspectives on Tropical Cyclones. Tech. Doc.* No. 693; World Meteor. Organiz.: Geneva, Switzerland, 1995; 289 pp.
35. Zipser, E.J. Some views on "hot towers" after 50 years of tropical field programs and two years of TRMM data. Meteorol. Monogr. **2003**, *29*, 49–49.
36. Fierro, A.O.; Simpson, J.; LeMone, M.A.; Straka, J.M.; Smull, B.F. On how hot towers fuel the Hadley cell: an observational and modeling study of line-organized convection in the equatorial trough from TOGA COARE. J. Atmos. Sci. **2009**, *66*, 2730–2746.
37. Montgomery, M.T.; Nicholls, M.E.; Cram, T.A.; Saunders, A.B. A vortical hot tower route to tropical cyclogenesis. J. Atmos. Sci. **2006**, *63*, 355–386.

Bibliography

Madden, R.A.; Julian, P.R. Observations of the 40–50–day tropical oscillation – A review. Mon. Wea. Rev. **1994**, *122*, 814–837.

Montgomery, M.T.; Smith, R.K. Paradigms for tropical-cyclone intensification. Q. J. R. Meteorol. Soc. **2011**, *137*, 1–31.

Philander, S.G.H. *El Niño, La Niña, and the Southern Oscillation.* Academic Press, San Diego, CA, 1990; 289 pp.

Trenberth, K.E.; Branstator, G.W.; Karoly, D.; Kumar, A.; Lau, N.; Ropelewski, C. Progress during TOGA in understanding and modeling global teleconnections associated with tropical sea surface temperatures. J. Geophys. Res. **1998**; *103*, 14291–14324.

Wang, B., Eds.; *The Asian Monsoon;* Springer/Praxis Publishing Co.: New York, 2006; 787 pp.

28

Ocean–Atmosphere Interactions

Vasubandhu Misra
Florida State University

Introduction

Ocean–atmosphere interaction commonly refers to the exchange of energy at the interface of the ocean and atmosphere. However, besides energy, there is also exchange of mass including that of fresh water (precipitation, runoff, melting of sea ice, evaporation) and of inert and sparingly soluble gases in seawater [e.g., oxygen (O_2), methane (CH_4), nitrous oxide (N_2O), carbon dioxide (CO_2)], which are quite significant even if they seem small on a unit area basis because of the extent of the ocean surface. We will, however, limit our discussion to the exchange of energy and freshwater. The energy exchange usually takes the form of exchange of heat (sensible heat), moisture (latent heat), and momentum (wind stress) at the overlapping boundary of the atmosphere and the ocean. The exchange of heat, moisture, and momentum are represented usually as fluxes, which are defined as the rate of exchange of energy per unit surface area of ocean–atmosphere interface. So the wind stress is the flux of horizontal momentum imparted by the atmospheric wind to the ocean. Similarly the latent heat flux refers to the rate at which energy associated with phase change occurs from the ocean to the atmosphere. Likewise the sensible heat flux refers to exchange of heat (other than that due to phase change of water) by conduction and/or convection. There is also heat flux from precipitation, which comes about as a result of the difference in temperature of the raindrops and the ocean surface. The heat flux from precipitation could become important in a relatively wet climate.

In the following sub-sections we discuss air-sea feedback, their global distribution, the common practices to diagnose this feedback, and the dynamical implications of ocean–atmosphere interactions, with concluding remarks in the final section.

Air–Sea Feedback

The surface energy budget of the ocean mixed layer, of which sensible heat flux, latent heat flux, precipitation heat flux are a part, dictates the temperature of the sea surface, which by virtue of its role in determining the stability of the lower atmosphere and the upper ocean dictates the fluxes across the interface. It may be noted that the surface mixed layer of the ocean refers to the depth up to which surface turbulence plays an important role in mixing so that the density is approximately the same as the surface. The entire mixed layer is active in transferring heat to the ocean–atmosphere interface. This forms a feedback loop between sea surface temperature (SST), stability of the lower atmosphere and the upper ocean, and the air–sea fluxes. Therefore determination of air–sea fluxes is a form of diagnosis of the coupled ocean–atmosphere processes. Similarly, the fresh water flux is part of the ocean surface salinity budget that dictates the stability of the upper ocean, which feeds back to the air–sea fluxes. For example, the fresh water flux from river discharge into the coastal ocean sometimes results in forming a relatively thin fresh water "lens" or an ocean barrier layer[1] that stabilizes the vertical column and inhibits vertical mixing in the upper ocean. This ocean barrier layer is then relatively easily modulated by the atmospheric variations, thus modulating the SST and, therefore, the air–sea fluxes. The influence of cloud radiation interaction on SST is another form of air–sea feedback. Several studies,[2–4] for example, note the existence of positive cloud feedback on SST in the summer time over the North Pacific (~35°N). This feedback refers to the reduction of SST under enhanced maritime stratiform clouds in the boreal summer season that reduces the downwelling shortwave flux and cools the mixed layer. In turn, the cooler SST favors further enhancement of the stratiform clouds affecting the stability of the atmospheric boundary layer. Wind Induced Surface Heat Exchange (WISHE) phenomenon is another form of air–sea feedback, which refers to the positive feedback between the surface (sensible and latent) heat fluxes with wind speed that results in increased fluxes leading to cooling of the SST as the wind speed increases. These negative correlations are apparent at large spatial scales across the subtropical oceans[5] as well as in mesoscale features (e.g., hurricanes).[6] The asymmetry of the inter-tropical convergence zone residing largely in the northern hemisphere in the eastern Pacific and in the eastern Atlantic Ocean was explained in terms of WISHE.[7]

Distribution of Air–Sea Fluxes

Direct measurements of air–sea flux are few, limited both in time and in space. These measurements of air–sea fluxes are important, however, for developing, calibrating, and verifying the estimated air–sea flux from parameterization schemes.[8,9] The parameterization schemes for air–sea fluxes use state variables of the atmosphere and the ocean (e.g., wind speed, temperatures, humidity) to estimate the fluxes. A commonly used parameterization scheme for air–sea fluxes is the "bulk-aerodynamic" formula. This is based on the premise that wind stress is proportional to the mean wind shear computed between surface and 10 m above surface, and sensible heat flux and latent heat flux are proportional to the vertical temperature and moisture gradients computed between surface and 2 m above surface. As a result, air–sea fluxes have been computed globally and regionally from a variety of analyzed and regionally observed or analyzed atmospheric and oceanic states leading to a number of air–sea flux intercomparison studies.[10–12] These intercomparisons provide insight into the uncertainty of estimating air–sea fluxes as well as revealing salient differences in the state variables used in the parameterization scheme. For example, Smith et al.[10] found that in many regions of the planet the differences in surface air temperature and humidity amongst nine different products of air–sea fluxes had a more significant impact than the differences in the surface air wind speed (at 10 m) on the differences in the air–sea fluxes. Climatologically, large values of sensible heat flux are observed in the winter along the western boundary currents of the middle-latitude oceans, when cold continental air passes over the warm ocean currents (e.g., Gulf Stream). In the tropics and in the sub-tropical eastern oceans, the sensible heat flux is usually small. In the former region, climatologically, the wind speeds and the vertical temperature gradients between the surface and 2 m above the surface are weak. In the sub-tropical eastern oceans, with prevalence of upwelling, the SSTs are relatively cold leading to generally smaller sensible heat flux. The latent heat flux is observed climatologically to be large everywhere in the global oceans relative to the sensible heat flux, with exceptions over polar oceans in the winter season. The ratio of sensible to latent heat flux, called the Bowen ratio, has a latitudinal gradient with a higher (smaller) ratio displayed in the polar (tropical) latitudes. This gradient of the Bowen ratio largely stems from the decreasing SST with latitude that affects the moisture-holding capacity of the overlying atmosphere and thereby affecting the latent heat flux.

The higher Bowen ratio may lead one to believe that air–sea coupling is weak in regions of cold SST. However, several studies[13,14] reveal that this is not the case. Using satellite-based (scatterometer) winds and SST that affords high space–time resolution, a robust relationship (positive correlation) between surface wind stress and SST is revealed along major ocean fronts that display strong SST gradients. Such a positive correlation between SST and surface wind speed is suggestive of a vertical momentum-mixing mechanism, which allows atmospheric wind to adjust to SST changes across major ocean fronts.[15,16] In other words, static stability of the lower atmosphere is reduced as the SST warms, which results in intensifying the turbulent mixing of the fast moving air aloft that causes the surface wind to accelerate. Such positive correlations have been observed in many regions of the global oceans at relatively small spatial scales.[14] On the other hand, negative correlations of wind speed with SST is suggestive of the so-called WISHE phenomenon discussed earlier.

Diagnosis of Air–Sea Feedback

Cayan[5] showed that vast regions of the middle-latitude ocean surface temperature variability is forced by the atmospheric variations. He showed this by comparing local, simultaneous correlations of the monthly mean values of latent heat and sensible heat fluxes with time tendency of SST that capture this interaction at spatial scales, which span a major portion of the North Atlantic and Pacific Oceans. The simultaneous correlation of SST with heat flux is shown to operate at a more local scale. Similarly, Wu et al.[17] showed that the atmospheric variability in the eastern equatorial Pacific is forced largely by the underlying SST variations, where simultaneous correlations of latent heat flux with tendency of SST is positive. This is in contrast to the tropical western Pacific Ocean where the correlations of latent heat flux with time tendency of SST is negative, suggesting that atmospheric variability is forcing the SST variability.

Dynamical Interactions

Ocean–atmosphere interaction also happens through dynamics as illustrated in the case of Asian monsoon.[18] Here it is shown that in order to maintain the net heat balance of the coupled ocean–land–atmosphere Asian monsoon system, the ocean transports heat to the winter hemisphere of the Indian Ocean as comparable heat is released in the atmosphere of the summer hemisphere from the monsoonal convection. Similarly, Czaja and Marshall[19] demonstrate that atmospheric heat transport from tropical to higher latitudes is comparably higher to that by the oceans to maintain the observed latitudinal gradients of temperature and hence the general circulation of the Earth. In other words, in both these examples, the ocean and the atmosphere is acting in tandem to maintain the energy balance of the climate system of the planet, exhibiting a manifestation of the coupled ocean–atmosphere processes. Another prime example of ocean-atmosphere interaction is the natural variability of the El Niño and the Southern Oscillation (ENSO), one of the largest and most well-known interannual variations of the Earth's climate system, manifesting most prominently as the SST variability in the eastern equatorial Pacific Ocean. ENSO is a result of ocean dynamics and air–sea interaction processes,[20,21] which has a quasi-periodic oscillation with a period in the range of 2–7 years that affects the global climate variations.

Conclusions

Ocean–atmosphere interaction comprises a critical part of the Earth's physical climate system. ENSO variations that affect the global climate variability was first explained in terms of air–sea feedback in the equatorial eastern Pacific Ocean in the late 1960s. The first forays of predicting seasonal climate in the

1970s came from the hypothesis that through air–sea interaction, the persistent tropical SST anomalies would impart climate memory to the overlying atmosphere. Ever since, steady progress has been made in modeling (or parameterizing) and observing air–sea fluxes. Our growing understanding of the importance of air–sea interaction on observed climate variability has also led us to wean away slowly from a reductionist approach of developing component (such as atmosphere, land, ocean) models in isolation to coupled climate numerical models. The air–sea interaction happens not only at the interface of the atmosphere–ocean boundary through fluxes (defined as the amount of exchange of energy or mass per unit area) but also through compensatory dynamical circulations to maintain the observed climate of the planet.

Acknowledgments

This work is supported by grants from NOAA, USGS, and CDC. However, the contents of the paper are solely the responsibility of the author and do not represent the views of the granting agencies.

References

1. Lukas, R.; Lindstrom, E. The mixed layer of the western equatorial Pacific Ocean. J. Geophys. Res. **1991**, *96*, 3343–3457.
2. Norris, J.E.; Zhang, Y.; Wallace, J.M. Role of low clouds in summertime atmosphere-ocean interactions over the North Pacific. J. Clim. **1998**, *11*, 2482–2490.
3. Park, S.; Deser, C.; Alexander, M.A. Estimation of the surface heat flux response to sea surface temperature anomalies over the global oceans. J. Clim. **2005**, *18*, 4582–4599, DOI: 10.1175/JCLI3521.1.
4. Wu, R.; Kinter III, J.L. Atmosphere-ocean relationship in the midlatitude North Pacific: Seasonal dependence and east-west contrast. J. Geophys. Res. **2010**, *115*, D06101, DOI: 10.1029/2009JD012579.
5. Cayan, D.R. Latent and sensible hat flux anomalies over the northern oceans: Driving the sea surface temperature. J. Phys. Oceanogr. **1992**, *22*, 859–881.
6. Emanuel, K.A. An air-sea interaction theory for tropical cyclones. Part I. J. Atmos. Sci. **1958**, *43*, 585–604.
7. Xie, S.-P.; Philander, S.G.H. A coupled ocean-atmosphere model of relevance to the ITCZ in the eastern Pacific. Tellus. **1994**, *46A*, 340–350.
8. Weller, R.A.; Anderson, S.P. Temporal variability and mean values of the surface meteorology and air–sea fluxes in the western equatorial Pacific warm pool during TOGA COARE. J. Clim. **1996**, *9*, 1959–1990.
9. Fairall, C.W.; Bradley, E.F.; Rogers, D.P.; Edson, J.B.; Young, G.S. The parameterization of air–sea fluxes for Tropical Ocean-Global Atmosphere Coupled-Ocean Atmosphere Response Experiment. J. Geophys. Res. **1996**, *101*, 3734–3764.
10. Smith, S.R.; Hughes, P.J.; Bourassa, M.A. A comparison of nine monthly air-sea flux products. Int. J. Climatol. **2010**, *31*, 1002–1027.
11. Bourras, D. Comparison of five satellite-derived latent heat flux products to moored buoy data. J. Clim. **2006**, *19*, 6291–6313.
12. Chou, S.-H. A comparison of airborne eddy correlation and bulk aerodynamic methods for ocean-air turbulent fluxes during cold-air outbreaks. Bound-Lay. Meteorol. **1993**, *64*, 75–100.
13. Xie, S.-P. Satellite observation of cool ocean-atmosphere interaction. Bull. Amer. Soc. **2004**, 195–208
14. Chelton, D.B., Schlax, M.G.; Freilich, M.H.; Milliff, R.F. Satellite radar measurements reveal short-scale features in the wind stress field over the world ocean. Science **2004**, *303*, 978–983.
15. Wallace, J.M.; Mitchell, T.P.; Deser, C. The influence of sea surface temperature on surface wind in the eastern equatorial Pacific: Seasonal and interannual variability. J. Climat. **1989**, *2*, 1492–1499.

16. Hayes, S.P.; McPhaden, M.J.; Wallace, J.M. The influence of sea-surface temperature on surface wind in the eastern equatorial Pacific: Weekly to monthly variability. J. Clim. **1989,** *2,* 1500–1506.

17. Wu, R.; Kirtman, B.P.; Pegion, K. Local air–sea relationship in observations and model simulations. J. Clim. **2006,** *19,* 4914–4932.

18. Loschnigg, J.; Webster, P.J. A coupled ocean-atmosphere system of SST regulation for the Indian Ocean. J. Clim. **2000,** *13,* 3342–3360.

19. Czaja, A.; Marshall, J. The partitioning of poleward heat transport between the atmosphere and ocean. J. Atmos. Sci. **2006,** *63,* 1498–1511.

20. Cane, M.A.; Zebiak, S.E. A theory for El Niño and the Southern Oscillation. Science **1985,** *228,* 1085–1087

21. Philander, S.G. *El Niño, La Niña, and the Southern Oscillation;* Academic Press: 1990; 289pp.

29

Oceans: Observation and Prediction

Oscar Schofield,
Josh Kohut, Grace
Saba, Xu Yi,
John Wilkin, and
Scott Glenn
Rutgers University

Introduction

The oceans cover the majority of the Earth's surface and despite centuries of exploration they remain relatively unexplored. This gap of knowledge reflects the difficulty of collecting physical, chemical, and biological data in the ocean, as it is a harsh and unforgiving environment in which to operate. Despite centuries of ship-based exploration, the immense size and hazards associated with wind, waves, and storms limit the ability of humans to sustain a coherent global sampling network. Satellite and aircraft remote sensing approaches provide powerful tools to map global synoptic properties (Figure 29.1); however, satellite systems largely provide information on the surface ocean. Fixed and mobile sensors deployed in the ocean can provide subsurface data, however, their numbers, while expanding, are limited and the technology still struggles with issues related to the onboard power availability and the number of available robust sensors. These sampling shortcomings have significant implications for human society especially as there is increasing evidence that the physics, chemistry, and biology of the ocean have changed over the last few decades. These changes reflect both natural cycles and the anthropogenic forcing from human activity.

Quantitatively understanding the relative importance of the natural and anthropogenic forcing in the ocean remains an open question, which needs to be resolved as the environmental impacts associated with human activity will increase, reflecting the growth of human populations.[1] The current projections suggest that human population growth at coastlines will be the most rapid and largest on the planet.[2] This will increase the importance of marine systems in national economies around the world making managing coastal systems critical. The close proximity of large populations will expose them to potential natural and man-made disasters associated with the oceans. These disasters include tsunamis, hurricanes, offshore industrial accidents, and human health issues such as outbreaks of waterborne disease. Our current capabilities to predict, respond to, manage, and mitigate these events is astonishingly poor. Improving our ability to observe and predict changes in the ocean will require technical

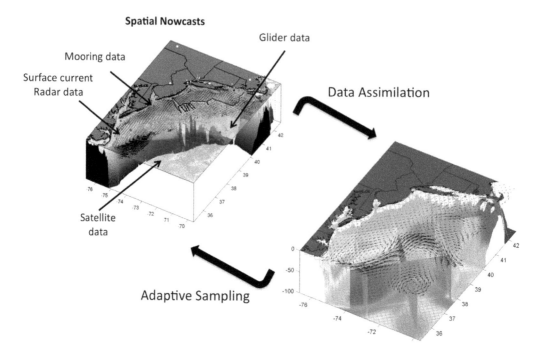

FIGURE 29.1 (See color insert.) An example of a coupled ocean observation and modelling network along the East coast of the United States. The spatial nowcast consists of observational data delivered in near real-time to shore. The nowcast consists of satellite data (sea surface temperature is shown), surface current radar (black arrows on the ocean), mooring data (black dots), subsurface data collected from gliders (water temperature shown here), and weather data collected by shore based stations (small white dots). The nowcast provides data to simulation models via data assimilation. The simulation model provides a 48-hour forecast, which is used to redistribute observational assets to provide an improved data set before the next forecast cycle.

improvements combined with an increased fundamental understanding of physical, chemical, and biological processes.

What Is the Path Forward?

Improving our understanding and management of the ocean system will require an improved ability to map ocean properties in the present and improving our ability to forecast future ocean conditions. The ability to map ocean properties will require a distributed portfolio of ocean infrastructure that will be linked together through an increasing number of ocean models.[3] The observation networks will collect quantitative data about the current status of the ocean. The forecasts are driven by numerical models that use current scientific understanding to project how the ocean will evolve. The combined observatory and numerical modelling capacity will improve our fundamental understanding and ability to respond to changes in the physics, chemistry, and biology of the marine systems.

The observations will assist the modelling efforts in several ways. Observations will provide data required to parameterize processes within the models. If the data are delivered in real-time they will be assimilated into the model to improve the predictive skill of the forecast. Finally, as new data are collected, they will be used to validate the predictive skill of the model. In turn the model forecasts will assist the observational efforts by providing forecasts that will allow scientists to adjust the spatial configuration and sampling rates of sensors to better sample future ocean conditions. These coupled systems are a rapidly maturing technology and builds off the more mature science of weather forecasting, which has its roots in the early 19th century. The fundamental approaches are based on the seminal work of Lewis Fry Richardson who is considered the father of numerical weather prediction in the 1920s. These approaches are computationally intensive and it was not until the advent of electronic computers that the science moved forward to become an indispensable tool for humanity. Modern computer models use data as inputs collected from automated weather stations and weather buoys at sea. These instruments, observing practices and timing are standardized through the World Meteorological Organization.

Oceanographic efforts are evolving in a similar fashion where observations inform operational models. The weather and ocean models, most often run by federal agencies, provide forecasts that are used by scientists, the maritime industry, state and local communities. Most often they are used to issue warnings of unsafe conditions due to storms and high waves. The motivation for global standardized ocean forecast systems can be traced back to the sinking of the Titanic in 1912, which prompted the international community to call for development of systems to improve safety at sea. Modem approaches and forecasting tools for the ocean did not mature until the 1980s and are rapidly evolving as the computing and ocean observation technologies are rapidly improving.

How Are Observations Made in the Oceans?

Many platforms are available for making ocean measurements and, although the list below is not exhaustive, it provides a snapshot of the major platforms. The platforms carry sensors that can measure physical, chemical, and biological properties of the sea; however, most new novel sensors can only be carried on ships. A smaller number of sensors can be deployed on autonomous platforms and the discussion in this entry is focused on those sensors that can be deployed on a variety of ocean platforms.

The most mature sensors are those that measure physical and geophysical variables, such as temperature, salinity, pressure, currents, waves, and seismic activity. Except for seismic variables, most of the physical sensors can be deployed on most of the platforms listed below. Many of the physical properties are the key variables in ocean numerical models. Currently, chemical sensors can measure dissolved gases (primarily oxygen) and dissolved organic material; however, recently the sensors capable of measuring nutrients (primarily nitrogen) are becoming commercially available. Biological sensors currently consist of optical and acoustic sensors. The optical sensors are used to provide information on the concentration, composition, and physiological state of the phytoplankton. Acoustic sensors can provide information on zooplankton to fish depending on the acoustic frequency band that is chosen.

Ships

The primary tool for oceanographers for centuries has been ships and will remain a central piece of infrastructure for the foreseeable future.[4] Ships are ideal as they are extremely flexible and allow teams to conduct experiments at sea. Ships are expensive to operate and must avoid hazardous conditions, such as storms, which limit the ability to make sustained measurements.

Satellites

Satellites are the most important oceanographic technology in modern times (beyond the ships).[5] Satellite observations have resulted in numerous advances in our fundamental understanding of the oceans[6] by resolving both global features associated with the mesoscale circulation of physical and biological properties. Satellite datum is fundamental to weather and ocean state prediction. Physical parameters available from space-based sensors include ocean surface temperature, wind speed and direction, sea surface height and topography, and sea ice distribution and thickness. Biological and chemical parameters can be derived from ocean color radiometers.

High-Frequency Radar

High-frequency radar measures ocean surface current velocities over hundreds of square miles simultaneously. Each site measures the radial components of the ocean surface velocity directed towards or away from the site[7,8] and the estimated velocity components allow surface currents (upper meter of water column) to be estimated.[9] These systems are cost effective and have many applied uses.

Ocean Moorings

The modern ocean moorings grew out of the weather stations established in the 1940s. Since the 1960s, modern buoys have enabled a wide range of studies addressing the ocean's role in climate, weather, as well as providing insight into the biogeochemistry of the sea. Moorings provide the backbone to many of the global ocean networks studying ocean–atmosphere interactions and are the foundation for the global tsunami warning system network. They will continue to be a key element of ocean observing infrastructure for the foreseeable future.

Seafloor Cables

Scientists often require high bandwidth and power for sustained periods of time. Seafloor electro-optic cables provide a means for maintaining a sustained presence in the ocean. Cables have been deployed off the east and west coasts of the United States, Canada, Japan, and Europe. Many other countries are planning to deploy seafloor cables.

Drifters and Floats

Passive Lagrangian platforms are tools for creating surface and subsurface maps of ocean properties. These platforms are relatively inexpensive and thus allow thousands of these platforms to be deployed. Drifters have historically been a key tool for oceanography as evidenced by the important works by Benjamin Franklin [10] and Irving Langmuir.[11] The drifters can carry numerous sensors to create global maps of surface circulation. The first neutrally buoyant drifters were designed to observe subsurface currents.[12] The subsurface profiling drifters were enabled in the early 1990s with communication capabilities[13] and now anchor the international ARGO program, which has over 3000 floats deployed in the ocean.

Gliders are a type of autonomous underwater vehicle (Figure 29.2) that use small changes in buoyancy in conjunction with wings to convert vertical motion to horizontal motion, and thereby propel itself forward with very low-power consumption.[14] Gliders follow a saw-tooth path through the water, providing data on large temporal and spatial scales. They navigate with the help of periodic surfacing for Global Positioning System (GPS) fixes, pressure sensors, tilt sensors, and magnetic compasses. Using buoyancy-based propulsion, gliders have a significant range and duration, with missions lasting up to a year and covering over 3500 km of range.[15–17]

Propeller-driven autonomous underwater vehicles (AUVs) are powered by batteries or fuel cells and can operate in water as deep as 6000 m. Similar to gliders, AUVs relay data and mission information to shore via satellite. Between position fixes and for precise maneuvering, inertial navigation systems are often available onboard the AUV to measure the acceleration of the vehicle, and combined with Doppler velocity measurements, it is used to measure the rate of travel. A pressure sensor measures the vertical

FIGURE 29.2 (See color insert.) A Webb underwater glider being deployed in the Ross Sea Antarctica in February 2011.
Source: Photo credit Chris Linder.

position. AUVs, unlike gliders, can move against most currents nominally at 3–5 knots, and can therefore systematically and synoptically survey a particular line, area, and/or volume.

What Numerical Ocean Models Are Available?

Over the last 30 years, there have been significant developments in three-dimensional numerical models for the ocean.[18] Many models exist spanning from global ocean scales down to the scale of individual estuaries. Models vary in their coordinate system (linear, spherical, and others), resolution in space and time, complexity (i.e., number of state variables), and the parameterization of key processes within the model. There are several excellent texts that outline many of the details of numerical ocean modeling[19,20] and one key lesson is that the choice of particular modelling approach depends on its intended application and on the available computational resources. Although an exhaustive list of the ocean models is beyond the scope of this text, several classes are described in the following paragraphs.

Mechanistic models are simplified models used to study a specific process, and are used to provide insight into the underlying processes influencing the physics, chemistry, and biology of the ocean. These models are most often constructed as a learning tool in order to assess processes and feedbacks within marine systems.

Simulation models are complex and describe three dimensional (3-D) ocean processes using the continuity and momentum equations. For this reason they are called the primitive equation models. These models can be used to simulate many processes, including ocean circulation, mixing, waves, and responses to external forces (such as storms). All these models are constructed using different assumptions. Additionally, the resolution of the models requires that the trade-offs of the computation burdens be measured against the processes that need to be simulated. For example, if the model must resolve mesoscale eddies, it will require the resolution of a few tenths of a degree of latitude and longitude. In contrast, most primitive equation climate models have much coarser horizontal resolution as they were designed to study large-scale hydrographic structure, climate dynamics, and water-mass formation over decadal time scales; however, for a specific question, there are climate models with sufficient resolution to resolve mesoscale eddies if one is willing to accept the computation cost. Simulations are also constructed for coastal systems and can resolve coastal currents, tides, and storm surges. Increasingly the biological and ocean chemistry models are being coupled into these 3-D simulation models. Although these biogeochemical models are rapidly improving, there is unfortunately no set of "primitive" equations yet capable for describing biological and chemical systems in the ocean; however, as these models evolve, they will be increasingly useful tools for managing living resources and water quality in the ocean.

Often models of varying resolutions are combined. Coarser-scale global or basin-scale models provide outer boundary inputs to higher resolution nested models, which allows a myriad of processes to be modeled with a lower computation burden but allow a range of processes to be simulated even if they require high resolution. This is often the case for coastal and continental shelf models. The advantage of this downscaling approach is that it allows basin scale models to resolve large-scale forcing that drives the regional to local-scale processes that are effectively modeled by a higher resolution model. The approach by which one links these models is a difficult problem and remains an area of active research.

Data assimilation is an approach by which model simulations are constrained by observations. For example, model calculations and observations of temperature and salinity can be compared, and then the model can be "adjusted" based on the mismatch. This is a difficult problem as 1) it represents an inverse problem (where a finite number of observations are used to estimate a continuous field), 2) many of the ocean processes of interest are non-linear, and 3) the observations and models both have unknown errors. Descriptions of data assimilation approaches for oceanography have been reviewed.[21,22] These approaches allow modelers to increase the forecast skill of their models by essentially keeping the models "on track" if the observations and data assimilation approaches can be provided in a timely fashion. Many in the ocean modelling community are focusing on using these approaches to increase model forecast skill as it determines how well these approaches will serve a wide range of science, commercial, and government needs.[23]

Conclusions

Ocean observation and modelling capabilities are rapidly diversifying and improving. These systems are increasingly linked by data assimilation approaches that when combined, provide a coupled observing and forecasting network. These approaches will increase the predictive skill of forecast models that in turn can serve a wide range of applications spanning from basic research to improving the efficiency of the maritime industry. The combined technologies will be critical to improving our understanding of the ocean today and the potential trajectory in the future.

Acknowledgments

We acknowledge the support of the Office of Naval Research's Major University Research Program, the National Oceanic and Atmospheric Administration's Integrated Ocean Observing System, and the National Science Foundation's Ocean Observing Initiative.

References

1. De Souza, R.; Williams, J.; Meyerson, F.A.B. Critical links: Population, health, and the environment. Popul. Bull. **2003**, *58* (3), 3–43.
2. Vitousek, P.M.; Mooney, H.A.; Lubchenco, J.; Melillo, J.M. Human domination of Earth's ecosystems. Science **1977**, *277*, 494–499, DOI: 10.1126/science.277.5325.494.
3. National Research Council. *Critical Infrastructure for Ocean Research and Societal Needs in 2030;* National Academy Press: Washington, D.C., 2011.

4. National Research Council. *Science at Sea: Meeting Future Oceanographic Goals with a Robust Academic Research Fleet*; National Academy Press: Washington, D.C., 2009.

5. Munk, W. Oceanography before, and after, the advent of satellites. In *Satellites, Oceanography and Society*, Halpern, D., Ed.; Elsevier Science: Amsterdam, 2000; 1–5.

6. Halpern, D., Ed. *Satellites, Oceanography and Society;* Elsevier Science: Amsterdam, 2000; 361 pp.

7. Crombie D.D. Doppler spectrum of sea echo at 13.56 Mc/s. Nature **1955**, *175*, 681–682.

8. Barrick, D.E.; Evens, M.W.; Weber, B.L. Ocean surface currents mapped by radar. Science **1977**, *198*, 138–144.

9. Stewart R.H.; Joy J.W. HF radio measurements of ocean surface currents. Deep Sea Res. **1974**, *21*, 1039–1049.

10. Franklin, B. Sundry marine observations. Trans. Am. Philos. Soc. **1785**, *1* (2), 294–329.

11. Langmuir, I. Surface motion of water induced by wind. Science **1938**, *87*, 119–123.

12. Swallow, J.C. A neutral-buoyancy float for measuring deep currents. Deep Sea Res. **1955**, *3*, 74–81.

13. Davis, R.E.; Webb, D.C.; Regier, L.A.; Dufour, J. The Autonomous Lagrangian Circulation Explorer (ALACE). J. Atmos. Ocean. Tech. **1992**, *9*, 264–285.

14. Rudnick, D.L.; Davis, R.E.; Eriksen, C.C.; Fratantoni, D.M.; Perry, M.J. Underwater gliders for ocean research. Mar. Technol. Soc. J. **2004**, *38*, 73–84.

15. Sherman, J.; Davis, R.E.; Owens, W.B.; Valdes, J. The autonomous underwater glider "Spray." IEEE J. Ocean. Eng. **2001**, *26*, 437–446.

16. Eriksen, C.C.; Osse, T.J.; Light, R.D.; Wen, T.; Lehman, T.W.; Sabin, P.L.; Ballard, J.W.; Chiodi, A.M. Seaglider: A long-range autonomous underwater vehicle for oceanographic research. IEEE J. Ocean. Eng. **2001**, *26*, 424–436.

17. Webb, D.C.; Simonetti, PJ.; Jones, C.P SLOCUM: An underwater glider propelled by environmental energy. IEEE J. Ocean. Eng. **2001**, *26*, 447–452.

18. Bryan, K. A numerical method for the study of the circulation of the world ocean. J. Comput. Phys. **1969**, *4* (3), 347–376, DOI: 10.1016/0021-9991(69)90004-7.

19. Kantha, L.H.; Clayson, C.A. *Numerical Models of Oceans and Oceanic Processes;* International Geophysics Series, Elsevier: Amsterdam, 1995; 750 pp.

20. Haidvogel, D.; Beckmann, A. *Numerical Ocean Circulation Modelling;* Imperial College Press: London, 1999; 318 pp.

21. Wunsch, C. *The Ocean Circulation Inverse Problem;* Cambridge University Press: Cambridge, 1996; 442 pp.

22. Malanotte-Rizzoli P. *Modern Approaches to Data Assimilation in Ocean Modelling;* Elsevier Science: Amsterdam, Netherlands, 1996; 455 pp.

23. Pinardi, N.; Woods, J. *Ocean Forecasting: Conceptual Basis and Applications;* Springer Verlag: Berlin, Germany, 2002; 472 pp.

30

Tropical Cyclones

Patrick J. Fitzpatrick
Mississippi State University

Introduction Using a Western Hemisphere Perspective

A tropical cyclone is a rotating, organized system of clouds and thunderstorms that originates over tropical or subtropical waters and has a closed low-level cyclonic circulation. These systems form and intensify from ocean heat extraction through complex processes, with a warm upper-level center which distinguishes it from other windy weather systems. Due to the Earth's rotation, these storms spin counterclockwise in the Northern Hemisphere, and clockwise in the Southern Hemisphere. Both hemispheric spins are referred to as *cyclonic rotation* because the sense of spin about a local vertical axis is the same as the Earth's rotation when viewed from above.

Classifications for tropical cyclones vary globally. In this initial discussion, we use Western Hemisphere terminology, which covers the North Atlantic, Northeast Pacific, and North Central Pacific basins. Classifications in the Eastern Hemisphere are different compared with the Western Hemisphere. Furthermore, classifications vary between the Northwest Pacific, the North Indian Ocean, the Southwest Indian Ocean, the Southeast Indian Ocean, and the Southwest Pacific Ocean. For example, the term *hurricane* in the Western Hemisphere is the equivalent to the term *typhoon* in the Northwest Pacific Ocean. Their origins are locally derived. Hurricane is derived from the Caribbean Indian word hurican meaning "evil spirit," who in turn translated it from the Mayan storm god Hurakan. Typhoon is derived probably from the Chinese words tung fung ("a terrible storm of east winds"), ta fung (a "great wind"), or t'ai fung (either "eminent wind" or "wind of Taiwan"). A Greek god is also called Typhon (a storm giant and father of all "monsters") but it is unknown if there is a connection to the word typhoon or China. The global classifications will be discussed in the last section.

Wind speed is the fundamental demarcation of tropical cyclone classifications. International units are in ms^{-1}, and most countries have adopted the metric standard. However, some regions, countries, and professions prefer other wind units. The United States uses mph, where $1\ ms^{-1} = 2.2$ mph. Other countries prefer kmh^{-1}, where $1\ ms^{-1} = 3.6\ kmh^{-1}$. Mariners prefer knots, where $1\ ms^{-1} = 1.9$ knots.

This text uses the metric standard of ms^{-1} for wind speed. Because the fastest winds in tropical cyclones occur near the ground, a standard height of 10 m (33 ft) is used. Winds are also averaged by either 1 or 10 minutes (depending on global region), and referred to as *sustained winds*.

A tropical cyclone does not form instantaneously. Initially a tropical cyclone begins as a *tropical disturbance* when a mass of organized, oceanic thunderstorms persists for 24 hours. The tropical disturbance becomes a *tropical depression* when a closed circulation is first observed and all sustained winds are less than 17 ms^{-1}. When these sustained winds increase to 17 ms^{-1} somewhere in the circulation, it is then classified as a *tropical storm*. A tropical cyclone becomes a *hurricane* when sustained surface winds are 33 ms^{-1} or more somewhere in the storm.

To clarify the life cycle, this entry discusses tropical cyclone formation in phases. The first phase is the *genesis stage*, and includes tropical disturbances and tropical depressions. The second phase includes tropical storms and hurricanes, called the *mature stage*. These phases are separated because most disturbances and a few depressions never reach tropical storm intensity, eventually dissipating, whereas mature systems may intensify, remain steady-state, or weaken. About 25% of mature systems evolve into a non-tropical system as they move out of the tropics; these systems may even re-intensify, becoming a storm with sustained winds up to 38 ms^{-1} and impacting regions outside the tropics. The genesis and mature stages will now be described.

Life Cycle

Tropical disturbances form where there is a net inflow of air at the surface, known as *convergence*, resulting in ascent to compensate. As air rises, it will often saturate and form a cloud base. Once the air is saturated, ascent may be enhanced where the atmosphere is in a state of *static instability*. In a *statically unstable* atmosphere, ascending saturated air is less dense than surrounding unsaturated air. As a result, it accelerates upward because the air is buoyant relative to its environment, forming a broad spectrum of vertically growing cumulus clouds.

However, thunderstorm formation is common in the tropics, and is only a prerequisite. Several conditions must simultaneously exist for a tropical disturbance to develop a closed circulation and become a tropical depression. First, the disturbance must be in a *trough*, defined as an elongated area of low *atmospheric pressure*. Atmospheric pressure is the weight of a column of air per area. All troughs away from the equator contain a weak, partial cyclonic rotation. These troughs can develop from a variety of mechanisms related to certain temperature and wind patterns beyond the scope of this entry. Some disturbances undergo the transition to tropical depression directly inside troughs. However, others experience this transition as *tropical waves* (called inappropriately *easterly waves* in the Western Hemisphere), which form when a trough "breaks down" into a cyclonic wave-like pattern in the wind field and travels westward away from the source region. About 60% of the Atlantic tropical cyclones actually originate from tropical waves, which form over Africa and propagate into the Atlantic. Tropical waves are fairly persistent features, and can propagate long distances. In the Atlantic about 55–75 tropical waves are observed, but only 10–25% of these develop into a tropical depression or beyond.

The second condition required for genesis is a water temperature of generally at least 27°C (80°F). Heat and water vapor transferred from the ocean to the air generates and sustains static instability (and therefore thunderstorms) in the disturbance. The third genesis condition is minimal *vertical wind shear*, defined as the difference between wind speed and direction generally at 12 km (40,000 ft) aloft and near the surface. In other words, for genesis to occur, the wind must be roughly the same speed and from the same direction above the surface at all height levels in the atmosphere. This allows thunderstorms and the wind structure to grow unimpeded. The three conditions—warm water, surface trough, and weak vertical wind shear—are generally necessary but insufficient conditions to develop closed rotation (and, by definition, a tropical depression). Furthermore, a few exceptions exist with strong vertical wind shear or water temperature less than 27°C, but a depression still forms. Because of insufficient understanding on the genesis stage, much research is currently underway on this vexing forecast problem.

Once a tropical depression forms, the favorable conditions of wind shear, warm water, and complete cyclonic rotation provide the "ignition" process for further development. The ascending air, developing warm-core aloft, and decreasing surface pressure in the depression stimulates low-level inflow towards the center. The cyclonic circulation of the disturbance increases. Typically, the genesis timeframe of both disturbance and depression lasts for several days or longer. However, under ideal conditions, they can evolve much quicker. When the cyclonic sustained winds increase to 17 ms^{-1} somewhere in the depression, the system is upgraded to a tropical storm. At this point, the mature stage begins.

For tropical storm intensification into a hurricane, the same conditions that allowed its initial development (warm water, moist air, and weak wind shear) must continue. However, a system with closed rotation develops faster compared with the genesis stage because the fluxes of heat and moisture from the ocean become more efficient at faster winds, and because a larger percentage of the heat is retained in the storm center. The column of air begins to warm, which decreases atmospheric pressure. More air will flow toward the lower surface pressure, trying to redistribute the atmosphere's weight, resulting in faster winds. The faster cyclonic winds also enhance convergence. Both factors increase thunderstorm production and low-level inflow. A feedback mechanism now occurs in which faster cyclonic winds maintain thunderstorms by fluxes and convergence, dropping surface central surface pressure more, creating stronger inflow and faster cyclonic winds, and so on. When sustained winds reach 33 ms^{-1} somewhere in the storm, it is classified as a hurricane.

Water temperature is unquestionably linked to these storms' development. Tropical cyclones rarely form over water colder than 27°C. They also weaken dramatically if a mature system moves over water colder than 27°C, or if they make landfall, because their heat and moisture source has been removed. The warmer the water, the greater are the chances for genesis, the faster is the rate of development, and the stronger these storms can become. Because tropical cyclones mix the ocean column, it is also important the warm surface water is at least 60 m (200 ft) thick. Under conditions of prolonged weak shear and an ocean surface temperature greater than 29°C (85°F), sustained winds may reach 89 ms^{-1}.

Fortunately, few hurricanes reach their maximum potential because of some inhibiting factor. Conditions that stop intensification include wind shear, landfall, dry air intrusion, storm-induced ocean mixing or upwelling, and movement over colder water. Occurrence of any (or a combination) of these influences will stall development or cause weakening. Furthermore, even with no inhibiting factors, strong hurricanes rarely maintain their intensity. The internal physics of a hurricane preclude a strong steady-state storm for more than 1–2 days. Instead, strong wind conditions promote interior adjustments near the storm's center, weakening it (see discussion in next section on concentric eyewalls).

Persistent occurrence of any or several inhibiting factors will cause disintegration of a tropical cyclone. Of these possibilities, most dissipating cases occur due to landfall or movement over colder water. Tropical cyclones making landfall rapidly decay. If the storm remains over land, its maximum sustained winds will decrease on average 20 ms^{-1}/day and the rate of dissipation is even faster for initially strong storms. Thirty-six hours after landfall, inland tropical cyclones rarely contain winds above tropical depression strength.

Tropical Cyclone Structure

The structure of a tropical cyclone is one of the most fascinating features in meteorology (Figure 30.1). Distinct cloud patterns exist for each stage of a tropical cyclone's life cycle. These distinct patterns allow meteorology centers worldwide to classify these systems as a depression, tropical storm, hurricane, or major hurricane based on cloud organization. During genesis, typically a mass of thunderstorms with a weak rotation is first observed. Usually these thunderstorms will temporarily dissipate or weaken, leaving a residual circulation. Often, no further development occurs. But should thunderstorms return (sometimes within 12 hours, but more often taking several days), and a closed circulation develops, a depression has formed. A dominant band of clouds gradually takes on more curvature around a cloud minimum center. When the band curves at least one-half distance around

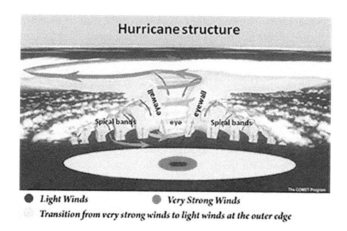

Hurricane structure

● *Light Winds*   ● *Very Strong Winds*

Transition from very strong winds to light winds at the outer edge

FIGURE 30.1 Cross-section of a mature hurricane showing the eye in the center surrounded by the eyewall and spiral bands circling its environment. Arrows depict a surface cyclonic airflow, air rising in the eyewall and spiral bands, air descending between the spiral bands in the eye, and anticyclonic outflow aloft. The circular colors depict the horizontal surface wind structure, in which winds are relatively calm in the eye, fastest in the eyewall, and decrease away from the storm center. [The source of this material is the COMET® Website at http://meted.ucar. edu/ of the University Corporation for Atmospheric Research (UCAR), sponsored in part through cooperative agreement(s) with the National Oceanic and Atmospheric Administration (NOAA), U.S. Department of Commerce (DOC). © 1997–2013 University Corporation for Atmospheric Research. All Rights Reserved.]
Source: Adapted from COMET MetEd.[1]

the storm center, typically tropical storm intensity has been achieved. A cloud shield extends further out with squall lines of thunderstorms of less vertical growth than near the storm center. These thunderstorm bands are known as *spiral bands*. Between the bands is light-to-moderate rain or areas of sinking air (downdrafts).

As the tropical storm strengthens, the dominant cloud band continues coiling around the center. When the band completely coils around the center, hurricane intensity usually has been reached. At this point, a clear region devoid of clouds forms in the center known as the *eye,* surrounded by a ring of thunderstorms known as the *eyewall.* The eyewall contains the fiercest winds and often the heaviest rainfall, making this feature the most dangerous part of a hurricane. The eyewall slants outward with height giving the eye a "coliseum" appearance, as if one is in a giant football stadium made of clouds. In the eye, winds become weak, even calm! This transition from hurricane force winds to calm is rather sudden (often within minutes). The average eye size diameter is 32–64 km (20–40 miles). Typically an eye starts at about 56 km wide during the transition from tropical storm to hurricane. As the hurricane intensifies, the eye usually contracts. Small eyes correlate to intense hurricanes, with diameters as little as 14 km (9 miles). However, intense hurricanes with large eyes also occur, and there is considerable variance with these numbers.

Sometimes a second eyewall forms outside the original eyewall about 64–96 km (40–60 miles) from the center. This outer eyewall "cuts-off" off the inflow to the inner eyewall, causing the inner one to weaken and dissipate. Because the eye is wider, temporary weakening occurs. The outer eyewall will begin to contract inward to replace the inner eyewall, and approximately 12–24 hours later, intensification resumes. This internal adjustment process, known as the *concentric eyewall cycle*, is one reason strong hurricanes experience intensity fluctuations in otherwise favorable conditions.

Outside the eyewall, weaker spiral bands accompanying the hurricane typically affect a large area. The average width of a hurricane's cloud shield is 800 km (500 miles), but varies tremendously. However, cloud size is not an accurate indication of strong wind coverage. One criterion for size is the radial extent of tropical storm-force winds (17 ms^{-1}). Mariners and navy fleets typically avoid winds stronger than 17 ms^{-1}. Furthermore, many hurricane preparedness exercises require completion before tropical storm-force winds begin; for example, bridges are often closed if winds exceed 17 ms^{-1}, impeding last-minute evacuees.

Damage from Tropical Cyclones

Coastal communities devastated by strong hurricanes usually take years to recover. Many forces of nature contribute to the destruction. Obviously, tropical cyclone winds are a source of structural damage. As winds increase, stress against objects increases at a disproportionate rate. For example, a 22-ms^{-1} wind causes a stress of 36 kg m^{-2} (7.5 lb ft^{-2}). In 45-ms^{-1} winds, stress becomes 147 kg m^{-2} (30 lb ft^{-2}). Although sustained wind is used as a reference standard, the actual wind will be 20–30% faster or slower than the sustained wind at any instantaneous period. These gusts often initiate structure damage, typically beginning at the roof. The removal of roof coverings often occurs with wind gusts at 36 ms^{-1}, roof decks at 45 ms^{-1}, and roof structure at 54 ms^{-1}. Wind interacting with a building is deflected over and around it. Positive (inward) pressures are applied to the windward walls and try to "push" them down. Negative (outward) pressures are applied to the side and leeward walls. The resulting "suction" force can peel away any exterior covering (siding, shingles, and so on). Negative (uplift) pressures are applied to the roof, especially along windward eaves, roof corners, and leeward ridges, similar to the lifting aerodynamics on aircraft wings, and cause roof failure.

Building damage occurs in other ways. Wind damage begins with exterior items such as television antennas, satellite dishes, unanchored air conditioners, wooden fences, gutters, storage sheds, carports, and yard items. As the wind speed increases, cladding items on buildings become susceptible to wind damage including vinyl siding, gutters, roof coverings, windows, and doors. Debris is also propelled by strong winds, compounding the damage. The weak points to internal building exposure are usually through roof, window, door, or garage damage. Any cause which compromises the inside of a building usually results in extreme damage to that structure from rain and wind. Damage patterns tend to be uneven due to the streakiness of wind gusts; building construction quality; building age; wind interference, wind acceleration, or vortex shedding from other buildings; proximity to open fields, trees, or vegetation. In addition to structure damage, large tree branches will snap, trees will be uprooted, and power poles will be toppled (or power lines snapped), resulting in power outages which can last weeks.

Floods produced by the rainfall can also be quite destructive, and are a leading cause of tropical cyclone-related fatalities. A majority (57%) of the 600 U.S. tropical cyclone-related deaths between 1970 and 1999 were associated with inland flooding. Tropical cyclone rainfall averages 150–300 mm (6–12 in.) at landfall regions, but varies tremendously. Heavy rainfall is not just confined to the coast. The remnants of tropical cyclones can bring heavy rain far inland, and is particularly dangerous in hills and mountains where acute concentrations of rain turn tranquil streams into raging rivers in a matter of minutes or cause mudslides. In addition, mountains "lift" air in tropical cyclones, increasing cloud formation and rainfall. Rainfall rates of 300–600 mm (1–2 ft) per day are not uncommon in mountainous regions when tropical cyclones pass through.

Although all these elements (wind, rain, floods, and mudslides) are dangerous, historically most people have been killed in the *storm surge*, defined as an abnormal rise of the sea along the shore associated with cyclonic wind storms such as tropical cyclones. Death tolls in coastal regions can be terrible for those who do not evacuate inundation zones. Most storm surge fatalities are associated with structural collapse or drowning. The worst natural disaster in the United States history occurred in 1900 when a hurricane-related 5 m (16 ft) storm surge inundated Galveston Island, TX, U.S.A., and claimed over 6000 lives. In 1893, nearly 2000 were killed in Louisiana and 1000 in South Carolina by two separate hurricanes. Camille in 1969 (181 fatalities), Katrina in 2005 (1836 fatalities), and Ike in 2008 (20–40 fatalities) are the latest U.S. examples.

However, these U.S. statistics pale compared with the lives taken globally. Regions around the North Indian Ocean, Japan, and China have experienced fatalities by storm surge ranging from 10,000 to 300,000 in the last few centuries, with the latest tragedy of 138,000 killed in Myanmar by Cyclone Nargis in 2008. In contrast, the 1900 Galveston hurricane ranks 16th globally in terms of storm surge fatalities based on the latest research.

Storm surge is caused by several factors. The main contribution to the storm surge results from the interaction of coastal water with wind-driven water at landfall known as the *wind stress effect*. In deep offshore water, a vertical ocean circulation occurs, which compensates surface water wind transport, and no storm surge from wind forcing exists. But as a tropical cyclone approaches the shoreline and begins to interact with the ocean floor and land boundaries, this circulation is disrupted and water levels must rise to compensate. As a tropical cyclone approaches a region, coastal waters begin to rise gradually from the wind stress effect, then quickly at landfall.

A current also develops parallel to the shoreline ahead of the tropical cyclone, causing water to rise in response to the Earth's rotation, known as a *surge forerunner*. The forerunner effect is dangerous because it peaks before landfall, sometimes trapping residents before they complete evacuation. For example, many of the Hurricane Ike fatalities occurred when roads on the Bolivar Peninsula flooded one day before landfall, cutting off the only escape route.

The final factor to storm surge is a minor contribution associated with the low pressure of a hurricane, which causes a bulge of water known as the *inverse barometer effect*. For every 10 mb pressure drop, water rises 10 cm (3.9 in.). The contribution ranges from 0.3–1 m (1–3 ft) elevated water depending on intensity, peaking at landfall.

Other factors that determine a surge's height include coastal bathymetry, storm intensity, storm size, and storm translation speed (Figure 30.2). All other factors being equal, the most intense tropical cyclones produce higher storm surge. However, for a given intensity, bathymetry causes tremendous variation in surge heights. A coastline with a shallow sloping ocean floor is prone to higher storm surge than a coastline with a steep ocean shelf nearby. For example, a Category 3 produces a surge of 1.5 m (5 ft) in very deep bathymetries but 6.7 m (22 ft) in very shallow, with most values in between for a given coastline. Low-lying regions adjacent to extended shallow seas, such as the Gulf of Mexico in the southern United States, and the Bay of Bengal bordering Bangladesh and India, are particularly vulnerable to the storm surge. In shallow bathymetries, large, slow-moving tropical cyclones produce more surge than average-sized, average speed hurricanes for the same intensity (Figure 30.2). The storm surge is always highest on the side of the eye corresponding to onshore winds, which is usually the right side of the point of landfall in the Northern Hemisphere, called the *right front quadrant*.

The total elevated water includes two additional components—the astronomical tide and ocean waves. The astronomical tide results from gravitational interactions between the Earth, moon, and sun, producing high and low tides every 12–24 hours. Should the storm surge coincide with high tide, additional inundation occurs. The total water elevation due to surge and tides is known as the *storm tide* (Figure 30.2), but in practice it is difficult to distinguish from storm surge during post-storm inspections. Waves are superimposed on top of the storm surge, running up past inundation boundaries, spilling over structures, sloshing against and inside structures, and enhancing damage. Waves may also contribute to inundation through a process called *wave setup* in which incoming waves exceed retreating waves, and is currently the subject of considerable research.

Water is very heavy, weighing 1025 kg m^{-3} (64 lb ft^{-3}), and dense. Therefore, waves possess much greater force than wind, and can undermine foundations and destroy support walls. Furthermore, if a wave is breaking, the force is 4–7 times greater than moving water. In addition, waves also cause an "uplift" force if the waves are under a horizontal structure, displacing the object and even toppling it, including bridge decks. Even without wave forces, water is very damaging. Buoyancy causes a lifting force on submerged wood buildings and beam foundations, damaging structures. Storm surge currents contain forces equal to or slightly greater than wind. Destruction can be thorough, leaving only slabs

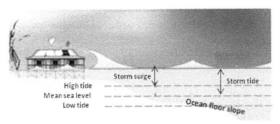

Storm surge at high tide

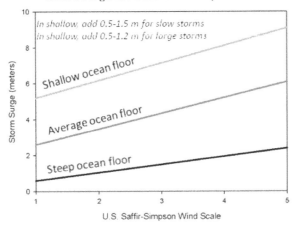

Storm surge for different bathymetries

FIGURE 30.2 (Top) Graphical portrayal of the storm surge at high tide impacting a structure along a shoreline. The definition of storm tide includes the storm surge plus water elevation departures from mean sea level due to the tide cycle. Waves are superimposed on the surge. (Bottom) Storm surge relationship to ocean floor slope and hurricane intensity using the U.S. wind scale. Note the large differences between shallow and steep bathymetry for a given intensity. Storm surge is 0.5–1.5 m higher for slow storms in shallow water, and is 0.5–1.2 m higher for large storms in shallow water. Storm size and speed only marginally modifies surge elevation in average and steep ocean floors, and is neglected.
Source: Top figure adapted from Australian Bureau of Meteorology.[2] Bottom figure adapted from Fitzpatrick et al.[3]

behind near the coast. Erosion of beaches and islands also is severe, and coastal highways can be devastated. Further inland where currents are weaker with little wave activity, inundation is still damaging. Most possessions, floors, and drywall that are submerged will be ruined, plus mold will grow on the items when the water retreats.

The United States has developed the *Saffir Simpson hurricane wind scale,* which describes the expected level of wind damage for a given hurricane intensity. This scale classifies hurricanes into five categories according to maximum sustained wind and escalating potential property damage. Categories 1–5 correspond to a lower threshold of 33, 43, 50, 58, and 70 ms^{-1}, respectively. Although all categories are dangerous, categories 3, 4, and 5 are considered *major hurricanes,* with the potential for widespread devastation and loss of life. Whereas only 24% of U.S. landfalling tropical cyclones are major hurricanes, they historically account for 85% of the damage. However, while low-intensity hurricanes cause less physical devastation, preventative measures are still required during a landfall threat. Low-intensity hurricanes also disrupt regional economic activity for several days to several weeks through business interruption and evacuation costs. Additionally, freshwater flood threat is present for any tropical storm or hurricane regardless of intensity.

It should be noted that storm surge ranges used to be included in the Saffir–Simpson scale, but were removed in 2010 since surge depends not only on wind but bathymetry, storm size, and translation speed. Central pressure was also removed because it does not correspond to eyewall winds but instead correlates to wind structure. The wind scale also does not include damage from floods, and small-scale intense features such as imbedded tornadoes. Because damage is dependent on so many factors, preparations for a tropical cyclone impact should be thorough.

Global Tropical Cyclone Prediction, Climatology, and Terminology

Different naming nomenclatures related to intensity classifications are used worldwide, and can be confusing. A global context of tropical cyclones will now be discussed. Tropical cyclones generally occur in every tropical ocean except the South Atlantic and eastern South Pacific (Figure 30.3). Locations include the tropical North Atlantic Ocean (including the Caribbean Sea and Gulf of

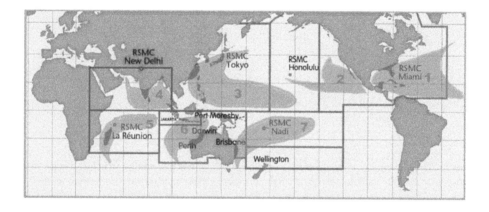

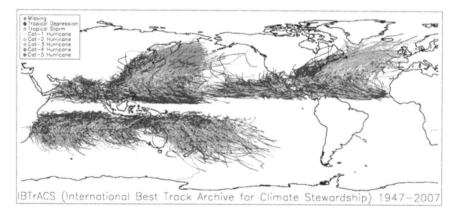

FIGURE 30.3 (Top) Global genesis regions (in light gray), the Regional Specialized Meteorological Centers (RSMCs), and Tropical Cyclone Warning Centers (TCWCs; Wellington, Brisbane, Perth, Darwin, Port Moresby, and Jakarta). For example, the National Weather Service National Hurricane Center and the Japan Meteorological Agency Typhoon Center are RSMCs. (Bottom) Global tropical depression, tropical storm, and hurricane tracks by U.S. Saffir–Simpson wind categories for 1947–2007. "Missing" indicates a position was available but intensity data was missing.
Source: Top figure adapted from World Meteorological Organization.[4] Bottom figure courtesy of Knapp et al. adapted from.[5] Bottom figure is © Copyright March 2010 AMS.

Mexico); the Northeast Pacific (off the west coast of Mexico); the Central North Pacific (near Hawaii); the Northwest Pacific (including the China Sea, Philippine Sea, and Sea of Japan); the North Indian Ocean (including the Bay of Bengal and the Arabian Sea); the Southwest Indian Ocean (off the coasts of Madagascar and extending almost to Australia); the Southeast Indian Ocean (off the northwest coast of Australia); and the Southwest Pacific Ocean (from the east coast of Australia to about 140°W). In addition, with improved satellite monitoring capabilities, several (mostly) weak, short-lived tropical cyclones in the South Atlantic Ocean have been unofficially identified the last two decades. Cyclone Catarina (2004) is the only official South Atlantic tropical cyclone with category 2 winds, named after the state in Brazil it made a landfall.

Regional Specialized Meteorological Centers (RSMCs) and Tropical Cyclone Warning Centers (TCWCs) monitor and forecast tropical cyclones used by each country, territory, and national meteorological service in their region (Figure 30.3). RSMCs provide a spectrum of weather forecast products and responsibilities, which include hurricane forecasts, whereas TCWCs specialize in tropical cyclone services. Official warnings are the responsibility of each country/territory's national meteorological service. Monitoring is conducted with surface observations from land-based instruments; buoys and ships; upper-air balloon instruments, launched generally twice per day; satellite imagery; and satellite-derived diagnostics. Because these observations do not provide complete data coverage, the United States is the only nation to also use expensive reconnaissance aircraft.

Forecast agencies issue predictions for storm track, intensity, rainfall, storm surge, and other parameters. The priority is predicting tropical cyclone movement. To understand tropical cyclone motion, it is helpful to use the analogy of a wide river with a small eddy rotating in it. The river generally transports the eddy downstream, but the eddy will not move straight because the speed of the current varies horizontally, causing a slight left or right deviation, which also changes the eddy's speed of movement. Furthermore, this eddy's rotation may alter the current in its vicinity, which in turn will alter the motion of the eddy.

Likewise, one may think of a tropical cyclone as a vortex embedded in a river of air. The orientation and strength of large-scale pressure patterns generally dictate the storm's motion, except that the steering depends on both the horizontal and vertical distribution of the steering current. The tropical cyclone can also interact with steering currents, and even other nearby tropical cyclones, ultimately altering its own currents. Other factors impact track such as the Earth's rotation and storm asymmetry, and are beyond the scope of this entry. This is a difficult forecast problem, especially in data-void regions such as the ocean.

The complexity of track—as well as intensity and storm surge—predictions requires the use of computer models. Computer models ingest current weather observations and approximate solutions to complicated equations for future atmospheric values such as wind, temperature, and moisture. Forecast agencies use a suite of models that differ in their mathematical assumptions and complexities in describing atmospheric processes. The most complex models must be run on the fastest computers in the world, known as supercomputers.

Many U.S. residents perceive the North Atlantic Ocean basin as a proliferate producer of hurricanes due to the publicity these storms generate. In reality, several oceans produce more hurricanes annually than the North Atlantic. For example, the most active ocean basin in the world—the Northwest Pacific—averages 17 hurricanes per year. The second most active is the Northeast Pacific, which averages nine hurricanes. In contrast, the North Atlantic mean annual number of hurricanes is six. Table 30.1 summarizes each basin's average number of hurricanes and total tropical cyclones using U.S. definitions.

Tropical cyclone season is typically limited to the warm seasons. In the Atlantic, the official tropical cyclone season begins June 1 and ends November 30, although activity has been observed outside this timeframe. However, tropical cyclones are most numerous and strongest in late summer and early fall. Exceptions to this late summer/early fall peak in tropical cyclone activity occur in certain parts of the world such as India since their monsoon trough moves inland during the summer. In addition, while activity does peak in late summer, the Northwest Pacific tropical cyclone season lasts all year.

TABLE 30.1 The Mean Number of Total Tropical Cyclones (Hurricanes and Tropical Storms), Hurricanes, and Major Hurricanes Per Year in All Tropical Ocean Basins Using U.S. Definitions

Tropical Ocean Basin	Mean Annual Tropical Storms and Hurricanes	Mean Annual Hurricanes	Mean Major Hurricanes
Northwest Pacific	26	17	8
Central North Pacific and Northeast Pacific	17	9	5
East Coast Australia and Southwest Pacific	10	5	2
West Coast Australia and Southeast Indian	8	4	1
North Atlantic	12	6	2
Southwest Indian	9	5	2
North Indian	5	2	Between 0 and 1
South Atlantic	0	0	0
Southeast Pacific	0	0	0
Global	86	47	20

Improved satellite monitoring capabilities have also unofficially identified several (mostly) weak, short-lived tropical cyclones in the South Atlantic Ocean the last two decades. Cyclone Catarina (2004) is the only official South Atlantic tropical cyclone with category 2 winds, named after the state in Brazil it made landfall. The official average is zero in the South Atlantic, but apparently a tropical cyclone forms every few year.

The classification schemes vary around the globe by assorted names, different time averaging for sustained winds, and diverse wind thresholds. Figure 30.4 summarizes global classifications. They all recognize classes from genesis to strong storms with consistent thresholds of 17 and 33 ms^{-1}, but use their own terms and wind thresholds for other delineations. For example, in the Northwest Pacific Ocean, the designation at 33 ms^{-1} is "typhoon," in India is "very severe cyclonic storm," and off Australia is "severe tropical cyclone." The rough equivalent term for major hurricanes in the North India Ocean is "super cyclonic storm," and in the Southwest Indian Ocean is "very intense tropical cyclone," although the wind thresholds are different. Finally, in the most active global region, the Northwest Pacific Ocean, storms with winds greater than 67 ms^{-1} are common enough that the Joint Typhoon Warning Center uses a special category exists for them—"supertyphoon."

Australia also uses a different wind damage scale than the United States (Figure 30.4). Their category 1 begins at 17 ms^{-1} for minor damage, escalating to "some roof and structure damage" (Australian category 3), "significant damage" (Australian category 4), and "widespread damage" (Australian category 5). The U.S. Saffir–Simpson wind scale does not include tropical storms, the Australian wind scale of 3 is approximately the same as category 1 in America, Australian category 4 is approximately category 2–3 in America, and Australian category 5 is approximately category 4–5 in America.

When 17 ms^{-1} is reached, a name is assigned in most basins. However, the number of lists varies, as does whether a revolving cycle is used. For example, the Atlantic basin storms use a 6-year repeating list. However, the Northwest Pacific uses five lists, which do not rotate annually, but simply goes to the next list when the last name is reached. Whenever a storm has had a major impact in terms of damage and/or fatalities, any country affected by the storm can request the name be "retired" by agreement of the World Meteorological Organization.

All names are determined at international meetings, and reflect the regional culture. For example, Northeast Pacific storms tend to have Hispanic names, whereas central North Pacific storms have Hawaiian names. In regions with multiple countries, names will vary reflecting the different cultures. Southwest Indian tropical cyclone names are based on French, Madagascar, and many African countries. Northwest Pacific uses Asian nomenclature that do not reflect personal names but instead natural objects (such as trees, flowers, rivers, or animals), "stormy" adjectives (such as swift, strong, sharp, fast),

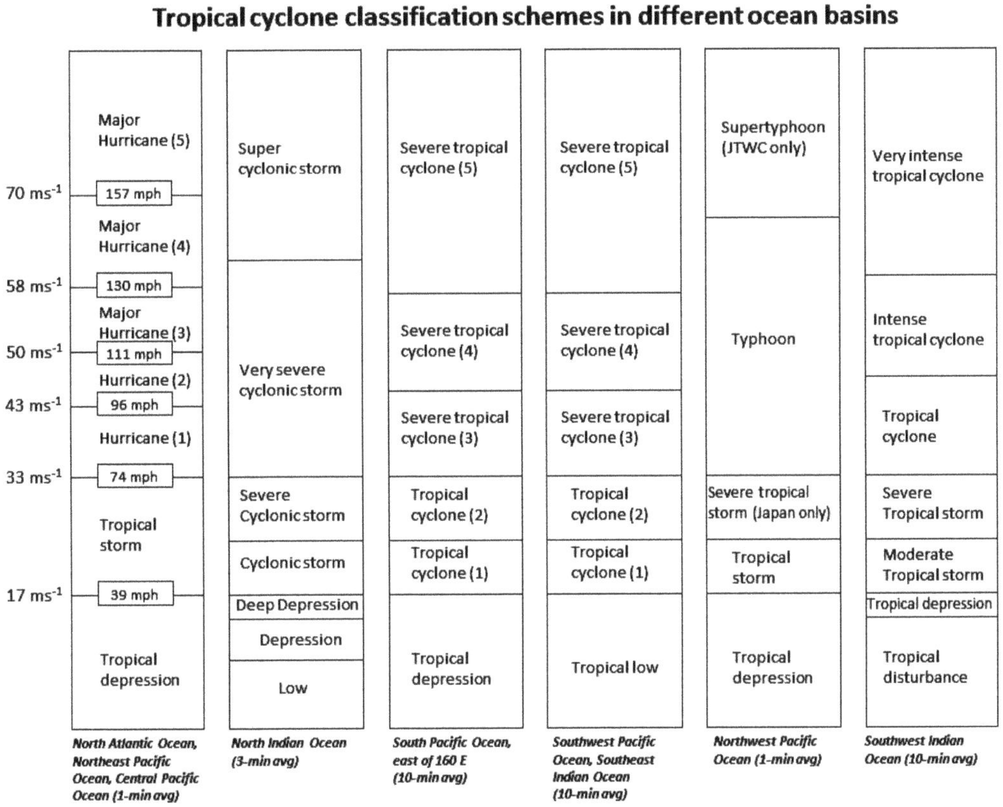

Tropical cyclone classification schemes in different ocean basins

Wind	North Atlantic Ocean, Northeast Pacific Ocean, Central Pacific Ocean (1-min avg)	North Indian Ocean (3-min avg)	South Pacific Ocean, east of 160 E (10-min avg)	Southwest Pacific Ocean, Southeast Indian Ocean (10-min avg)	Northwest Pacific Ocean (1-min avg)	Southwest Indian Ocean (10-min avg)
	Major Hurricane (5)	Super cyclonic storm	Severe tropical cyclone (5)	Severe tropical cyclone (5)	Supertyphoon (JTWC only)	Very intense tropical cyclone
70 ms⁻¹ — 157 mph						
	Major Hurricane (4)					
58 ms⁻¹ — 130 mph						
	Major Hurricane (3)	Very severe cyclonic storm	Severe tropical cyclone (4)	Severe tropical cyclone (4)	Typhoon	Intense tropical cyclone
50 ms⁻¹ — 111 mph						
	Hurricane (2)					
43 ms⁻¹ — 96 mph			Severe tropical cyclone (3)	Severe tropical cyclone (3)		Tropical cyclone
	Hurricane (1)					
33 ms⁻¹ — 74 mph						
	Tropical storm	Severe Cyclonic storm	Tropical cyclone (2)	Tropical cyclone (2)	Severe tropical storm (Japan only)	Severe Tropical storm
		Cyclonic storm	Tropical cyclone (1)	Tropical cyclone (1)	Tropical storm	Moderate Tropical storm
17 ms⁻¹ — 39 mph		Deep Depression				Tropical depression
	Tropical depression	Depression	Tropical depression	Tropical low	Tropical depression	Tropical disturbance
		Low				

$$ $$

FIGURE 30.4 Classification schemes for the six ocean regions where tropical cyclone terminology varies from genesis to intense categories. The U.S. definitions are used for reference in the leftmost column. Also shown are the wind damage scales used in the United States and in Australia. Wind averaging schemes are shown for each basin. One-minute averaging results in winds that are approximately 14% more than 10-minute average winds (1-minute winds = 1.14 times 10-minute winds). Wind units are provided in ms⁻¹ and mph on the left side.

and gods or goddesses. In addition, the Philippines assign local names to tropical cyclones in their region, even though it will also have an "official" international name, resulting in two names for the same storm. The two purposes for names are to facilitate easy regional communication of a storm between forecasters and the general public, and for historical context.

Excellent Background Information and Imbedded Links on Hurricanes Are Available at the Following Websites (Accessed September 2, 2013)

1. The National Hurricane Center: http://www.nhc.noaa.gov
2. Australian Bureau of Meteorology: http://www.bom.gov.au/cyclone/
3. India Meteorological Department: http://www.imd.gov.in/section/nhac/dynamic/cyclone.htm
4. World Meteorological Organization Tropical Cyclone Programme: http://www.wmo.int/pages/prog/www/tcp/index_en.html
5. Frequently Asked Questions About Hurricanes: http://www.aoml.noaa.gov/hrd/tcfaq/tcfaqHED.html

6. International Best Track Archive for Climate Stewardship: http://www.ncdc.noaa.gov/oa/ibtracs/index.php
7. Historical Hurricane Tracks: http://www.csc.noaa.gov/hurricanes/

References

1. COMET MetEd. Tropical cyclones. In *Introduction to Tropical Meteorology, 2nd edition*; Chapter 8, 2013. Available at https://www.meted.ucar.edu/training_module.php?id=868
2. Australian Bureau of Meteorology. Storm surge preparedness and safety, 2013. Available at http://www.bom.gov.au/cyclone/about/stormsurge.shtml
3. Fitzpatrick, P.J.; Lau, Y.; Tran, N.; Li, Y.; Hill, C.M. The role of bathymetry in local storm surge potential. Nat. Hazards Earth Syst. Sci. **2013**. In revision.
4. World Meteorological Organization. Organizational structure, 2013. Available at http://www.wmo.int/pages/prog/www/tcp/organization.html. Also see http://www.wmo.int/pages/prog/www/tcp/Advisories-RSMCs.html.
5. Knapp, K.R.; Kruk, M.C.; Levinson, D.H.; Diamond, H.J.; Neumann, C.J. The International Best Track Archive for Climate Stewardship (IBTrACS): Unifying tropical cyclone best track data. Bull. American Meteor. Soc. **2010**, *91*, 363–376.

Bibliography

Fitzpatrick, P. J.; *Hurricanes – A Reference Handbook*; ABC-CLIO: Santa Barbara, CA, 2005; 412 pp.

31

Urban Heat Islands

James A. Voogt
*University of
Western Ontario*

Introduction

The urban heat island (UHI) represents the relative warmth of the air, surface, and subsurface in urban areas compared to their nonurbanized surroundings. It arises from the consequences of urban development on the surface and atmospheric characteristics in urban areas and represents an unintentional change to the climate of cities. The magnitude (or intensity) of an UHI is defined by the difference between urban and nonurban temperatures.

Types of UHIs and Their Characteristics

The relative warmth of the urban atmosphere, surface, and substrate materials leads to several distinct heat islands (Figure 31.1, Table 31.1).[1,2] The *urban canopy-layer (UCL) heat island* is the best-known heat island.[3] It describes the warmth of a near-surface air layer that extends from the ground to the mean height of the buildings and vegetation. It is the most commonly measured layer of the atmosphere in cities for air temperatures because it is easily accessed by ground-based instruments that are either fixed or mobile (e.g., on vehicles). Canopy-layer heat islands are typically the largest at night (Figure 31.2) during weather conditions in which winds are calm and skies are clear. Heat island intensities under such conditions are typically a few degrees Celsius in large parts of most cities and may exceed 10°C for the most densely developed parts of a large city. They are smaller and sometimes may even slightly negative (representing an urban "cool island") during clear daytime conditions; together, these indicate a strong temporal variability of the heat island over the course of a day under calm, clear weather conditions.[3] When averaged over all weather conditions, the heat island magnitude in the UCL is typically 1°C–3°C.

The spatial structure of the heat island shows a pattern of isotherms—isolines of equal temperature—that follow the border of the city with a relatively tight spacing when winds are calm, and hence the topographic analogy with an island (Figure 31.2). The spatial temperature gradient, the change of temperature over space, is typically large near the edge of the "island," forming the so-called cliff in response to the large relative change in surface characteristics in the region of rural-to-urban transition. Within the urban area, temperatures in the UCL are strongly controlled by the local characteristics of the urban surface. They may show substantial spatial variability (on the order of several °C) when weather conditions are favorable. The highest nighttime temperatures, the peak of the heat island, are

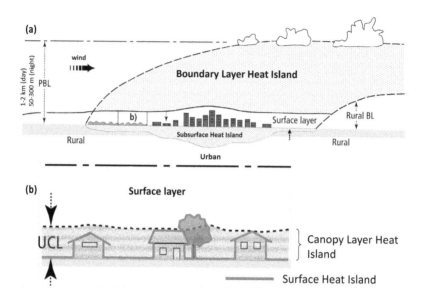

FIGURE 31.1 Schematic diagram showing the different heat island types and their locations within an urban area. Shading represents the affected volume, but it does not show the expected spatial variations. (a) Under fair weather conditions, within the first 1–2 km of the atmosphere, known as the planetary boundary layer (PBL), a boundary-layer heat island exists with a characteristic plume shape extending in the downwind direction. (b) Within the UCL air volume, a canopy-layer heat island exists (warmer air temperatures). The surface heat island consists of all surfaces, but usually only a subset of these surfaces can be seen from any one observation point. Below the surface, a subsurface heat island exists. (Adapted from Oke 1997.)

usually associated with the area of most intense urban development, and some warmer air is often transported horizontally downwind of the city.

Above the UCL, the urban *boundary-layer heat island* represents an urban-scale warming through the depth of the urban boundary layer (up to 1–2 km during daytime with clear skies and a few 10s to 100s of meters at night) that has a smaller magnitude and is much less spatially and temporally variable than that of the underlying UCL heat island. The heat island magnitude here is positive both day and night. The warmed boundary-layer air above the city is often transported downwind by the mean wind leading to a plume of warmer air above downwind non-rural areas (Figure 31.1). Measurements of the urban boundary-layer heat island are relatively rare as access to this layer is difficult. Thermometers must be mounted on tall towers, balloons, or aircraft; or temperatures can be observed by remote sensing techniques using ground-based instruments (Table 31.1).

The *surface UHI* represents the relative warmth of the surfaces in cities compared to that in rural areas. Surface temperatures can be observed using remote sensing techniques from satellite or aircraft-mounted instruments. These instruments provide a spatially continuous view of the urban area but tend to be biased toward viewing the highest upward-facing unobstructed surfaces such as roofs, open spaces, and the tops of trees.[4] The implications of this method of observing the heat island must be borne in mind when interpreting remotely sensed images of the surface UHI. Under clear skies and light winds, daytime surface temperatures seen from above are much warmer than air temperatures in the canopy layer and show much more spatial variability (Figure 31.2 and Table 31.1) due to the juxtaposition of surfaces with contrasting properties in urban areas (e.g., hot dry rooftops adjacent to an irrigated lawn). However, the variability is not always easily observed—for example, when the sensors have spatial resolutions that are substantially larger than the scale of the surface structure that provides the temperature variability. This can be the case when using satellite observations of surface temperature. The daytime surface heat island is positive (Figure 31.2) and is controlled by the amounts of impervious

TABLE 31.1 Heat Island Types and Their Spatial and Temporal Characteristics

Heat Island Type	Variable Affected	How Measured	Spatial Characteristics	Temporal Characteristics
Canopy layer heat island	Air temperature	Fixed or mobile (vehicle-based) measurements using thermometers.	Shows significant spatial variability associated with important elements of surface structure: building height-to-width ratio, amount of vegetation, topographic features.	Largest at night. Magnitude grows rapidly in the late afternoon and early evening. May be negative (a cool island) during the daytime. Highly sensitive to wind and cloud.
Boundary-layer heat island	Air temperature	Thermometers mounted on very tall towers, balloons, kites, or aircraft. Remote-sensing from ground-based instruments.	Exhibits a "domed" structure in near-calm conditions and a distinct downwind "plume" as winds increase. Boundary-layer depth is 1–1.5 km by day, but only 50–300 m at night. Heat island magnitude decreases with height in the boundary layer by night and is approximately constant by day.	Shows relatively small diurnal variation. Sensitive to wind and cloud.
Surface heat island	Surface temperature	Remote sensing from towers, aircraft, or satellites.	Significant spatial variability associated with variations in surface characteristics, including shading, surface orientation, moisture status, thermal properties, surface reflectivity, and vegetation coverage. Variability of rural surface temperature is also large and affects heat island magnitude.	Largest during daytime in the summer season. Nighttime value is also positive and largest in summer. Highly sensitive to weather conditions. Varies with season especially if there are significant changes to moisture or vegetation characteristics or where substantial winter heating is used.
Subsurface heat island	Ground temperature	Ground or borehole temperature measurements.	Spatial variability occurs due to variations in surface characteristics and subsurface heat sources from urban infrastructure. Urban areas show a greater depth of temperature decrease from the surface before temperatures reverse to show the geothermal gradient.	Temporal response is increasingly lagged with depth so that subsurface patterns reflect past conditions at the surface. Temporal influences below the first meter are typically only seasonal or longer.

and vegetated surfaces as well as the surface characteristics of the rural reference. At night, the overall spatial structure and magnitude of the surface heat island are similar to those of the canopy-layer heat island, reflecting the importance of surface controls (Table 31.2) on the near-surface air temperature. The spatially averaged surface heat island magnitude under favorable weather conditions is larger by day than at night,[5] making it the reverse of that in the canopy-layer air. The use of simple correlations to relate surface and air temperatures and their heat islands is problematic without the use of fully coupled surface–atmosphere energy budget models that can represent the physical processes that govern the exchanges of energy between the surface and the air.

The *subsurface UHI* is influenced by the relative warmth of the surface and the atmosphere above it as well as by contributions of heat from the basements of buildings and subsurface infrastructure. It

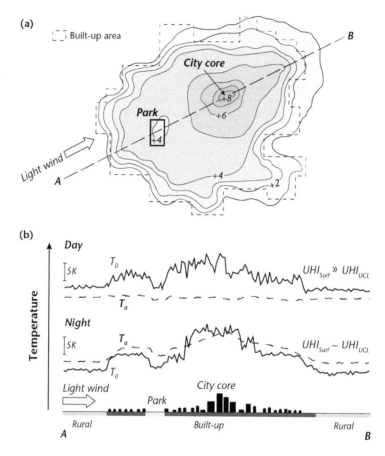

FIGURE 31.2   Top: A plan view of the nighttime canopy-layer heat island. Bottom: Conceptual cross sections of the canopy layer and surface UHIs by day and night. (Oke et al. 2017, with permission.)

TABLE 31.2   Factors That Lead to the Formation of and/or Influence the Magnitude of UHIs

Heat Island Influence	Effect on UHI Magnitude
Surface geometry	UHI magnitude increases as the ratio of building height to street width increases and view of the nighttime sky is obstructed.
Surface thermal properties	UHI magnitude increases with materials that make the city a better storer of heat; those with higher heat capacity and/or thermal conductivity relative to rural materials. Variations in rural moisture can influence UHI magnitude.
Anthropogenic heat input	UHI magnitude increases as anthropogenic heat increases. This input can have large seasonal variations in some climates as well as intraurban spatial variability related to the density of development and magnitude of energy use.
City size	UHI magnitude tends to increase with city size up to a limiting amount.
Wind speed	UHI magnitude decreases rapidly as wind speed increases.
Cloud cover	UHI magnitude decreases as cloud cover increases.
Season	UHI magnitude is typically the largest in the warm season in mid-latitudes. In high latitudes, the UHI is the largest in the winter due to anthropogenic heat input. In tropical cities with distinct wet and dry seasons, the UHI is typically largest in the dry season.
Time of day	The canopy-layer UHI is the largest at night (air temperatures). The surface UHI magnitude is larger during the day (clear, sunny conditions).

Source: Adapted from Arnfield.[6]
These factors are most directly related to the canopy-layer UHI but also affect other UHI types.

too shows spatial variability, although much less than that of the surface, and it decreases with depth. It can be measured from temperature measurements deep in the soil or from boreholes or wells. Where substrate materials are water-saturated, subsurface flows may also yield a warm plume associated with the subsurface heat island.

The exact spatial configuration and behavior of each heat island type in a given city depend on the layout of the city and its topographic setting; cities in coastal or mountainous areas will be influenced by sea–valley or mountain–valley breeze systems that can significantly impact the spatial pattern of the heat islands and their temporal development.[1]

Causes of the UHI: The Urban Energy Budget

The UHI arises from the fact that urban areas modify their surface and atmospheric characteristics relative to the surrounding rural regions. These alterations result in a modified *surface energy budget* in cities. The energy budget describes the exchanges of energy at the Earth's surface: how the radiation energy arriving at the city surface from the Sun and atmosphere is absorbed and partitioned into heat energy that is used to warm the air through mixing (convective sensible heat), heat energy used to evaporate water from the surface or to transpire water from vegetation (convective latent heat), and heat that is conducted and stored in the substrate materials. As a general rule, urban areas tend to increase both the relative amount of energy used to warm the air and that which is stored in the urban substrate materials, and to decrease the relative amount of energy directed toward evapotranspiration (the combined evaporation and transpiration of water) compared to the energy budget of surrounding nonurban areas. In addition, cities also directly add heat to their atmosphere through the use of energy. Building heating and cooling, electricity use, and transportation are important contributors to this anthropogenic energy component.[7] Therefore, we can say that altered energy budgets underlie UHI formation. To better understand the reasons for urban energy budget changes, we turn to an examination of the surface characteristics of urban areas.

Causes of the UHI: Characteristics of the Urban Surface

The surface characteristics of urban areas provide important controls on the surface energy budget and therefore on the UHI. Three groups of urban surface characteristics are important. First, the surface coverage and relative fractions of buildings, vegetation, and impervious surfaces are important. Buildings and other impervious surfaces such as roads are designed to shed water so that energy absorbed by these surfaces is directed only toward heating the material and the air above it. As the relative fraction of buildings and impervious surfaces increases and vegetation decreases, greater fractions of energy will be directed toward warming the air and substrate because evapotranspiration from vegetation becomes more limited.

Second, the structure or form of the surface as determined by the dimensions and spacing of buildings and trees is important to UHI formation. By day, the urban surface is better at absorbing solar radiation than most rural surfaces. Absorption is enhanced by several factors: the reflectivity of the surface to solar radiation (known as the albedo)—which may be less (darker) for some surfaces (such as asphalt roads and dark-colored roofs) compared to the rural surroundings—and the three-dimensional structure of the urban surface, which serves to increase the effective area of the surface that can absorb sunlight and which also serves to "trap" some radiation that is initially reflected off of individual surfaces. At night, the three-dimensional structure of the surface serves to effectively block the view of the sky for many parts of the urban surface, especially within the UCL (e.g., the roadway separating adjacent tall buildings), and thereby reduces the loss of radiative heat energy. The sky view factor that describes the relative obstruction of the sky to the surface is strongly related to the UHI magnitude of the UCL air.[1] The structure of the urban surface is also responsible for altering winds near the surface; this provides both sheltering effects and increased turbulence (mixing). During the day, the effective mixing of the atmosphere assisted by the rough surface helps to warm the urban boundary layer with heat from the surface. At night, the sheltering of air between buildings contributes to the increased warmth of the nighttime UCL heat island.

Third, the properties of materials that affect temperature, such as radiative, thermal, and moisture characteristics, differ in urban areas from their rural surroundings. Many urban materials are drier, have higher heat capacities, and may be darker (lower albedo). This allows them to more readily absorb solar energy, to store it, and to have higher temperatures because no energy is used to evaporate water. Urban materials thus tend to act as a sink for heat during daytime and a source of heat at night. The material properties contribute to a temporal lag in the warming of urban areas in the morning and a similar lag in their cooling at night.

The widely used Land Cover Zone classification provides a climatically relevant land-cover classification scheme for cities and their surroundings.[8] It explicitly recognizes the importance of surface characteristics to variations in urban-scale climate and is a highly useful tool for planning and assessing UHI studies, especially those related to the UCL.

Weather and Climate Influences

Surface and atmospheric heat islands are significantly impacted by weather conditions. All heat islands are best expressed under clear and calm conditions when surface controls that affect microclimates have their maximum effect.[3] As winds increase, the heat island magnitude in the UCL decreases exponentially, and as cloud cover increases, it decreases linearly. Winds also transport warmer urban boundary-layer air downwind and affect the diurnal development of the UHI. Seasonally, heat island magnitudes are typically largest in the summer or dry season when the input of solar energy to the surface is high.[6] In some special cases, such as in very high-latitude settlements, the heat island may be largest in winter when there is a large anthropogenic heat input to the atmosphere from space heating requirements. Table 31.2 summarizes the factors that affect heat island magnitude. These apply most directly to the canopy-layer heat island, but also affect the other heat island types.

Heat Island Impacts

UHIs have a range of impacts. Of particular concern are health impacts arising from warmer urban temperatures, especially in the context of combined large-scale climate warming, which is expected to lead to more frequent, longer, and more intense heat wave episodes. The additional warmth of the UHI exacerbates temperatures in cities, and moreover, the UHI magnitude is often larger during heatwaves.[9] The UCL warmth is of particular concern during hot summer nights, because it is then when the heat island effects are largest, and the physiological impacts on the body become important when it is unable to achieve a period of cooling at night during sleep. Prolonged exposure to such conditions can be a serious health threat, and there have been several incidents of significant urban-enhanced mortality during heat waves, for example, in Paris in the summer of 2003 and Russia in 2010, both associated with mortalities in the tens of thousands.[10]

Daytime boundary-layer heat islands can also impact air quality through an enhancement of smog (ozone) formation rates within the boundary layer. These formation rates are temperature dependent and are more efficient at higher temperatures. Higher temperatures also provide two positive feedbacks that can further exacerbate air quality. Warmer temperatures drive a greater demand for electricity to power cooling needs that can lead to both emissions of more pollutants and greenhouse gases from power plants that use fossil fuels and at the same time increase natural (biogenic) emissions of smog-forming pollutants. Increased demand for space cooling at these times is one synergistic factor as additional anthropogenic heat flux exacerbates the atmospheric heat island. Summer heat islands can also exacerbate the demand for water use in cities.

More positively, the UHI can reduce the incidence of cold exposure to urban inhabitants in winter. It also increases the length of the growing season and alters the timing associated with the stages of plant development. At the same time, the length of the frost season is reduced. Associated with the relative warmth is a decrease in the need for energy required for space heating of homes and businesses.

The UHI plays a role in modifying the weather a city experiences. The relative warmth of the UHI can help induce a local circulation pattern, known as a city breeze, analogous to a sea breeze, in which cooler rural air is transported horizontally into the city. This wind system can have consequences for pollutant dispersion and urban design for pathways for the cleaner, cooler rural air. The relatively warm urban boundary-layer air provides a deeper layer into which pollutants may be mixed and can also provide for a higher base to convective clouds that may form over the urban area. The warmer urban atmosphere can, in select cases, provide a greater likelihood of warm-season convective cloud formation and possible increases in precipitation over and downwind of the city.[11] Impacts from high precipitation rate events can be magnified in urban areas because of the large impervious surface fraction that increases the runoff from such events and can lead to flash-flooding events. In the cold season, the heat island may influence the relative fraction of rain to snow events.

Future changes to UHIs are of interest because an increasing majority of the world's population is expected to live in urban areas. The associated increased conversion of land to urban use, intensification of the urban development, and increases in urban energy use are the factors that can promote an increase in UHI magnitude. Future climate change is expected to provide additional thermal stress to urban inhabitants because of the increased frequency of hot summer nights in urban areas relative to rural areas.[12] Under global climate warming forced by increases in greenhouse gas concentrations, the absolute value of urban (and rural) temperatures is expected to increase, but UHI magnitudes may not necessarily increase. This situation occurs because changes to the frequency of weather conditions that favor heat island formation and to the moisture conditions of surrounding rural areas may occur that can favor either increased or decreased heat island magnitude, depending on the location of the city in question and the regional impacts of large-scale climate change. When the expansion of urban areas is incorporated in the assessment, the combined warming due to climate change and that from the urban expansion is less than the sum of their individual contributions, but overall, there is a large projected urban warming.[13]

Urban-scale modifications to the characteristics of the urban surface, which involve so-called cool technologies such as high-reflectivity surfaces for reducing the absorption of the Sun's energy, light-weight materials that reduce energy storage, thermochromic materials, and increases in vegetative cover, are increasingly being considered in many cities to ameliorate the negative impacts of warmer urban temperatures associated with both the UHI and the large-scale climate change. However, many of these infrastructure-based adaptations may be unable to fully offset urban nighttime warming with climate change reinforcing the need for mitigation of greenhouse gas emissions to address climate change.[13] The optimal design of cities with respect to climate and the UHI is a challenge because of the dichotomy that arises from compact cities and their efficient greenhouse gas emissions versus greater potential heat stress and air pollutant densities.[14]

References

1. Oke, T.R.; Mills, G.; Christen, A.; Voogt, J.A. *Urban Climates*; Cambridge University Press: Cambridge, 2017.
2. Oke, T.R. Urban Environments. In The Surface Climates of Canada; Bailey, W.G.; Oke, T.R.; Rouse, W.R., Eds.; McGill-Queen's University Press: Montreal, 1997; 303–327.
3. Oke, T.R. The energetic basis of the urban heat island. *Q. J. R. Meteorol. Soc.* **1982**, *108*, 1–24.
4. Voogt, J.A.; Oke, T.R. Thermal remote sensing of urban climates. *Rem. Sens. Environ.* **2003**, *86*, 370–384.

5. Imhoff, M.L.; Zhang, P.; Wolfe, R.E.; Bounoua, L. Remote sensing of the urban heat island effect across biomes in the continental USA. *Rem. Sens. Environ.* **2010**, *114*, 504–513.

6. Arnfield, A.J. Two decades of urban climate research: A review of turbulence, exchanges of energy and water, and the urban heat island. *Int. J. Climatol.* **2003**, *23*, 1–26.

7. Sailor, D. A review of methods for estimating anthropogenic heat and moisture emissions in the urban environment. *Int. J. Climatol.* **2011**, *31*, 189–199.

8. Stewart, I.D.; Oke, T.R. Local climate zones for urban temperature studies. *Bull. Am. Meteorol. Soc.* **2012**, *93*, 1879–1900.

9. Zhao, L.; Oppenheimer, M.; Zhu, Q.; Baldwin, J.W.; Ebi, K.L.; Bou-Zeid, E.; Guan, K.; Liu, X. Interactions between urban heat islands and heat waves. *Environ. Res. Lett.* **2018**, *13*(3), 034003.

10. Barriopedro, D.; Fischer, E.M.; Luterbacher, J.; Trigo, R.M.; Garcia-Herrera, R. The hot summer of 2010: Redrawing the temperature record map of Europe. *Science*, **2011**, *332*(6026), 220–224.

11. Shepherd, J.M.; Stallins, J.A.; Jin, M.L.; Mote, T.L. Urbanization: Impacts on Clouds, Precipitation and Lightning. In *Urban Ecosystem Ecology*; Aitkenhead-Peterson, J.; Volder, A. Eds.; Agron. Monogr. 55., ASA, CSSA, and SSA: Madison, WI, 2010; 1–28.

12. McCarthy, M.P.; Best, M.J.; Betts, R.A. Climate change in cities due to global warming and urban effects. *Geophys. Res.Lett.* 2010, 37, L09705. doi: 10.1029/2010GL042845.

13. Krayenhoff, E.S.; Moustaoui, M.; Broadbent, A.; Gupta, V.; Georgescu, M. Diurnal interaction between urban expansion, climate change and adaptation in US cities. *Nat. Clim. Change* **2018**, *8*, 1097–1103.

14. Martilli, A. An idealized study of city structure, urban climate, energy consumption, and air quality. *Urban Clim.* **2014**, *10*, 430–446.

32

Wind Speed Probability Distribution

Adam H. Monahan
University of Victoria

Introduction

The fact that the atmosphere is in ceaseless motion across the Earth's surface is of fundamental importance to a broad range of physical, biological, and chemical processes in the Earth system, as well as being of societal and economic significance. Lower atmospheric winds exercise an important control on the fluxes of mass, momentum, and energy between the atmosphere and the underlying surface. Extreme surface winds represent hazards to both the built and natural environments. Furthermore, the moving air is a reservoir of energy, which has been used by humankind for millennia and is a central element in the portfolio of energy resources alternative to fossil fuels. An elementary fact about surface winds is that they are not steady: the motion of the air changes on timescales from seconds through to days, years, and beyond. This variability is a key aspect of many of the contexts in which winds are of are importance. For example, as the wind power density is proportional to the cube of the wind speed, the mean power density is not that produced by the mean wind speed: higher order statistical moments of the wind speed influence the mean power. It is therefore important to develop mathematical tools to characterize this variability in the wind.

The most natural of such tools is the probability distribution, which describes the range of values a quantity takes and their relative likelihood. As the velocity of the air is a continuous variable (rather than one taking discrete values), we can represent the probability distribution in terms of the associated probability density function (pdf). In particular, denoting the wind speed as w and its pdf as $p_w(w)$, the probability of observing a speed w between the values w_1 and w_2 is

$$\text{prob}(w_1 \leq w \leq w_2) = \int_{w_1}^{w_2} p_w(w)\,dw. \qquad (32.1)$$

The pdf carries all information about the range and relative likelihood of wind speeds. In many contexts, the statistical moments are required. These are defined by integrals over $p_w(w)$, and include the mean, standard deviation, and skewness:

$$\text{mean}(w) = \int_0^\infty w p_w(w)\, dw \tag{32.2}$$

$$\text{std}(w) = \text{mean}\left\{ (w - \text{mean}(w))^2 \right\} \tag{32.3}$$

$$\text{skew}(w) = \text{mean}\left\{ \left(\frac{(w - \text{mean}(w))}{\text{std}(w)} \right)^3 \right\}. \tag{32.4}$$

Different moments carry information about different aspects of wind variability. The mean and standard deviation are measures of magnitude: they represent respectively the overall magnitude of the wind speed and the strength of the variability around it. In contrast, the skewness is a measure of the shape of the distribution. A skewness value of zero indicates that the pdf is symmetric around its mean, so anomalies (relative to the mean) of equal size and opposite sign are equally likely. A positive skewness indicates that positive anomalies of a given size are more likely than negative values of the same magnitude; for negative skewness, the asymmetry favors negative anomalies. Other measures of shape exist, such as kurtosis (a measure of the flatness of the distribution); as these are more sensitive to sampling variability than lower-order moments, they are less often used in practical applications.[1] As noted earlier, the moments of w enter into the calculation of mean wind power density Π: assuming a constant air density ρ,

$$\text{mean}(\Pi) = \frac{1}{2}\rho \int_0^\infty w^3 p_w(w)\, dw = \frac{1}{2}\rho \left(\text{mean}^3(w) + 3\,\text{mean}(w)\text{std}^2(w) + \text{skew}(w)\text{std}^3(w) \right). \tag{32.5}$$

Computations of the mean power R captured by a turbine of cross-sectional area A,

$$R = \frac{1}{2}A\rho \int_0^\infty P(w) p_w(w)\, dw, \tag{32.6}$$

account for the physical response of the turbine through the power curve $P(w)$.[2] The integral in Equation (32.6) cannot generally be expressed as a simple function of the moments of w: computation of R requires knowledge of the full distribution of wind speeds.

Physical controls on wind variability involve processes on scales from global-scale differential heating to the millimeter-scale dissipation of turbulence kinetic energy by viscosity. These processes are reviewed in standard treatments of dynamic meteorology[3,4] or boundary layer meteorology.[5,6] While energy is present across all timescales from seconds through years to beyond, it is not evenly distributed across scales. There are astronomically forced peaks in variability at the daily and annual scales: winds often display marked diurnal and seasonal variations. As well, in many settings it is observed that there are concentrations of energy on timescales less than several minutes and greater than an hour or two. Between these is a "spectral gap" that allows us to distinguish between faster, smaller-scale "turbulent" winds and slower, larger-scale "eddy averaged" winds.[6] It is the second of these, which is generally defined in terms of winds averaged on timescales from 10 minutes to an hour and that is the focus of this review.

There is an extensive literature on the probability distribution of velocities in turbulent flow.[6] For eddy-averaged winds, the vertical component of the wind (relative to the local surface) is generally much smaller than the horizontal components. As such, we will neglect the vertical wind component and write the wind speed in terms of the orthogonal vector wind components u and v as $w = \sqrt{u^2 + v^2}$. Velocity is a dimensional quantity, and can be reported in a range of different units. For concreteness, we will refer to velocities in ms^{-1}. The vector components have their own pdfs, which are closely related to that of wind speed (as will be discussed in the later section, "Relation of Wind Speed and Vector Wind Component pdfs"). Because it is of importance to a broader range of applications, the primary focus of this entry will be the pdf of wind speed. Furthermore, this entry will consider the entire probability distribution of surface winds without specific discussion of extreme events in the tail of the pdf. A general discussion of extreme events is presented in the entry "Climate: Extreme Events" in this encyclopedia.[7]

General Features of the Wind Speed Pdf

There are a number of basic constraints that must be satisfied by any wind speed pdf. These are as follows:

1. $p_w(w) = 0$ for $w < 0$ ms^{-1} (wind speeds must be positive).
2. $p_w(w) \geq 0$ for $w \geq 0$ ms^{-1} (any pdf must be nonnegative).
3. $p_w(w)$ must be normalized to unity (the wind speed must take some value):

$$\int_0^\infty p_w(w)\,dw = 1 \qquad (32.7)$$

4. $p_w(0) = 0$ unless the joint pdf of the vector wind components is infinite at $(u, v) = (0, 0)$ ms^{-1}. This point is addressed in more detail in the section, "Relation of Wind Speed and Vector Wind Component pdfs."

From the normalization constraint (Equation 32.7), it follows that the units of the pdf $p_w(w)$ are the inverse of the units of w. That is, for w measured in ms^{-1}, the units of $p_w(w)$ are m^{-1}s.

Beyond these general requirements of any surface wind speed pdf, there are a number of other features, which are generally observed. Mean wind speeds over the oceans are generally larger than those over land, with the largest values found in the middle latitudes and the subtropics. Variability in w is generally largest in the midlatitude storm tracks.[4] Over land, $p_w(w)$ is generally positively skewed;[8] over the oceans, positive skewness is found in the Northern Hemisphere middle latitudes, near to zero skewness over the Southern Ocean, and negative skewness in much of the tropics and subtropics.[9] Over land, mean(w) and std(w) are generally larger during the day than at night, as a result of the influence of surface stratification on the downward transport of momentum.[5,10] In coastal areas, this diurnal cycle can be reversed over water as a result of changes in stratification due to horizontal temperature advection.[11] Land surface winds tend to be more strongly skewed at night than during the day.[8] Wind speed pdfs generally have a single maximum value (i.e., there is a single most likely speed), although in regions of complex topography multimodal wind speed pdfs are observed.[12,13]

Wind observations often include a nonzero frequency of calms (times when $w = 0$ ms^{-1}) resulting from a finite cut-off speed below which w cannot be measured by the anemometer. To account for this, hybrid pdfs in which a standard parametric distribution is augmented with a delta function at $w = 0$ are sometimes used.[14] As these calms result from limited instrumental precision rather than a true nonzero probability of $w = 0$ ms^{-1}, such hybrid distributions will not be further considered in this review.

Empirical Models of the Wind Speed Probability Distribution

A number of parametric distributions have been suggested for the empirical modelling of the wind speed probability distribution.[14–18] The most common of these in present use is the two-parameter Weibull distribution:

$$p_w^{wbl}(w) = \frac{b}{a}\left(\frac{w}{a}\right)^{b-1}\exp\left[-\left(\frac{w}{a}\right)^b\right] \quad w \geq 0. \tag{32.8}$$

The parameters a and b denote, respectively, the scale and shape. We can use the notation w ~ W(a, b) to indicate that w is Weibull with parameters a and b. Explicit formulas for the moments of a Weibull distributed variable follow from the general result:

$$\text{mean}(w^k) = a^k\Gamma\left(1+\frac{k}{b}\right), \tag{32.9}$$

where $\Gamma(x)$ is the gamma function. In particular:

$$\text{mean}(w) = a\Gamma\left(1+\frac{1}{b}\right) \tag{32.10}$$

$$\text{std}(w) = a\left[\Gamma\left(1+\frac{2}{b}\right) - \Gamma^2\left(1+\frac{1}{b}\right)\right]^{1/2} \tag{32.11}$$

$$\text{skew}(w) = \frac{\Gamma\left(1+\frac{3}{b}\right) - 3\Gamma\left(1+\frac{1}{b}\right)\Gamma\left(1+\frac{2}{b}\right) + 2\Gamma^3\left(1+\frac{1}{b}\right)}{\left[\Gamma\left(1+\frac{2}{b}\right) - \Gamma^2\left(1+\frac{1}{b}\right)\right]^{1/2}}. \tag{32.12}$$

Both skew(w) and the ratio mean(w)/std(w) are independent of the scale parameter a: for a Weibull distributed variable, the skewness is a unique function of the ratio of the mean to the standard deviation. To a good approximation,

$$b = \left(\frac{\text{mean}(w)}{\text{std}(w)}\right)^{1.091}. \tag{32.13}$$

A number of methods exist for estimating the parameters of the Weibull distribution from observations.[14,19] Among these, the most straightforward is the method of moments, which makes use of Equations (32.10) and (32.13).

Observations indicate that $b \simeq 2$ is the most common value of the shape parameter over land.[8] The special case of $p_w^{wbl}(w)$ with $b = 2$ is the Rayleigh distribution:

$$p_w^{ral} = \frac{w}{\sigma^2}\exp\left(-\frac{w}{2\sigma^2}\right). \tag{32.14}$$

While the Weibull pdf is often a useful model for $p_w(w)$, it is not an exact characterization and deviations of various degrees from Weibull behavior are common.[14,20]

It is worth noting that if w is Weibull, so is w^k: w ~ W(a, b) implies w^k ~ W(a^k, b/k). In particular, if the speed is Weibull, then the wind power density is as well. Furthermore, if w is Rayleigh: w ~ W(a, 2) then

$w^{1/2} \sim W(a^{1/2}, 4)$ has a skewness near to zero. This result is consistent with the fact that over land surfaces, $w^{1/2}$ can often be approximated as Gaussian.[21]

It is also possible to model the joint distribution of wind speed and direction, $p_{w\theta}(w, \theta)$, with specified marginal distributions $p_w(w)$ and $p_\theta(\theta)$:

$$P_{w\theta}(w, \theta) = 2\pi g \left(2\pi F_w(w) - 2\pi F_\theta(\theta) \right) p_w(w) p_\theta(\theta). \tag{32.15}$$

In Equation (32.15), $F_w(w)$ and $F_\theta(\theta)$ are respectively the cumulative distribution functions of w and θ, and $g(\zeta)$ is a pdf on $0 \le \zeta < 2\pi$, which specifies the dependence of w and θ. [22,23]

Relation of Wind Speed with Vector Wind Component Pdfs

A natural question in the discussion of the wind speed pdf is its relation to the pdf of vector wind components, as was noted in many of the earliest studies of $p_w(w)$.[24–26] Vector winds can be described in terms of either their components u and v (e.g., eastward and northward) or in their speed w and direction θ:

$$(u, w) = (w\cos\theta, w\sin\theta). \tag{32.16}$$

(where the orientation of $\theta = 0$ is arbitrary and can be chosen for convenience). We will denote the joint pdf of (u, v) by $p_{uv}(u, v)$. While $p_{uv}(u, v)$ and $p_{w\theta}(w,\theta)$ are different functions (with different units), these must be related as the description of vector winds in terms of components is equivalent to that in terms of speed and direction. The probability that the vector winds fall within a given range should be independent of the coordinate system in which we choose to describe them, so

$$p_{uv}(u, v) du\, dv = p_{w\theta}(w, \theta) dw\, d\theta, \tag{32.17}$$

where the infinitesimal areas $du\, dv$ and $dw\, d\theta$ are related through the standard change of variables from Cartesian to plane polar coordinates: $du\, dv = w\, dw\, d\theta$. It follows that the relationship between the joint distributions is

$$p_{w\theta}(w, \theta) = w p_{uv}(w\cos\theta, w\sin\theta). \tag{32.18}$$

To obtain the marginal pdf for w alone, we integrate $p_{w\theta}(w, \theta)$ over θ to obtain:

$$p_w(w, \theta) = w \int_{-\pi}^{\pi} p_{uv}(w\cos\theta, w\sin\theta) d\theta. \tag{32.19}$$

From this result, we can relate any expression for the pdf of the vector wind components to that of the wind speed. As well, we see that $p_w(0) \ne 0$ only if $p_{uv}(0, 0)$ is infinite (which is not observed).

If we make the approximation that the vector winds are Gaussian with mean $(\bar{u}, \bar{v})$ and isotropic, uncorrelated variability of standard deviations σ, then

$$p_{uv}^{gau}(u, v) = \frac{1}{2\pi\sigma^2} \exp\left(-\frac{(u-\bar{u})^2}{2\sigma^2} - \frac{(v-\bar{v})^2}{2\sigma^2} \right). \tag{32.20}$$

Evaluation of the integral in Equation (32.19) yields the Rice distribution

$$p_w^{rice}(w) = \frac{w}{\sigma^2} \exp\left(-\frac{(\bar{U}^2 - w^2)}{2\sigma^2} \right) \left(\frac{(\bar{U}w)}{\sigma^2} \right), \tag{32.21}$$

Where $\bar{U}^2 = \bar{u}^2 + \bar{v}^2$ and $I_k(z)$ are the associated Bessel function of the first kind of order k.[27] In this case, the mean and standard deviation of the wind speed are given by

$$\text{mean}(w) = \sqrt{\frac{\pi}{2}} \sigma \exp\left(\frac{-\bar{U}^2}{4\sigma^2}\right)\left[\left(1 + \frac{\bar{U}^2}{2\sigma^2}\right)\right.$$

$$\left. I_0\left(\frac{\bar{U}^2}{4\sigma^2}\right) + \frac{\bar{U}^2}{2\sigma^2} I_1\left(\frac{\bar{U}^2}{4\sigma^2}\right)\right] \tag{32.22}$$

$$\text{std}(w) = \sqrt{2\sigma^2 + \bar{U}^2 - \text{mean}^2(w)}. \tag{32.23}$$

It is possible to generalize Equation (32.21) to allow for unequal variances in the vector wind components, but the resulting expressions (involving infinite series of Bessel functions) are awkward.[15,28] Non-Gaussianity in the vector wind components can also be accounted for, although the resulting expressions are either analytically intractable or violate the positivity constraint of the section, "General Features of the Wind Speed pdf." Nevertheless, it has been shown that non-Gaussianity in the vector winds must be accounted for to accurately model wind speed skewness over the oceans.[15]

Mechanistic Models of the Wind Speed Probability Distribution

While most studies of the wind speed pdf have been empirical, recent efforts have been made to develop physically based models of $p_{uv}(u, v)$ and $p_w(w)$. Through this approach, the pdfs are expressed in terms of dynamically meaningful parameters rather than abstract statistical parameters. For example, Monahan used an idealized slab model of the sea surface boundary layer momentum budget driven by fluctuating large-scale pressure gradients[9] to express $p_w(w)$ in terms of the boundary layer depth, boundary layer top entrainment velocity, surface drag coefficient, and pressure gradient statistics.[9] He et al. used a generalized version of this model[8] to demonstrate that surface buoyancy fluxes and the character of the land surface have important influences on the pdf of wind speed.[8] A subsequent analysis suggested that the long positive tail of the nighttime wind speed pdf is a result of intermittent turbulent mixing at the top of the normally quiescent nocturnal boundary layer.[20] Much more work remains to be done on this problem; the mechanistic study of surface wind pdfs remains in its infancy.

Conclusions

The wind speed probability distribution is a central ingredient in studies of wind hazards, wind energy assessment, and fluxes between the atmosphere and the underlying surface. While the structure of $p_w(w)$ can be constrained by some fundamental general requirements, it is not known to be fully characterized by any single family of distributions. Characterizations of $p_w(w)$ can be empirical (by parametric distributions such as the Weibull), or derived from representations of the joint distribution of the vector wind components. An emerging area of study seeks to develop physically based models of $p_w(w)$. These approaches offer distinct and complementary insights into the pdf of surface winds.

References

1. Wilks, D.S. *Statistical Methods in the Atmospheric Sciences; Academic Press*: 2005.
2. Burton, T.; Jenkins, N.; Sharpe, D.; Bossanyi, E. *Wind Energy Handbook*; Wiley: Chichester, UK, 2011.
3. Holton, J.R. *An Introduction to Dynamic Meteorology*; Academic Press: 2004.
4. Peixoto, J.P.; Oort, A.H. *Physics of Climate*; American Institute of Physics: New York, 1992.

5. Arya, S.P. *Introduction to Micrometeorology*; Academic Press: 2001.
6. Stull, R.B. *An Introduction to Boundary Layer Meteorology*; Kluwer: Dordrecht, 1997.
7. Sura, P. Extreme climate events. In *Encyclopedia of Natural Resources;* Wang, Y., Ed.; Taylor and Francis: 2012.
8. He, Y.; Monahan, A.H.; Jones, C.G.; Dai, A.; Biner, S.; Caya, D.; Winger, K. Land surface wind speed probability distributions in North America: Observations, theory, and regional climate model simulations. J. Geophys. Res. **2010**, *115*, doi: 10.1029/2008JD010708.
9. Monahan, A.H. The probability distribution of sea surface wind speeds. Part I: Theory and sea winds observations. J. Clim. **2006**, *19*, 497–520.
10. Dai, A.; Deser, C. Diurnal and semidiurnal variations in global surface wind and divergence fields. J. Geophys. Res. **1999**, *104*, 31109–31125.
11. Barthelmie, R.J.; Grisogono, B.; Pryor, S.C. Observations and simulations of diurnal cycles of near-surface wind speeds over land and sea. J. Geophys. Res. **1996**, *101*, 21327–21337.
12. Romero-Centeno, R.; Zavala-Hidalgo, J.; Gallegos, A.; O'Brien, J.J. Isthmus of Tehuantepec wind climatology and ENSO signal. J. Clim. **2003**, *16*, 2628–2639.
13. Jiménez, P.A.; Dudhia, J.; Navarro, J. On the surface wind probability density function over complex terrain. Geophys. Res. Lett. **2011**, *38*, L22803, doi:10.1029/2011GL049669.
14. Carta, J.A.; Ramírez, P.; Velázquez, S. A review of wind speed probability distributions used in wind energy analysis. Case studies in the Canary Islands. Ren. Sust. Energy Rev. **2009**, *13*, 933–955.
15. Monahan, A.H. Empirical models of the probability distribution of sea surface wind speeds. J. Clim. **2007**, *20*, 5798–5814.
16. Hennessey, J.P. Some aspects of wind power statistics. J. Appl. Meteor. **1977**, *16*, 119–128.
17. Li, M.; Li, X. MEP-type distribution function: a better alternative to Weibull function for wind speed distributions. Ren. Energy **2005**, *30*, 1221–1240.
18. Morrissey, M.L.; Greene, J.S. Tractable analytic expressions for the wind speed probability density functions using expansions of orthogonal polynomials. J. App. Meteor. Clim. **2012**, *51*, 1310–1320.
19. Justus, C.G.; Hargraves, W.R.; Mikhail, A.; Graber, D. Methods for estimating wind speed frequency distributions. J. Appl. Meteor. **1978**, *17*, 350–353.
20. Monahan, A.H.; He, Y.; McFarlane, N.; Dai, A. The probability distribution of land surface wind speeds. J. Clim. **2011**, *24*, 3892–3909.
21. Brown, B.J.; Katz, R.W.; Murphy, A.H. Time series models to simulate and forecast wind speed and wind power. J. Clim. Appl. Meteor. **1984**, *23*, 1184–1195.
22. Johnson, R.A.; Wehrly, T.E. Some angular-linear distributions and related regression models. J. Am. Stat. Assoc. **1978**, 73, 602–606.
23. Carta, J.A.; Ramírez, P.; Bueno, C. A joint probability density function of wind speed and direction for wind energy analysis. Energy Conversation. Manag. **2008**, *49*, 1309–1320.
24. Brooks, C.E.P.; Durst, C.S.; Carruthers, N. Upper winds over the world. Part I, The frequency distribution of winds at a point in the free air. Q. J. Roy. Met. Soc. **1946**, *72*, 55–73.
25. Crutcher, H.L. On the standard vector-deviation wind rose. J. Meteor. **1957**, *14*, 28–33.
26. Davies, M. Non-circular normal wind distributions. Quart. J. Roy. Meteor. Soc. **1958**, *84*, 277–279.
27. Rice, S.O. Mathematical analysis of random noise (part 2). Bell Syst. Tech. J. **1945**, 24, 46–156.
28. Weil, H. The distribution of radial error. Ann. Math. Stat. **1954**, 25, 168–170.

III

Climate Change

33

Climate Change

Robert L. Wilby
University of Loughborough

Introduction

Climate change is an abstract concept in the sense that no one can feel a 30-year average rainfall total or global mean temperature. Yet these types of indexes are widely used by scientists to describe the state of the climate system (Figure 33.1). However, climate averages are accumulated from day-to-day weather, which is perceptible, as are meteorological variations from pole to the equator or from sea level to summit. Retreating ice sheets, increasing ocean heat content, and rising sea levels are further manifestations of a warming planet.

Climate change is also a plastic concept, because the term means very different things to different people.[1] More than a century ago, atmospheric warming linked to "carbonic acid" (now termed carbon

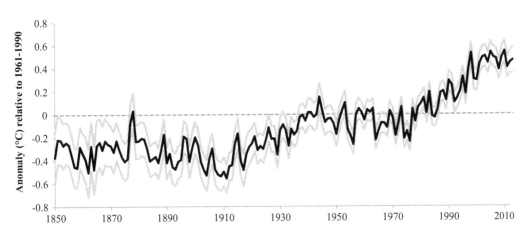

FIGURE 33.1 Global annual mean temperature anomalies (black line) based on analyses of surface data (HadCRUT4) showing lower and upper bounds of the 95% confidence interval of the combined effects of measurement and sampling, bias, and coverage uncertainties (gray lines). Annual anomalies are calculated with respect to the 1961–1990 mean.
Source: UK Met Office, http://www.metoffice.gov.uk/hadobs/hadcrut4/.

dioxide or CO_2) was regarded by Arrhenius as beneficial by delaying the onset of the next ice age.[2] In contrast, many modern-day governments and their citizens view climate change as a present and/ or imminent threat to be managed. Others do not see the same urgency or question whether current (policy and economic) responses are proportionate to the established risk.

This entry begins with an overview of the natural and human drivers of climate variability and change over decades to millennia. Global climate models and their projections are then described. The final section considers the technical challenges surrounding formal detection and attribution of climate change. These are all areas of rapid scientific development and vigorous debate.

Natural Causes of Climate Change

Natural causes of climate change are conventionally described as either external or internal to the climate system. External forces affect the amount of solar radiation entering the upper atmosphere; internal forces determine the amount of radiation that actually reaches the surface of the Earth. This is distinct from internal natural climate variability, which arises from the interplay between ocean and atmosphere processes in response to radiative forcing.

Even minor changes in solar output and receipt of radiation by the Earth are sufficient to trigger large responses in the planet's ocean–atmosphere system. Variations in solar radiation output are linked to levels of sunspot activity. The number of sunspots varies, on an average, over 11-year, 22-year, 80–90-year, and longer timescales. However, the length of individual sunspot cycles varies between 10 and 12 years. During periods of high sunspot activity and rapid cycles (such as the 1950s), solar output increases by about one-tenth of a percent relative to the long-term average of 1370 Wm^{-2}. Conversely, periods with fewer sunspots and slower cycles have coincided with notable cool interludes, such as the Little Ice Age, which was harshest during the Maunder minimum of 1645 to 1715.

Solar activity is not the only determinant of the Earth's solar energy balance. By the mid-19th century, Agassiz recognized that Alpine glaciers had once been much more extensive; Croll's astronomical theory of Ice Ages provided an explanation. Calculations by Milanković subsequently showed how variations in the eccentricity (stretch of the Earth's elliptical orbit), obliquity (tilt of the Earth's axis), and precession of the equinoxes (wobble of the North Pole orientation) affect receipt of solar radiation by the upper atmosphere. The rhythm of these cycles has since been corroborated by oxygen isotope analysis of deep sediment cores as 23,000 years for precession, 41,000 years for obliquity, and 96,000 years for eccentricity. Rapid uplift of mountain chains and/or changes in ocean currents due to continental drift can trigger glacial epochs over even longer periods.

The amount of solar radiation received at the Earth's surface depends partly on the composition of the atmosphere. Violent volcanic eruptions like Tambora (1815), Krakatoa (1883), Santa Maria (1902), Novarupta (1912), Mount St. Helens (1980), and Pinatubo (1991) inject huge quantities of sulfur dioxide into the stratosphere. Atmospheric processes convert the gas into aerosols that can linger and reflect radiation for several years. For example, the eruption of Mount Pinatubo added an estimated 10 teragrams of sulfur (TgS) to the stratosphere and caused a global cooling of ~0.5°C up to 18 months later.[3] Volcanic eruptions also provide opportunities to observe the effect of changing atmospheric temperatures on water vapor content—a potent climate feedback mechanism represented in "climate models." Moreover, data from eruptions have been used to show the potential efficacy and side effects of solar radiation management via artificial seeding of the atmosphere with sulfate aerosols (a controversial geoengineering option).

Internal climate variability is generated by major transfers of energy, mass, and momentum within the climate system. Some exchanges exhibit organization at the hemispheric or even planetary scales. For example, the El Niño is characterized by warm waters and high rainfall extending westward across the Pacific from Peru and Ecuador. Quasiperiodic oscillations between warm and cool (La Niña) phases occur every 4 to 7 years. The influence of El Niño is expressed most forcefully by drought in northeast Brazil, eastern Australia, Indonesia, and India, but greater-than-average rainfall over southwest and southeast United States. The North Atlantic Oscillation (NAO) and Pacific Decadal Oscillation (PDO) are further ocean–atmosphere variations that modulate regional rainfall anomalies on interannual to interdecadal timescales. Positive NAO is generally associated with strong westerly airflows, warmer- and wetter-than-average winter rainfall, and flood-rich episodes across northwest Europe. When a positive PDO and El Niño coincide, the Pacific coast dipole (dry northwest–wet southwest) is amplified.

Other harmonics of natural climate variability such as the Atlantic Multidecadal Oscillation (AMO) occur over even longer timescales. Long instrumental and paleoclimate records show that sea surface temperatures fluctuate through anomalously warm and cool phases, each lasting several decades. The recent warming of the North Atlantic to a condition not seen since the 1950s is thought to explain the spate of unusually wet summers in northern Europe, and hot, dry summers in southern Europe since the 1990s.[4] Similarly, very long reconstructions of rainfall and river flow indicate decades—even centuries—of drought in North America. For example, drought-sensitive tree-ring chronologies suggest four "mega droughts" in the Great Plains between AD 1 and 1200. These events have no modern counterpart, and their origins are poorly understood.[5] Some explanations refer to changes in the strength and location of the North Atlantic thermohaline circulation; others theories suggest that El Niño/La Niña effects may have been amplified by external climate forcing.

Human Causes of Climate Change

Anthropogenic climate change arises from changes in the composition of the atmosphere and/or modifications to surface energy, moisture, carbon fluxes, and stores.

By the 1850s and 1860s, Tyndall had determined that CO_2 (and other greenhouse gases [GHGs] such as nitrous oxide, methane, and water vapor) selectively absorbs solar radiation in the infrared band. The atmospheric concentration of CO_2 was about 280 ppm prior to the Industrial Revolution; by 2013, the concentration had exceeded 400 ppm for the first time at the Mauna Loa Observatory in Hawaii. Emissions from fossil fuel combustion and land use changes are adding to the CO_2 concentration and thereby strengthening radiative forcing. What remains less clear is how laboratory measurements scale up to a global warming response, once feedbacks are taken into account (e.g., from changing cloud, snow, or ice cover). Climate sensitivity is typically defined as the expected global warming by doubling the effective CO_2 concentration. The term "effective" is used because the sensitivity implicitly captures forcing by other GHGs. The value of the sensitivity parameter is important because it strongly determines the amount of climate change projected by models and is a point of contention.

Arrhenius calculated that either halving or doubling the concentration of CO_2 would change global mean surface temperatures by 4°C to 5°C. Over a century later, the Intergovernmental Panel on Climate Change (IPCC) Fifth Assessment Report (AR5) stated that the equilibrium climate sensitivity is likely in the range 1.5°C to 4.5°C with high confidence, extremely unlikely less than 1°C (high confidence), and very unlikely greater than 6°C (medium confidence) (Figure 33.2). As climate sensitivity cannot be measured directly, the range reflects different interpretations of transient changes in surface, upper air, and ocean temperature measurements, estimated radiative forcing, satellite data, and environmental reconstructions for the last millennium. Upper bound sensitivities are difficult to constrain because of non-linear feedbacks in the climate system and large uncertainty about aerosols and ocean heat uptake. One study suggests that the reduced rate of global mean warming observed over the past decade supports the case for low-end climate sensitivity: somewhere in the region of 1.2°C to 3.9°C.[6]

Human modifications of land cover contribute to climate change in three main ways. First, through forest clearance, crop cultivation, and soil degradation, carbon stores are converted to carbon sources. During the period 1990 to 2005, net emissions of carbon from land use change were 1.5 ± 0.7 Pg Cyr^{-1} (compared with 8.7 ± 0.7 Pg Cyr^{-1} from fossil fuel combustion in 2008).[7] Replacement of natural vegetation by pasture or arable landscapes may also contribute emissions of methane from cattle or submerged fields and nitrous oxide from fertilizers. Permafrost thawing and forest dieback through drought and fire driven by human-induced regional rainfall and temperature changes could further enhance GHG emissions.

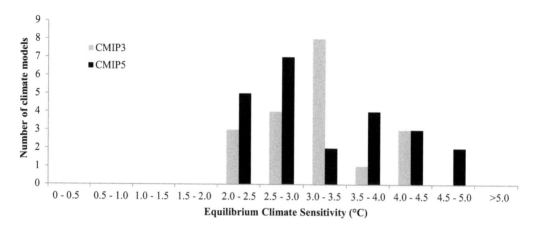

FIGURE 33.2   Equilibrium Climate Sensitivity (ECS) (°C) based on the CMIP3 and CMIP5 ensembles. The mean ECS (3.2°C) is approximately the same for CMIP3 and CMIP5.
Source: IPCC (2007) Fourth Assessment Report and IPCC (2013) Fifth Assessment Report.

Second, destruction or degradation of terrestrial and marine environments affects their capacity to act as carbon sinks. It is estimated that nearly half of all CO_2 emitted from burning fossil fuels and land use change between 1959 and 2008 has remained in the atmosphere.[7] This fraction exhibits much year-to-year variability and is thought to be increasing as the effectiveness of natural carbon sinks declines. However, considerable uncertainty surrounds the future behavior of the carbon cycle under a changed climate. For example, changes in ocean currents or acidity could affect the rate of carbon sequestration by the marine ecosystem.

Third, changes in land use affect regional energy and moisture budgets, an effect that is most pronounced in built environments. It has long been recognized that urban centers can be several degrees warmer than the surrounding countryside.[8] Compared with vegetated surfaces, building materials retain more solar energy during the day and have lower rates of radiant cooling during the night. Urban areas also have lower wind speeds, less convective heat losses, and less evapotranspiration, yielding more energy for surface warming. Artificial space heating, air conditioning, transportation, cooking, and industrial processes further add to the heat load. Other regional climate changes have been reported following reservoir creation or widespread irrigation: the former stimulating more intense precipitation events and the latter local cooling.

Climate Models

The climate system is represented in models by equations for radiation transfer, momentum, mass, and water vapor, solved at regular grid points on the Earth's surface, heights in the atmosphere, and depths in the ocean. The sophistication and number of climate model experiments have been rising since the first double CO_2 experiments in the 1970s. Early models had simplified "slab" oceans, fixed cloudiness, and no heat transport by oceans, idealized topography, and coarse grid resolution. State-of-the-art global climate models now incorporate many more physical elements including atmospheric chemistry, cloud convection, snow and ice stores, carbon cycling, rudimentary biomes, and hydrology, ocean salinity, and circulation (Figure 33.3). They also include important mechanisms that either amplify or dampen radiative forcing via positive (e.g., water vapor, ice-albedo, carbon and methane release, ocean acidity) or negative (e.g., low cloud, lapse rate) feedbacks. The number of calculations involved is so immense that climate model simulations can only be performed on supercomputers or via massively distributed networks of computers (such as the public participation ClimatePrediction.net experiment).

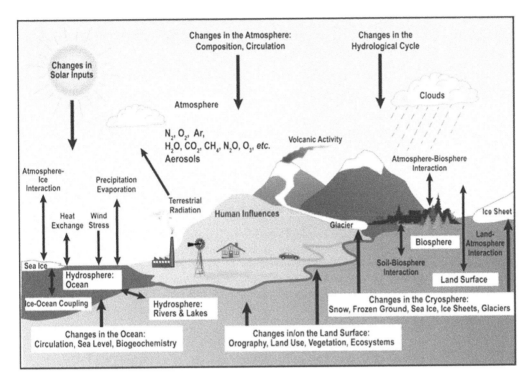

FIGURE 33.3 The major constituents of a global climate model.
Source: Intergovernmental Panel on Climate Change 2007.

Despite rapid developments in computing hardware, global climate models still operate at comparatively coarse spatial resolutions, relative to some of the phenomena that they seek to represent. The Japanese Earth Simulator runs at 20 km globally but most global climate models have resolutions of 100 km. This means that important features of the climate system such as clouds and convective storms must be described statistically (parameterized). Notwithstanding these constraints, climate models yield convincing representations of zonal average temperature trends;[9] they are less skillful at replicating regional rainfall patterns such as the South Asia monsoon.[10] Climate models are also known to produce precipitation more often and more lightly than is observed and to exaggerate soil moisture feedbacks in convection schemes. Although other techniques can be used to "downscale" to finer spatial scales, these are entirely dependent on the quality of climate information transferred from the coarser resolution global model.[11]

Future rates and patterns of climate change are uncertain because of natural forcing (by variations in solar output and volcanic aerosols), natural climate variability (due to ocean-atmosphere patterns such as El Niño), and socioeconomic and demographic trends (translated into GHG emissions, then atmospheric concentrations). Further uncertainty arises from parameter and process representations within climate models themselves. At global scales, natural climate variability and climate model uncertainty dominate the total uncertainty for the next few decades; emission scenario uncertainty increases in importance with time and is the major contributor to total uncertainty beyond the mid-21st century (Figure 33.4). Uncertainty in aerosol emissions is important much sooner but mainly on regional scales.

These elements of uncertainty can be quantified by running large numbers of models (ensembles) with different initial conditions for oceans and atmosphere and different emission scenarios. For example, the climate model ensemble used in the IPCC Fifth Assessment Report projects a likely global mean temperature change of 0.3°C to 4.8°C by 2081–2100 relative to 1986–2005 (Table 33.1). Some climate

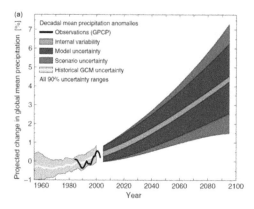 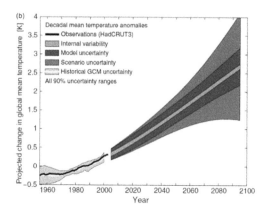

FIGURE 33.4 The total uncertainty in an ensemble (CMIP3) of global mean, decadal mean, and model projections for the 21st-century, separated into three components: internal variability (thick white line), model uncertainty (shadow gray), and scenario uncertainty (dark gray). The gray regions show the uncertainty in the 20th century integrations of the same climate models, with the mean in white. The black lines show an estimate of the observed historical changes. (a) Precipitation, with observations (GPCP). (b) Temperature, with observations (HadCRUT3). All anomalies are calculated relative to the 1971–2000 mean, except for the precipitation observations, for which a 1979–2000 mean is used. **Source:** Hawkins and Sutton 2011.

TABLE 33.1   Projected Global Mean Surface Temperature Change and Global Mean Sea Level Rise by 2046–2065 and 2081–2100

Variable	Scenario	2046–2065		2081–2100	
		Mean	Likely Range[c]	Mean	Likely Range[c]
Global Mean Surface Temperature Change (°C)[a]	RCP2.6	1.0	0.4 to 1.6	1.0	0.3 to 1.7
	RCP4.5	1.4	0.9 to 2.0	1.8	1.1 to 2.6
	RCP6.0	1.3	0.8 to 1.8	2.2	1.4 to 3.1
	RCP8.5	2.0	1.4 to 2.6	3.7	2.6 to 4.8
Global Mean Sea Level Rise (m)[b]	RCP2.6	0.24	0.17 to 0.32	0.40	0.26 to 0.55
	RCP4.5	0.26	0.19 to 0.33	0.47	0.32 to 0.63
	RCP6.0	0.25	0.18 to 0.32	0.48	0.33 to 0.63
	RCP8.5	0.30	0.22 to 0.38	0.63	0.45 to 0.82

The AR5 applied four new scenarios, termed Representative Concentration Pathways (RCPs), identified by their approximate total radiative forcing in year 2100 relative to 1750. These are 2.6 Wm^{-2} (RCP2.6: a mitigation scenario leading to very low forcing); 4.5 Wm^{-2} (RCP4.5: a lower stabilization scenario); 6.0 Wm^{-2} (RCP6.0: a higher stabilization scenario); and 8.5 Wm^{-2} (RCP8.5: a very high emissions scenario)
Source: IPCC (2013) Fifth Assessment Report Summary for Policymakers.
[a] Based on the CMIP5 ensemble; anomalies calculated with respect to 1986–2005.
[b] Based on 21 CMIP5 models; anomalies calculated with respect to 1986–2005.
[c] Calculated from projections as 5–95% model ranges.

models have been run beyond year 2100 to quantify the multicentury commitment, even irreversible, global warming, and sea level rise. One study found thermal expansion of the ocean (an important component of sea level rise) of up to 1.9 m if 21st century CO_2 concentrations exceed 1000 ppmv.[12]

The above climate model results are not predictions. This is because other, first-order climate forcings beyond CO_2 (e.g., aerosol effects on clouds, black carbon deposition, reactive nitrogen, and changes in land use/cover) are not always included, yet are known to be significant over multidecadal timescales.[13] Nonetheless, models were originally conceived to aid policy decisions about the costs and benefits of cutting GHG emissions; increasingly, model outputs are applied to impact and adaptation

TABLE 33.2 Examples of National Climate Change Risk Assessments and Adaptation Programs

Country	Report	Year	Lead Agency
Australia	Climate Change Risks to Australia's Coast: A First Pass National Assessment	2009	Department of Climate Change
Belgium	Belgian National Climate Change Adaptation Strategy	2010	Flemish Environment Nature and Energy Department
Canada	From Impacts to Adaptation: Canada in a Changing Climate 2007	2008	Natural Resources Canada
Denmark	Danish strategy for Adaptation to a changing climate	2008	Danish Energy Agency
Finland	Evaluation of the Implementation of Finland's National Strategy for Adaptation to Climate Change 2009	2009	Ministry of Agriculture and Forestry
France	French National Climate Change Impact Adaptation Plan 2011–2015	2011	Ministry of Ecology, Sustainable Development, Transport and Housing
Germany	German Strategy for Adaptation to Climate Change	2008	Federal Ministry for the Environment, Nature Conservation and Nuclear Safety
Iceland	Iceland's Climate Change Strategy	2007	Ministry for the Environment
Ireland	Ireland National Climate Change Strategy 2007–2012	2007	Department of the Environment, Heritage and Local Government
Japan	Wise Adaptation to Climate Change	2008	Ministry of Environment
Netherlands	Working on the Delta: Acting Today, Preparing for Tomorrow	2012	Ministry of Infrastructure and the Environment; Ministry of Economic Affairs, Agriculture and Innovation
Spain	Evaluación Preliminar de los Impactos en España por Efecto del Cambio Climático	2005	Ministerio de Medio Ambiente
Sweden	Sweden Facing Climate Change – Threats and Opportunities	2007	Swedish Commission on Climate and Vulnerability
Switzerland	Stratégie Suisse d'adaptation aux changements climatiques: Rapport intermédiaire au Conseil federal	2010	Département fédéral de l'Environnement, des Transports, de l'Energie et de la Communication
UK	The UK Climate Change Risk Assessment 2012 Evidence Report	2012	Department for Environment, Food and Rural Affairs
USA	Scientific Assessment of the Effects of Global Change on the United States	2008	U.S. Global Change Research Program; Committee on Environment and Natural Resources, National Science and Technology Council

studies (Table 33.2). Climate models are also used to explore potential tipping points in the climate system; to set stabilization targets for emissions to avoid dangerous climate change; or to evaluate the consequences of more radical interventions such as geoengineering. For example, one climate model ensemble suggests that cumulative carbon emissions should not exceed 1 trillion tonnes if a peak warming of 2°C above preindustrial temperatures is to be avoided.[14] Other studies examine not *if* but *when* global warming could reach 4°C, and one found the best estimate to be some time in the 2070s.[15]

Detecting and Attributing Climate Change

Climate change detection and attribution is only possible when three ingredients are present: a coherent theory of causation; a physically plausible model of the climate system; and long, high-quality records of environmental change (e.g., air or ocean temperatures, atmospheric pressure, rainfall patterns).

Climate change *detection* involves demonstrating that environmental data have changed in some defined statistical sense, without having to offer any reason(s) for that change. Detection occurs when an event or trend lies beyond the range expected to occur by chance. Change need not occur in a linear

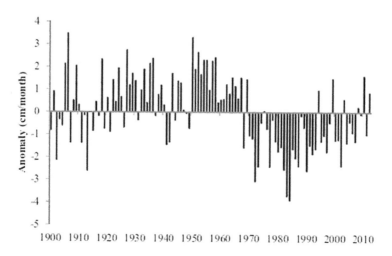

FIGURE 33.5   Sahel precipitation anomalies for the rainy season (June to October) over the domain 20–10°N, 20°W–10°E. Explanations for the abrupt transformation include the following: (1) changes in meteorological networks; (2) agricultural extensification and vegetation removal; (3) natural forcing by sea surface temperatures in the Atlantic, tropical Pacific, Indian Ocean, and Mediterranean; or (4) forcing by a combination of natural SST variability, anthropogenic greenhouse gases, and aerosols. Climate model experiments enable testing of the last three hypotheses.
Source: http://jisao.washington.edu/data/sahel/.

or in a monotonic way. There are plenty of cases where a record has changed abruptly, as with rainfall in southeast Australia in the 1980s[16] or Sahel in the late 1960s (Figure 33.5). Detection may be confounded if an underlying trend is weak compared with the "noise" of natural climate variability; on the other hand, there is always a chance of spurious detection even in random data. False trends can also emerge if records are not homogeneous or have been influenced by non-climatic influences such as expansion of urban areas, changes in observer practices, meteorological equipment, station location and surround-ings, or monitoring network density. Assessing trends in climate change *impacts* can be just as problem-atic. For example, much debate has surrounded trends detected in U.S. hurricane damages, a rise that is thought to be explained by increased prosperity and exposure of the population.[17]

Climate change *attribution* involves establishing the probable cause(s) of detected changes at a given level of statistical confidence. The difficulty for climate change studies is that there are no observational controls, as in epidemiological studies. Attribution, therefore, depends on comparing observations against climate models run with and without anthropogenic GHG emissions. In such experiments, attribution of observed climate change to human influence is accepted if there is consistency between the modeled response (to combined human-plus-natural forcing) and observed behavior. Alternatively, models can be used to compute the change in risk of a certain climate event with and without human influences. For instance, the chance of temperature anomalies like the European heat wave of the sum-mer of 2003 is believed to have doubled due to historic emissions.[18]

Callendar was the first to connect rising atmospheric concentrations of CO_2 with evidence of increasing global mean temperatures—the basis for climate change detection and attribution. Using a network of less than 150 stations and records from the 1890s to 1930s, he calculated that "excess" CO_2 was increasing world temperatures at a rate of 0.3°C per century.[19] Seventy-five years later, the IPCC declared that "it is extremely likely that human influence has been the dominant cause of the observed warming since the mid-20th century." However, it is now recognized that climate models reproduce global mean temperatures in different ways due to compensating effects between aerosols and climate sensitivity and that the trade-off between these parameters affects their respective future warming estimates.[20]

Evidence of change detection and attribution at global, hemispheric, and zonal scales has accumulated over the last two decades, but detection and attribution of changes in the intensity and frequency of extremes or attribution of individual extreme events are more demanding propositions.[21] Nonetheless, there are good reasons to expect robust detection of certain extreme events. For instance, regional climate models suggest that heavy rainfall could increase disproportionately more than average events under a warmer atmosphere with greater moisture content. However, precipitation observations show mixed results: evidence of increases in North America and Europe but no change for the Asia-Pacific or central African regions. This may be explained by the inability of climate models to adequately resolve extreme precipitation at local scales, the mismatch between point (observations) and grid (modeled) rainfall, natural climate variability, and by definition, the rarity of extreme events. Others focus on explaining the causes of specific extreme events. For example, analysis of the Thailand floods, East African and Texan droughts, and UK winter coldness in 2011 suggests that the human influence is making certain types of events more likely.[22] But these studies also demonstrate the importance of understanding the role of natural climate variability and non-climatic factors.

Conclusion

Climate change arises from three sets of forces: external variations in solar radiation receipt caused by periodicities in the earth's orbit and changes in solar output; internal variations in aerosol concentrations; and human-induced changes in atmospheric GHG concentrations due to fossil fuel combustion and modifications to land surface properties. Observed variations in global climate since the Industrial Revolution are best explained by the interplay of natural and human forces, along with natural variability in ocean–atmosphere systems. However, the climate sensitivity to rising concentrations of GHGs remains a highly contested figure that partly shapes attitudes to both the urgency and necessity of decarbonizing economies. Decadal climate projections also show divergent behavior because of large uncertainty in future socioeconomic drivers of GHG emissions and the nascent ability of climate models to represent aerosols, clouds, ice sheet dynamics, and ocean heat uptake.

References

1. Hulme, M. *Why We Disagree about Climate Change: Understanding Controversy, Inaction and Opportunity*; Cambridge University Press: 2009; 393 p.
2. Crawford, E. Arrhenius' 1896 model of the greenhouse effect in context. Ambio **1997**, *26*, 6–11.
3. Soden, B.J.; Wetherald, R.T.; Stenchikov, G.L.; Robock, A. Global cooling after the eruption of Mount Pinatubo: A test of climate feedback by water vapor. Science **2002**, *296*, 727–730.
4. Sutton, R.T.; Dong, B. Atlantic Ocean influence on a shift in European climate in the 1990s. Nat. Geosci. **2012**, *5*, 788–792.
5. Cook, E.R.; Seager, R.; Heim, R.R.; Vose, R.S.; Herweijer, C.; Woodhouse, C. Megadroughts in North America: Placing IPCC projections of hydroclimatic change in a long term palaeoclimate context. J. Q. Sci. **2009**, *25*, 48–61.
6. Otto, A.; Otto, F.E.L.; Boucher, O.; Chruch, J.; Hegerl, G.; Forster, P.M.; Gillett, N.P.; Gregory, J.; Johnson, G.C.; Knutti, R.; Lewis, N.; Lohmann, U.; Marcotzke, J.; Myhre, G.; Shindell, D.; Stevens, B.; Allen, M.R. Energy budget constraints on climate response. Nat. Geosci. **2013**, *6*, 415–416.
7. Le Quéré, C.; Raupach, M.R.; Canadell, J.G.; Marland, G.; et al. Trends in the sources and sinks of carbon dioxide. Nat. Geosci. **2009**, *2*, 831–836.
8. Wilby, R.L.; Jones, P.D.; Lister, D. Decadal variations in the nocturnal heat island of London. Weather **2011**, *66*, 59–64.
9. Sakaguchi, K.; Zeng, X.B.; Brunke, M.A. The hindcast skill of the CMIP ensembles for the surface air temperature trend. J. Geophys. Resear. Atmospheres **2012**, *117*, D16113.

10. Annamalai, H.; Hamilton, K.; Sperber, K.R. The South Asian summer monsoon and its relationship with ENSO in the IPCC AR4 simulations. J. Clim. **2007**, *20*, 1071–1092.

11. Wilby, R.L.; Wigley, T.M.L. Downscaling General Circulation Model output: A review of methods and limitations. Prog. Phys. Geogr. **1997**, *21*, 530–548.

12. Solomon, S.; Plattner, G-K.; Knutti, R.; Friendlingstein, P. Irreversible climate change due to carbon dioxide emissions. Proc. Nat. Acad. Sci. **2009**, *106*, 1704–1709.

13. Pielke Sr., R.; Beven, K.; Brasseur, G.; Calvert, J.; Chahine, Climate change: The need to consider human forcings besides greenhouse gases. Eos **2009**, *90*, 413–414.

14. Allen, M.R.; Frame, D.J.; Huntingford, C.; Jones, C.D.; Lowe, J.A.; Meinshausen, M.; Meinshausen, N. Warming caused by cumulative carbon emissions towards the trillionth tonne. Nature **2009**, *458*, 1163–1166.

15. Betts, R.A.; Collins, M.; Hemming, D.L.; Jones, C.D.; Lowe, J.A.; Sanderson, M.G. When could global warming reach 4 degrees C? Philos. Trans. R. Soc. A **2011**, *369*, 67–84.

16. Cai, W.; Cowan, T. Southeast Australia rainfall reduction: A climate-change-induced poleward shift of ocean-atmosphere circulation. J. Clim. **2013**, *26*, 189–205.

17. Pielke Jr., R.A.; Gratz, J.; Landsea, C.W.; Collins, D.; Saunders, M.A.; Musulin, R. Normalized hurricane damage in the United States: 1900–2005. Nat. Hazards Rev. **2008**, *13*, 29–41.

18. Stott, P.A.; Stone, D.A.; Allen, M.R. Human contribution to the European heatwave of 2003. Nature **2003**, *432*, 610–614.

19. Callendar, G.S. 1938. The artificial production of carbon dioxide and its influence on temperature. Q. J. R. Meteorol. Soc. **1938**, *64*, 223–240.

20. Knutti, R. Why are climate models reproducing the observed global surface warming so well? Geophys. Resear. Lett. **2008**, *35*, L18704.

21. Hegerl, G.C.; Karl, T.R.; Allen, M.; Bindoff, N.L.; Gillett, N.; Karoly, D.; Zhang, X.B.; Zwiers, F. Climate change detection and attribution: Beyond mean temperature signals. J. Clim. **2006**, *19*, 5058–5077.

22. Peterson, T.C.; Stott, P.A.; Herring, S. Explaining extreme events of 2011 from a climate perspective. Bull. Am. Meteorol. Soc. **2012**, *93*, 1041–1067.

Bibliography

Hawkins, E.; Sutton, R. The potential to narrow uncertainty in projections of regional precipitation change. Clim. Dyn. **2011**, *37*, 407–418.

Intergovernmental Panel on Climate Change (IPCC) Fifth Assessment Report, http://www.ipcc.ch/.

Mauna Loa Observatory, Earth System Research Laboratory, NOAA Global Monitoring Division, http://www.esrl.noaa.gov/gmd/obop/mlo/.

O'Hare, G.; Sweeney, J.; Wilby, R.L. *Weather, Climate and Climate Change: Human Perspectives*; Pearson Education Ltd: Harlow, 2005; 403.

Oppenheimer, M.; Petsonk, A. Article 2 of the UNFCCC: Historical origins, recent interpretations. Clim. Change **2005**, *73*, 195–226.

The Earth Simulator Center, Japanese Agency for Marine Earth and Technology http://www.jamstec.go.jp/esc/index.en.html.

34

Climate Change: Boreal Forests

Pertti Hari
University of Helsinki

Introduction

Coniferous forests at high latitudes are called boreal forests. They are mostly located between the latitudes 50°N and 70°N and they extend on the Eurasian and American continents from ocean to ocean. The stands are often rather evenly aged and the number of species is low, often one species dominating the stands. The mean biomass of trees is about 4.2×10^7 g ha^{-1}.[1] Dwarf shrubs dominate the field layer, but grasses and herbs also exist, especially in fertile soils. The carbon pool is often larger in the soil than in the living biomass.

Clear summer and winter seasons are characteristic of boreal forests. The mean July temperature is around 15–20°C and the mean January temperature is as low as −10 to −20°C. The annual variation is smaller near the oceans than in the middle of the continents. The climate in the boreal forests is rather humid because the rather low temperatures hinder evaporation, and the rainfall is usually small.

Human activities have increased the flow of several compounds into the atmosphere and the atmospheric composition reacts to these flows. The concentrations of several compounds are increasing in the air. The atmosphere hinders thermal radiation into the space more than previously, resulting in climate change. The boreal forests are reacting and will react to the changes in the environmental factors, especially atmospheric CO_2 concentration and temperature.

Climate Change

The air bubbles in ice in Antarctica and in Greenland have stored the climatic history on Earth for 700,000 years. Slow, in the time scale of 1000 years, and rather large changes characterize temperature and CO_2 concentration (Figure 34.1). These oscillations in temperature are evidently generated by small periodic changes in the spinning of Earth and in the circulation of Earth around the Sun. The climate is evidently rather stable in the time scale of 10,000 years and the temperature has varied very little after the ice age.

The human population on Earth and its need for food and energy before 1800 were so small that the human influence on the material and energy fluxes had only minor effects on

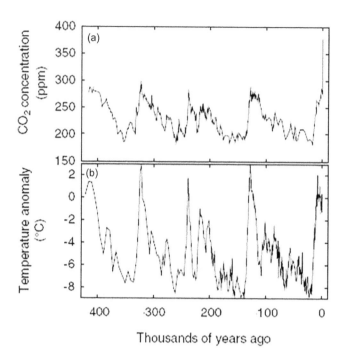

FIGURE 34.1 (a) Concentration of CO_2 and (b) change in temperature during the last 420,000 years. The CO_2 concentration data are compiled from Vostok,[2] Siple station ice core data,[3] and atmospheric measurements.[4] The temperatures are obtained from Vostok ice core isotopes as difference from present values.[5-8]
Source: Copyright by Springer.

the atmospheric concentrations. The rapid population growth and development of technology changed the situation, and human activities started to change the atmospheric concentrations and climate. Exponential growth has been characteristic for human-induced fluxes; thus, the changes in the atmospheric concentrations and climate have also been strongest during the twentieth century.

The need of new agricultural land for crops dominated the anthropogenic CO_2 fluxes into the atmosphere in the nineteenth century and the release of CO_2 from burning fossil fuels grew slowly in importance and now it covers about 90% of the anthropogenic emissions. Over two centuries, the atmospheric CO_2 concentration has increased from 280 ppm to nearly 400 ppm in the present.

Solar radiation is the source of energy for processes and transport phenomena taking place in the atmosphere. Several phenomena convert energy into other forms in the atmosphere. Gases and aerosols in the atmosphere absorb solar radiation and convert radiation energy into heat. Aerosols (clouds) reflect radiation into the space. Latent heat flux, i.e., flux of water vapor, is important for the atmospheric dynamics since evaporation consumes and condensation releases large amounts of energy.

All radiations have their spectrum, i.e., the radiation energy flux at each wavelength of the radiation. All bodies on earth emit thermal radiation. The spectrum of thermal radiation and the radiation flux depend on the temperature of the emitting body. Solar radiation occurs mainly within the range of 300–2000 nm, and half of its energy is within the visible light from 400 nm to 700 nm. The bodies on Earth emit thermal radiation at wavelengths of about 10,000 nm.

Gas molecules can absorb solar and thermal radiation. The ability to absorb, the absorbance, depends strongly on the wavelength of the radiation. The so-called greenhouse gases, i.e., CO_2, CH_4, and O_3, are transparent at visible light (0.3–0.7 nm), but they absorb the thermal radiation of several wavelengths. The greenhouse gases are able to partially absorb the thermal radiation from earth at several wavelengths.

Solar radiation generates flows of material and energy in the atmosphere, resulting in the climate at each location, warm and humid near the equator and cool and humid in the boreal region. The solar radiation has a very strong annual pattern in boreal areas due to the tilting position of the spinning axis of the Earth. This annual pattern is reflected in the climate, quite warm summers and very cold winters.

The physical phenomena, such as absorption and emission of radiation, take place in the atmosphere at a molecular level, and we have to convert our knowledge to deal with the global climate. Numeric simulation models enable the conversion of the knowledge dealing with different processes and transport phenomena in the atmosphere and in the oceans at the global scale. Several versions of global climatic models (GCMs), often also called global circulation models, have been constructed. The basic structure of the models is similar in these models.

The atmosphere is treated with volume elements, voxels, which are 200–300 km wide and the thickness varies from a few hundred meters to several kilometers at the top of the atmosphere. The temperature, water vapor concentration, three-dimensional wind velocities, etc., are treated in each voxel. The mass and energy fluxes change temperature and concentration in the voxels. The in and out fluxes of energy and material are combined with the conservation principle of mass, energy, and momentum, resulting in differential equations describing temperature, gas concentrations, and wind velocities.

The GCMs are able to simulate the climate at different locations rather well, utilizing physical knowledge and conservation principles. They are able to model the climate of boreal forests, rather warm or cool summer and cold winter, especially in the central parts of the continents. Although the GCMs simulate the global climate rather well, they are still under development.

Analysis and simulations of the present climate change are the most important applications of the GCMs. The absorption of the thermal radiation from earth in the atmosphere is increasing. This effect is built in the GCMs since they deal with all relevant aspects of energy flows in the atmosphere. There are two facts that generate the response of our climate to increasing greenhouse gas concentrations: 1) the atmosphere is quite transparent for the solar radiation, increasing greenhouse gas concentrations generate only a minor increase in the absorption of solar radiation, and 2) the greenhouse gases only partially absorb the thermal radiation of several wavelengths from Earth and any increase in concentrations of greenhouse gases results in enhanced absorption in the atmosphere. This fundamental physical basis of the present climate change is well understood and widely tested with measurements in the laboratory and in the field.

The development of the greenhouse concentrations in the atmosphere is well known during the last two centuries. After the ice age, the concentrations have been quite stable for about 10,000 years and the strong increase has occurred during the last 200 years. The temperature has been stable for the same 10,000 years, although some small variations have taken place (Figure 34.1).

The temperature measurements started in the eighteenth century and systematic monitoring began during the nineteenth century. The measurements in the boreal region indicate rather irregular fluctuations and a clear increasing trend of temperature starting around 1970. The phonological observations and dates of formation of ice in the Tornionjoki River show similar patterns.

The development of the atmospheric greenhouse gas concentrations during the twenty-first century is not clear. However, the coal resources are so large that the availability of coal will not limit the use of fossil fuels. Several countries have tried to reduce CO_2 emissions according to the Kyoto agreement with some success in the industrialized countries. However, the rapid development in China, India, and some other developing countries has overridden the reduction of CO_2 emissions and the emissions are growing very fast. The atmospheric carbon budget has been intensively studied and alternative developments of CO_2 concentration in the atmosphere can be found at http://www.globalcarbonproject.org/carbonbudget.

The expected concentration of the atmospheric CO_2 at the year 2100 varies greatly between scenarios from stabilization at 500 ppm to a strong increasing trend at 1000 ppm. The development of CO_2 concentration generates the largest uncertainty in the simulated global temperatures in the year 2100. The simulations of the climate change during the twentieth century show a clear increasing trend in temperature, especially in the boreal region.

The evidence of rapid temperature increase is piling up. The most important one is systematic temperature measurements during the last century. They indicate a clear increasing global trend and an especially strong increase at high latitudes (50–70°N). There has also been additional information from observations made in nature. The flowering of birch and bird cherry trees is taking place earlier than previously. The melting of ice cover on Tornionjoki River is occurring earlier than previously in the long time series covering the years 1693–2013. The polar ice on the Arctic Sea has been recorded low in the year 2012 and the glaciers are drawing back.

Boreal Forests

Boreal forest stands are often born after a catastrophe, forest fire, storm, or clear cut. The trees grow and the ecosystem develops until the next catastrophe and thereafter the stand starts again. The ground vegetation is rich after a catastrophe until the shading of the growing trees begins to reduce the growth of dwarf shrubs and herbs. The litter fall and root exudates from trees and ground vegetation feed the microbes in the soil.

Living organisms have long chains of enzymes, membrane pumps, and pigments for metabolic and specialized tissues for transport tasks, especially for transport of water and sugars within the individuals. The enzymes, membrane pumps, and pigment complexes are proteins. Proteins are amino acid polymers and their nitrogen content is as high as 15–17%.

Photosynthesis converts the solar radiation energy into chemical form as sugars, which are the main source of raw material of growth and the only source of energy for metabolism. The sugars are used for the synthesis of macromolecules, such as cellulose, lignin, and lipids, in the growing cells in needles, in woody components, and in fine roots. The macromolecules enter as litter into the organic material pool in the soil. The microbes decompose the macromolecules with extracellular enzymes and utilize the resulting small molecules in their metabolism. The metabolism of vegetation and microbes requires energy that is obtained from small carbon molecules in the process called respiration, then adenosine diphosphate is converted to energy-rich adenosine triphosphate (ATP) and CO_2 is released into the atmosphere. The flow of carbon compounds through a forest ecosystem is visualized in Figure 34.2.

Processes consume or produce materials giving rise to concentration and pressure differences that generate flows within the system. The carbon flows through the system via photosynthesis and respiration by either vegetation or microbes. The nitrogen fluxes in and out of the system are very small, and the nitrogen circulates effectively within the system. The effects of global change on processes are highlighted with dashed arrows in Figure 34.2.

The metabolism of vegetation and microbes is based on the action of enzymes, membrane pumps, and pigments. These proteins generate the important role of nitrogen as a nutrient, and the metabolism of a cell is impossible without proteins. We can say that photosynthetic products provide the structure and nitrogen uptake, together with photosynthesis, the tools in the structure.

Roots take up nitrogen as ions or amino acids. The ions are utilized in the synthesis of amino acids that are combined in the synthesis of proteins. When a cell dies, about half of the proteins in the cell are decomposed into amino acids for reuse in growing tissues and the other half enters an organic matter pool in the soil. The microbes decompose the proteins in the litter with extracellular enzymes to amino acids that can be used in the metabolism of microbes and vegetation. Microbes emit ammonium ions that are picked up by vegetation. Very small fluxes, i.e., deposition, biological nitrogen fixing, and leaching, connect the forest ecosystem with its surroundings, and thus the nitrogen circulates in a rather closed loop in the forest ecosystem.

The transpiration of vegetation is about hundredfold when compared with photosynthesis. Thus, trees need an effective transport system for water. We consider the woody structures in branches, stem, and coarse roots as water transport systems in the tree. We use sapwood area as a measure of the water transport capacity of wood. We assume that there is a balance between the water transport in branches, stem, and coarse roots and the needle mass in a whorl.

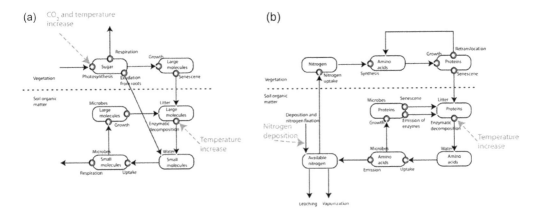

FIGURE 34.2 (a) Carbon and (b) nitrogen fluxes in a forest ecosystem. Boxes, arrows, and double circles denote amounts, flows, and processes, respectively.

A coniferous tree annually forms a whorl on the top of the stem. Thereafter, the whorl grows and finally dies. The trees have to solve the allocation problem, i.e., the use of available sugars and nitrogen for the growth of needles, water pipes, and fine roots. We assume that the trees solve the allocation problem annually at the level of whorls. The annual amount of sugars and nitrogen, i.e., the amount of sugars and nitrogen flown into the whorl during a year, is used for the growth of the whorl in such a way that: 1) all available sugars and nitrogen are used for growth and to obtain ATP for metabolism; 2) the roots can provide the nitrogen needed for the synthesis of proteins in the new tissues; and 3) the capacity of the water transport system meets the need of water in needles. The above three requirements result in carbon and nitrogen balance equations and the growths are obtained as the solution of the equations. These balance equations are very similar to those describing the behavior of heat and concentrations in a voxel in the GCMs.

When we describe all the carbon and energy fluxes in Figure 34.2, we obtain the forest ecosystem model MicroForest. The model describes cellulose, lignin, lipid, starch, and protein pools in the ecosystem in trees, in ground vegetation, and in soil. We have tested the model with measurements of tree diameter and height from very young to rather mature stands. We measured five stands in Estonia. The model estimated with measurement at station for measuring ecosystem-atmosphere relations II is able to predict the tree growths in Estonia without any changes in the parameter values. We demonstrate the fit between measured and predicted growths for the stands with best and weakest fit and the means of the five stands (Figure 34.3). The prediction of the mean growths is especially close to the measured one, indicating that random noise dominates the discrepancies at stand level. We have determined only the initial state of soil fertility from measurements in Estonia.

The Effect of Climate Change on Boreal Forests

The atmospheric CO_2 concentration is increasing and the climate is getting warmer in the boreal regions. In addition, the nitrogen deposition, although it is rather low, is high when compared to the flux a century ago and other fluxes between forest ecosystems and their surroundings. Boreal forests are reacting to the climate change, i.e., changes in carbon dioxide and nitrogen ion concentrations and in temperature.

The response of boreal forest to climate change takes place at the cell level in vegetation and in the soil. These small-scale reactions have to be converted to the annual ecosystem level. We had the same problem when several small-scale processes were combined to obtain the changes in the behavior of

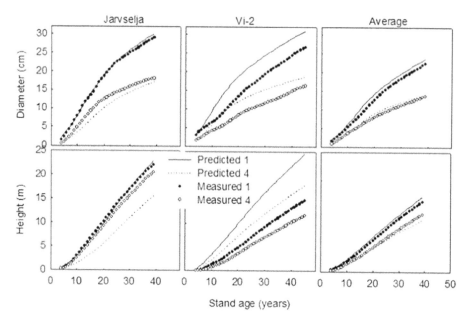

FIGURE 34.3 Predicted and measured diameters and heights at two sites in Estonia [Järvselja the best fit and Vihterpalu 2 (Vi-2) the poorest fit] and the average of all the sites in Estonia.
Source: Copyright by Springer.

the atmosphere. We used GCMs for the transition to the global level based on the material and energy flows in the atmosphere. Similarly, we can use MicroForest as a tool for the transition from cell level to ecosystems.

Availability of CO_2 limits photosynthesis, especially at high light intensities. Thus, the increase in the atmospheric CO_2 concentration has increased the formation of sugars, i.e., the flow of radiation energy into chemical energy. In addition, the temperature increase expands the photosynthetically active period and in this way also enhances the formation of sugars in photosynthesis.

The protein pool in the soils of boreal forests is large, about $0.5\,kg\,m^{-2}$. Microbes release ammonium ions from the protein pool in decomposition with extracellular enzymes. The annual release of NH_4 is small when compared with the amount of proteins, only about 2% of the protein pool in the soil. The rate of enzymatic reactions depends strongly on temperature; thus, the climate change accelerates the decomposition of proteins in the boreal forest soils, resulting in increased availability of nitrogen for vegetation.

The use of fossil fuels and nitrogen fertilizers in agriculture has increased the flux of reactive nitrogen into the atmosphere. The additional reactive nitrogen changes its chemical form in the reactions in the atmosphere and finally falls down as either wet or dry deposition within some thousand kilometers from the location of the emissions. The nitrogen deposition increased rapidly during the later part of the twentieth century to nearly 10-fold when compared with the pre-industrial values, providing an additional source of nitrogen for boreal forests.

Climate change has increased and will increase photosynthesis and has accelerated and will accelerate the decomposition of proteins in the soil, and nitrogen deposition has provided and will provide additional reactive nitrogen for vegetation. Thus, the metabolism of trees and microbes and the material fluxes are reacting to the climate change in boreal forests. These changes in the material fluxes are described in the ecosystem model called MicroForest.[9] The simulations with MicroForest indicate considerable changes in boreal forest and in photosynthesis, needle mass, and stem volume as a response to the changing climate (Figure 34.4).

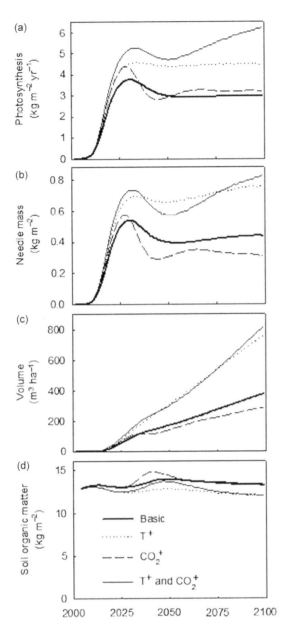

FIGURE 34.4 (a) Simulated photosynthesis (expressed in units of sugars), (b) needle mass, (c) volume, and (d) mass of soil organic matter (dry weight) by MicroForest for the years 2000–2100 with different scenarios: no change in temperature or atmospheric CO_2 concentration, increase in temperature (0.4°C/decade), increase in CO_2 concentration (1.5 ppm yr^{-1}), and increase in both temperature and CO_2 concentration.
Source: Copyright by Springer.

Feedback from Boreal Forests to Climate Change

The responses of boreal forests to climate change have generated and will generate feedback to the climate change. The carbon pool in the boreal forests will change, absorption of solar radiation into coniferous trees will enhance, and aerosol formation and growth will increase, resulting in increasing reflection of solar radiation.

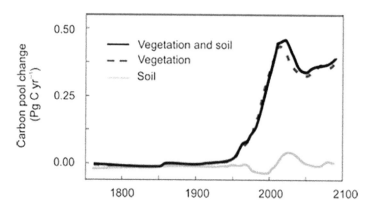

FIGURE 34.5 Simulated changes in carbon pools of vegetation, soils, and ecosystems in boreal forests during 1750–2100.

There are several changes in the carbon pools of a boreal forest due to the climatic change as seen in Figure 34.4. We can convert these changes to the level of boreal forests with MicroForest simulations. We introduce three additional assumptions: 1) even aged stands form the boreal forests; 2) the rotation time is 100 years; and 3) the age distribution of the stands is even. We divided the boreal forests into six areas, three on each continent, and we applied an area-specific development of temperature and nitrogen deposition during the period 1750–2100 in the simulations of carbon pools in the boreal forests.[10] The responses varied to some extent between areas, high values for densely populated areas of high nitrogen deposition and low ones for extremely remote areas. The obtained changes in the carbon pools in the boreal forests are rather large. The annual carbon uptake by boreal forests is about 0.5 Pg yr^{-1}, which is about 6% of the annual anthropogenic CO_2 emissions in the year 2000 (Figure 34.5).

The snow reflects nearly all solar radiation and, in contrast, the coniferous needles effectively absorb visible light. The boreal forests will expand to tundra and their needle mass will increase. The expansion of the area of boreal forests is a rather slow phenomenon, but the increase in needle mass takes place in the annual time scale. We simulated with GCM the effect of twofold needle mass on the monthly mean temperatures in boreal forests. The simulations indicated a clear increase in the spring, even 1.5°C in Central Asia and 1°C in Central North America, and in near coasts the increase was clearly smaller.[11] In midsummer the effects were clearly smaller.

The aerosols scatter solar radiation in the atmosphere and reduce the solar energy flux to the Earth's surface and in this way cool the climate. The vegetation, especially trees, emit volatile organic compounds (VOCs) that react in the atmosphere and lose their volatility. Thereafter, they condense on aerosol particles and the particles grow bigger. Riipinen et al.[12] roughly estimated the effects of doubling VOC emissions on the radiation energy reaching the earth's surface. This simulation procedure involves large uncertainties and the effect of aerosols in the changing climate is weakly known.

Conclusion

The physical basis of climate change is well understood and the simulation models (GCMs) are based on the physical knowledge of the phenomena involved and on conservation principles. The simulated changes are in a reasonable agreement with measurements and observations in nature. In addition, the responses of boreal forests are quite well understood: the photosynthesis is enhanced, decomposition of proteins in the soil is accelerated, and nitrogen deposition provides additional nitrogen for vegetation. The biomass in the boreal forests will increase and the boreal forests are able to bind about 6% of the anthropogenic emissions. In addition, the increasing needle mass is able to enhance the absorption of solar radiation in the spring. The role of changing aerosol concentrations increases the uncertainty in the analysis of climate change.

References

1. Botkin, D.B.; Simpson, L. The first statistically valid estimate of biomass for a large region. Biogeochemistry **1990**, *9* (9), 161–174.
2. Barnola, J.-M.; Raynaud, D.; Lorius, C.; Barkov, N.I. Historical CO_2 record from the Vostok ice core. In *Trends: A Compendium of Data on Global Change;* Carbon Dioxide Information Analysis Center, Oak Ridge National Laboratory, U.S. Department of Energy: Oak Ridge, Tennessee, USA, 2003.
3. Friedli, H.; Lötscher, H.; Oesschger, H.; Siegenthaler, U.; Stauffer, B. Ice core record of $^{13}C/^{12}C$ ratio of atmospheric CO_2 in the past two centuries. Nature **1986**, *324* (6094), 237–238.
4. Keeling, C.D.; Whorf, T.P. Atmospheric CO_2 records from sites in the SIO air sampling network. In *Trends: A Compendium of Data on Global Change*; Carbon Dioxide Information Analysis Center, Oak Ridge National Laboratory, U.S. Department of Energy: Oak Ridge, Tennessee, USA, 2005.
5. Jouzel, J.; Lorius, C.; Petit, J.R.; Genthon, C.; Barkov, N.I.; Kotlyakov, V.M.; Petrov, V.M. Vostok ice core: A continuous isotope temperature record over the last climatic cycle (160,000 years). Nature **1987**, *329*, 403–408.
6. Jouzel, J.; Barkov, N.I.; Barnola, J.M.; Bender, M.; Chappellaz, J.; Genthon, C.; Kotlyakov, V.M.; Lipenkov, V.; Lorius, C.; Petit, J.R.; Raynaud, D.; Raisbeck, G.; Ritz, C.; Sowers, T.; Stievenard, M.; Yiou, F.; Yiou, P. Extending the Vostok ice-core record of palaeoclimate to the penultimate glacial period. Nature **1993**, *364* (6436), 407–412.
7. Jouzel, J.; Waelbroeck, C.; Malaize, B.; Bender, M.; Petit, J.R.; Stievenard, M.; Barkov, N.I.; Barnola, J.M.; King, T.; Kotlyakov, V.M.; Lipenkov, V.; Lorius, C.; Raynaud, D.; Ritz, C.; Sowers, T. Climatic interpretation of the recently extended Vostok ice records. Clim. Dyn. **1996**, *12* (8), 513–521.
8. Petit, J.R.; Jouzel, J.; Raynaud, D.; Barkov, N.I.; Barnola, J.-M.; Basile, I.; Bender, M.; Chappellaz, J.; Davis, M.; Delayque, G.; Delmotte, M.; Kotlyakov, V.M.; Legrand, M.; Lipenkov, V.Y.; Lorius, C.; Pepin, L.; Ritz, C.; Saltzman, E.; Stievenard, M. Climate and atmospheric history of the past 420,000 years from the Vostok ice core, Antarctica. Nature **1999**, *399* (6735), 429–436.
9. Hari, P; Nikinmaa, E. Ecosystem responses to climate change. In *Boreal Forest and Climate Change*; Springer Dordrecht Heidelberg: New York, London, 2008; 499–503.
10. Hari, P.; Bäck, J.; Nikinmaa, E. Responses of forest ecosystems to climate change. In *Physical and Physiological Forest Ecology;* Springer Dordrecht Heidelberg: New York, London, 2013; 435–439.
11. Räisänen, J.; Smolander, S. Climatic effects of increased leaf area: Reduced surface albedo and increased transpiration. In *Physical and Physiological Forest Ecology;* Springer Dordrecht Heidelberg: New York, London, 2013; 446–454.
12. Riipinen, I.; Petäjä, T.; Hari, P.; dal Maso, M.; Bäck, J.; Kulmala, M. Forests, aerosols and climate change. In *Physical and Physiological Forest Ecology*; Springer Dordrecht Heidelberg: New York, London, 2013; 454–469.

35

Climate Change: Coastal Marine Ecosystems

Jennifer P. Jorve,
Rebecca L.
Kordas, Kathryn
M. Anderson,
Jocelyn C. Nelson,
Manon Picard,
and Christopher
D. G. Harley
*University of British
Columbia*

Introduction

Coastal marine ecosystems extend from the high tide mark to the edge of the continental shelf. They contain a staggering amount of life and range in form from tropical coral reefs to temperate-zone kelp forests to high-latitude sea ice habitats. Seventy-five percent of all countries are bordered by oceans or seas, and almost half of the world's population lives within 100 km of coasts. Coastal marine ecosystems supply an estimated 43% of the world's ecosystem goods (e.g., food and raw materials) and services (e.g., disturbance regulation and nutrient cycling), a contribution valued at over US$14 trillion per year.[1] However, the health of coastal ecosystems and the benefits associated with them are vulnerable to the effects of climate change.

Climate Change in the Ocean

Climate change is caused by anthropogenic emissions of greenhouse gases such as carbon dioxide (CO_2), which will soon reach the highest atmospheric concentration in tens of millions of years.[2] The rise in CO_2 concentrations since the industrial revolution are 100–1000 times faster than at any point in the past 420,000 years, and these rates of increase are expected to accelerate by a factor of 2–6 by the year 2100.[3] Although one of the most widely publicized repercussions of greenhouse gas emissions is the predicted 2–3°C increase in average air and ocean temperatures over the next 100 years,[2] several additional changes will occur in the ocean. For example, excess CO_2 reacts with seawater to increase the concentration of hydrogen ions (i.e., reduce pH) and reduce the concentration of carbonate ions, which are a key building block for calcified shells and skeletons. This suite of changes in ocean chemistry is known as "ocean acidification." The acidity of the surface ocean (the uppermost ocean layer where mixing due to a combination of temperature changes, waves, and currents acts to homogenize chemical conditions) has already increased by 30%, and may increase up to 150% by the end of this century.[2] Additional factors associated with climate change include sea level rise and changes in upwelling, storminess, and coastal salinity (see Solomon, Qin, et al.[2] for details). Sea level rise, caused by thermal expansion of water and melting of land-based ice sheets, is expected to occur at a magnitude of anywhere between 0.18 and 0.59 meters by the end of the twenty-first century. Uneven heating of the Earth's surface will affect wind patterns, which in turn determine storm frequency and the strength of upwelling along coastlines. Changes in precipitation will vary from place to place, resulting in increasing or decreasing river outflow and thus changes in coastal salinity and terrestrially derived nutrients and contaminants.

Biological Responses to Climate Change in Coastal Marine Ecosystems

Climate change will have a wide range of biological consequences, including changes in the rates and seasonal timing of growth and reproduction, the distribution of species, and the biodiversity of entire ecosystems.[4] Although the specific ecological impacts of climate change depend upon the system of interest (see Table 35.1), there are some generalities that are widely applicable. In general, changes in water temperature and CO_2 concentration are predicted to have the largest impacts on marine life. Experimental studies indicate that warming negatively impacts individuals living near their upper thermal tolerance limits, such as many corals,[20] while ocean acidification has a broadly negative effect on calcifying organisms but a positive effect on seaweeds.[21] While each abiotic change (e.g., warming or ocean acidification) may cause specific changes in an individual's physiology or growth rate, such direct effects are frequently modified by interactions with other organisms, such as competitors or consumers. For example, a small amount of warming has been shown to benefit the Atlantic population of the habitat-forming seaweed *Sargassum muticum* in the absence of herbivores. However, the higher energy demands of herbivores in warmer water increased the grazing pressure on *Sargassum,* and therefore the overall effect of warming on the seaweed was negative.[22]

The sum of all impacts of environmental change (from both direct and indirect effects) can create large changes in the abundance of species.[12] When the affected species are particularly ecologically important—e.g., habitat-forming species such as kelps and corals, and top predators such as sea stars and sharks—dramatic shifts in entire ecosystems may result. In coral reef ecosystems, warming and ocean acidification cause coral bleaching and mortality, promoting a transition to a system dominated by fleshy algae.[3] Following this shift, coral-associated fish and other reef residents are lost from the system, resulting in a decline in biodiversity. Dramatic ecological changes such as shifts from coral reefs to seaweed beds can occur rapidly when systems are pushed beyond some threshold or "tipping point."[23] Once an ecosystem has shifted into a new phase, it may take a long time to recover even if environmental conditions become more favorable. Because of these time lags, and because ecosystems respond to multiple climate stressors and other human pressures simultaneously, predicting the future state of marine ecosystems is no easy task.

Economic Implications

Fisheries and Aquaculture

Marine fisheries provide protein for approximately 1.5 billion people worldwide, and the demand for seafood continues to rise.[24] As with all marine species, the distribution and abundance of fished species are linked to changes in climate. Warming promotes poleward shifts of fish and plankton and may positively affect fisheries in higher latitudes as species migrate to track optimal environmental conditions, although this may occur at the expense of native, cold-water fish stocks.[25] The spread of disease also tracks temperature changes: historically limited by lower temperatures, diseases are spreading poleward,[26] and warming increases disease prevalence by increasing susceptibility to pathogens.[27] Although increasing

TABLE 35.1   Key Impacts by Habitat Type

Habitat	Key Stressors	Impacts	References
Coral reefs	Warming	Increased coral bleaching and mortality events	[3]
	Ocean acidification	Increased susceptibility to coral disease	
	Sea level rise	Reduced coral calcification and increased dissolution	
	Changing storminess	Reef drowning by sea level rise	
Mangroves	Warming	Slowed juvenile mangrove growth and eventual mortality	[3,5–7]
	Sea level rise	Habitat loss	
	Changing storminess	Reduced species richness	
		Loss of connectivity between populations and systems	
Sandy beaches	Sea level rise	Changes in beach composition and morphology	[8]
	Changing storminess	Beach erosion and habitat loss	
Estuaries	Warming	Eutrophication and changes in water clarity	[9,10]
	Ocean acidification	Changes in phytoplankton growth rates	
	Sea level rise	Reduction of calcifying species (e.g., oysters)	
	Altered run-off		
Seagrass beds	Warming	Changes in sedimentation	[11]
	Sea level rise	Reduced seagrass biomass and productivity	
	Altered run-off	Change in species distributions	
Rocky shores	Warming	Compressed vertical habitat space	[12]
	Ocean acidification	Localized species loss in "hot spots"	
		Poleward range shifts	
		Changes in species composition	
Salt marshes	Sea level rise	Habitat loss via marsh drowning	[9,13,14]
	Altered run-off	Reduced water filtration and carbon sequestration	
Kelp forests	Warming	Changes in kelp abundance, density, and productivity	[11,15,16]
	Ocean acidification	Community shift from kelp forests to algal turfs in some regions	
	Changing storminess		
Continental shelf habitats	Warming	Eutrophication, low oxygen levels in bottom waters	[17]
	Ocean acidification	Reduction of demersal fish and benthic invertebrate populations	
	Altered run-off	Biogeographic range shifts	
	Altered currents and upwelling		
Open water (pelagic) habitats	Warming	Shifts in latitudinal ranges and depth distributions	[6,11]
	Ocean acidification	Changes in primary and secondary productivity	
		Changes in biogeochemical cycling	
Sea ice habitats	Warming	Declines in phytoplankton body size and population size	[6,18,19]
		Poleward range shifts	
		Habitat loss	

Note that all impacts may include important changes in community structure, defined as the species present in the system, their abundance, and how they interact with one another (e.g., in a food web). Omission of a particular stressor does not mean that it will not be important in some situationsE

CO_2 boosts growth rates for few commercially important species such as seaweeds, lobsters, crabs, and shrimp,[28] the net effect of ocean acidification on most commercially important calcifying species, like oysters, clams, sea urchins, and abalone, is negative.[21] Aquaculture can provide an alternative to reliance on wild stocks, but the ocean is the primary source of water used in aquaculture facilities and farmed animals will face many of the same physiological problems with warming and acidification as wild stocks.[29] As with land crops, many of the areas currently used for farming marine animals will no longer be suitable as optimal thermal conditions shift poleward and local environments change.

Species Invasions

Along with climate change, species invasions are considered one of the greatest overall threats to marine biodiversity.[30] Species introductions can carry disease, increase competition for resources, reduce native species abundance, or result in the homogenization of ecosystems.[31] In 2005, damage and management due to species invasions cost nearly $120 billion in the USA alone.[32] Climate change, which disrupts local ecosystems and improves environmental conditions for many invaders, is predicted to increase the frequency and severity of invasions.[33] In general, invasive species tolerate significantly higher temperatures than native species, and as temperature regimes change, experimental evidence shows that the abundance of invasive species increases while native species abundance decreases.[34] Additionally, some invasive species recruit and grow earlier in the year than native species, which in turn shifts community dominance to invasive species.[30] Overall increases in invasive species numbers in the past decade, as in Southern California and Mexico, have been linked to climactic shifts promoting the spread of new invaders arriving via shipping, aquaculture, and ballast water vectors.[35]

Managing for Risk

As ecological shifts in coastal marine systems become more apparent, it may become advantageous to implement management strategies focused on mitigating climate change impacts. For example, managers can control several stressors at local scales and in the short term by reducing nutrient pollution in run-off and over-fishing.[11] Additionally, the control of fishing through management tools such as Marine Protected Areas (MPAs; see Caselle entry on "Marine Protected Areas") may enable ecosystems to retain enough resilience to weather some of the impacts of climate stressors such as warming and acidification. Although climate change will certainly impact the effectiveness with which MPAs meet specific conservation goals, it should be emphasized that the overall importance of MPAs—via their role in enhancing resilience—increases with climate change. Therefore, research is needed on assessment methods of ecosystem health and function, but also on the identification of key indicators of change in communities (e.g., the "canary in the coal mine"). Because many ecological responses to climate change only become apparent when they are combined with other stressors, such as over-fishing or pollution,[36] MPAs can act to reduce the total stress on ecosystems by managing those aspects that are under short-term regulatory control.

Conclusions

Nearshore marine systems supply economically and ecologically valuable goods and services to human communities worldwide.[1] As the global population size increases, the demand for these provisions will only increase.[24] Sadly, the recurring theme in scientific research addressing the future state of nearshore marine communities indicates that ecosystem function and health are in decline. Climate change stressors, such as warming and ocean acidification, combined with over-fishing and species invasions, continuously put pressure on the current state and shape of marine communities. Although it is difficult to precisely predict how whole ecosystems will respond to simultaneous climate stressors, the sum of negative impacts on key species implies that the overall abundance of many species and biodiversity as a whole will likely decline during the current century.[4] If we are going to mitigate the impacts of climate change on nearshore marine systems, we need to curb greenhouse gas emissions and implement effective management strategies that set realistic goals (e.g., fishing levels) that would allow species and ecosystems to recover and persist.

Acknowledgments

We thank Rebecca Gooding and Norah Brown for valuable contributions to the text, the Hakai Network for Coastal People, Ecosystems, and Management, and the Killam Trusts for support during the writing of this entry.

References

1. Costanza, R.; d'Arge, R.; de Groot, R.; Farber, S.; Brasso, M.; Hannon, B.; Limburg, K.; Naeem, S.; O'Neill, R.V.; Paruelo, J.; Raskin, R.G.; Sutton, P.; van den Belt, M. The value of the world's ecosystem services and natural capital. Nature **1997**, *387* (6630), 253–260.
2. Solomon, S.; Qin, D.; Manning, M.; Chen, Z.; Marquis, M.; Averyt, K.B.; Tignor M.; Miller, H.L., Eds.; IPCC. *Climate Change 2007: The Physical Science Basis, Contribution of Working Group I to the Fourth Assessment Report of the Intergovernmental Panel on Climate Change;* Cambridge University Press: Cambridge, UK, and New York, NY, USA, 2007.
3. Hoegh-Guldberg, O.; Mumby, P.J.; Hooten, A.J.; Steneck, R.S.; Greenfield, P.; Gomez, E.; Harvell, C.D.; Sale, P.F.; Edwards, A.J.; Caldeira, K.; Knowlton, N.; Eakin, C.M.; Iglesias-Prieto, R.; Muthiga, N.; Bradbury, R.H.; Dubi, A.; Hatziolos, M.E. Coral reefs under rapid climate change and ocean acidification. Science **2007**, *318* (5857), 1737–1742.
4. Munday, P.L.; Jones, G.P.; Pratchett, M.S.; Williams, A.J. Climate change and the future for coral reef fishes. Fish Fish. **2008**, *9* (3), 261–285.
5. Alongi, D. Mangrove forests: Resilience, protection from tsunamis, and responses to global climate change. Est. Coast. Shelf Sci. **2008**, *76* (1), 1–13.
6. Doney, S.C.; Ruckelshaus, M.; Duffy, J.E.; Barry, J.P.; Chan, F.; English, C.A.; Galindo, H.M.; Grebmeier, J.M.; Hollowed, A.B.; Knowlton, N.; Polovina, J.; Rabalais, N.N.; Sydeman, W.J.; Talley, J.D. Climate change impacts on marine ecosystems. Annu. Rev. Mar. Sci. **2012**, *4*, 11–37.
7. Gilman, E.L.; Ellison, J.; Duke, N.C.; Field, C. Threats to mangroves from climate change and adaptation options: A review. Aquat. Bot. **2008**, *89* (2), 237–250.
8. Brown, A.C.; McLachlan, A. Sandy shore ecosystems and the threats facing them: Some predictions for the year 2025. Environ. Conserv. **2002**, *29* (1), 62–77.
9. Scavia, D.; Field, J.C.; Boesch, D.F.; Buddemeier, R.W.; Burkett, V.; Cayan, D.R.; Fogarty, M.; Harwell, M.A.; Howarth, R.W.; Mason, C.; Reed, D.J.; Royer, T.C.; Sallenger, A.H.; Titus, J.G. Climate change impacts on U.S. coastal and marine ecosystems. Estuaries **2002**, *25* (2), 149–164.
10. Kennedy, V.S. Anticipated effects of climate change on estuarine and coastal fisheries. Fisheries **1990**, *15* (6), 16–24.
11. Hoegh-Guldberg, O.; Bruno, J.F. The impact of climate change on the world's marine ecosystems. Science **2010**, *328* (5985), 1523–1528.
12. Harley, C.D.G. Climate change, keystone predation, and biodiversity loss. Science **2011**, *334* (6059), 1124–1127.
13. Gedan, K.; Silliman, B.R.; Bertness, M.D. Centuries of human-driven change in salt marsh ecosystems. Annu. Rev. Mar. Sci. **2009**, *1*, 117–141.
14. Day, J.W.; Christian, R.R.; Boesch, D.M.; Arancibia, A.Y.; Morris, J.; Twilley, R.R.; Naylor, L.; Schaffner, L. Consequences of climate change on the ecogeomorphology of coastal wetlands. Est. Coast. **2008**, *31* (3), 477–491.
15. Harley, C.D.G.; Hughes, A.R.; Hultgren, K.M.; Miner, B.G.; Sorte, C.J.B.; Thornber, C.S.; Rodriguez, L.F.; Tomanek, L.; Williams, S.L. The impacts of climate change in coastal marine systems. Ecol. Lett. **2006**, *9* (2), 228–241.
16. Connell, S.D.; Russell, B.D. The direct effects of increasing CO_2 and temperature on non-calcifying organisms: Increasing the potential for phase shifts in kelp forests. Proc. R. Soc. B **2010**, *277* (1686), 1409–1415.

17. Hall, S.J. The continental shelf benthic ecosystem: Current status, agents for change and future prospects. Biol. Conserv. **2002**, *29* (3), 350–374.
18. Schofield, O.; Ducklow, H.W.; Martinson, D.G.; Meredith, M.P.; Moline, M.A.; Fraser, W.R. How do polar marine ecosystems respond to rapid climate change? Science **2010**, *328* (5985), 1520–1523.
19. Wassmann, P.; Duarte, C.M.; Agusti, S.; Sejr, M.K. Footprints of climate change in the Arctic marine ecosystem. Global Change Biol. **2011**, *17* (2), 1235–1249.
20. Hughes, T.P.; Baird, A.H.; Bellwood, D.R.; Card, M.; Connolly, S.R.; Folke, C.; Grosberg, R.; Hoegh-Guldberg, O.; Jackson, J.B.C.; Kleypas, J.; Lough, J.M.; Marshal, P.; Nystrom, M.; Palumbi, S.R.; Pandolfi, J.M.; Rosen, B.; Roughgarden, J. Climate change, human impacts, and the resilience of coral reefs. Science **2003**, *301* (5635), 929–933.
21. Kroeker, K.J.; Kordas, R.L.; Crim, R.N.; Singh, G.G. Meta-analysis reveals negative yet variable effects of ocean acidification on marine organisms. Ecol. Lett. **2010**, *13* (11), 1419–1434.
22. O'Connor, M.I. Warming strengthens an herbivore-plant interaction. Ecology **2009**, *90* (2), 388–398.
23. Mumby, P.J.; Hastings, A.; Edwards, H.J. Thresholds and the resilience of Caribbean coral reefs. Nature **2007**, *450* (7166), 98–101.
24. FAO. *The State of World Fisheries and Aquaculture 2008;* Food and Agriculture Organization of the United Nations: Rome, 2009.
25. Roessig, J.M.; Woodley, C.M.; Cech, J.J.; Hansen, L.J. Effects of global climate change on marine and estuarine fishes and fisheries. Rev. Fish Biol. Fish. **2004**, *14* (2), 251–275.
26. Hofmann, E.; Ford, S.; Powell, E.; Klinck, J. Modeling studies of the effect of climate variability on MSX disease in eastern oyster (*Crassostrea virginica*) populations. Hydrobiologia **2001**, *460* (1–3), 195–212.
27. Harvell, C.D.; Mitchell, C.E.; Ward, J.R.; Altizer, S.; Dobson, A.P.; Ostfeld, R.S.; Samuel, M.D. Climate warming and disease risks for terrestrial and marine biota. Science **2002**, *296* (5576), 2158–2162.
28. Ries, J.B.; Cohen, A.L.; McCorkle, D.C. Marine calcifiers exhibit mixed responses to CO_2-induced ocean acidification. Geology **2009**, *37* (12), 1131–1134.
29. Cooley, S.; Doney, S. Anticipating ocean acidification's economic consequences for commercial fisheries. Environ. Res. Lett. **2009**, *4* (2), 8 pp.
30. Stachowicz, J.; Terwin, J.; Whitlatch, R.; Osman, R. Linking climate change and biological invasions: Ocean warming facilitates nonindigenous species invasions. Proc. Natl. Acad. Sci. **2002**, *99* (24), 15497–15500.
31. Crooks, J.A. Characterizing ecosystem-level consequences of biological invasions: The role of ecosystem engineers. Oikos **2002**, *97* (2), 153–166.
32. Larson, D.L.; Phillips-Mao, L.; Quiram, G.; Sharpe, L.; Stark, R.; Sugita, S.; Weiler, A. A framework for sustainable invasive species management: Environmental, social, and economic objectives. J. Environ. Manag. **2011**, *92* (1), 14–22.
33. Hellmann, J.J.; Byers, J.E.; Bierwagen, B.G.; Dukes, J.S. Five potential consequences of climate change for invasive species. Conserv. Biol. **2008**, *22* (3), 534–543.
34. Sorte, C.J.B; Williams, S.L.; Zerebecki, R.A. Ocean warming increases threat of invasive species in a marine fouling community. Ecology **2010**, *91* (8), 2198–2204.
35. Miller, K.A.; Aguilar-Rosas, L.E.; Pedroche, F.F. A review of non-native seaweeds from California, USA Baja California, Mexico. Hidrobiologica, **2011**, *21* (3), 365–379.
36. McLeod, E.; Salm, R.; Green, A.; Almany, J. Designing marine protected area networks to address the impacts of climate change. Front. Ecol. Environ. **2009**, *7* (7), 362–370.

Climate Change: Polar Regions

Roger G. Barry
Cooperative Institute
for Research in
Environmental Sciences

Introduction

The two polar regions have undergone intervals in the past when their climatic conditions were broadly similar with extensive ice sheets and sea ice cover and other times, as today, when their climates are quite dissimilar. This entry begins by sketching the climatic history of past geological epochs and then treats in detail the climate of the two regions over the last century.

The Geological Record of the Cenozoic Era and the Quaternary Period

From approximately 65 to 50 million years ago (Ma) in the early Cenozoic (Greek for new life) era, global climatic conditions were warm, probably due to high levels of atmospheric carbon dioxide (CO_2) (up to 1600 ppmv). As a greenhouse gas, this traps infrared radiation emitted by the surface in the lower atmosphere raising the temperature. During this time there were no ice sheets, although the Antarctic continent was located in high southern latitudes. Now we discuss how world climate evolved during the Tertiary period (the start of the Cenozoic era; Cenozoic is Greek for new life) that began approximately 65 Ma, after the age of dinosaurs. Information about climatic conditions during this remote past is derived from sedimentary rocks that contain plant and animal fossils, and especially limestones that yield data on past CO_2 levels in the atmosphere. Conditions remained warm from 65 to 50 Ma with no ice sheets, perhaps due to high CO_2 concentrations. Between 55.5 and 52 Ma there were several abrupt, extreme global warming events. These have been linked to intervals with high eccentricity and high obliquity in the Earth's orbit that triggered massive releases of soil organic carbon that had been locked up in permafrost in the Arctic and Antarctic regions.[1]

Then, temperatures especially in the deep ocean began a steady decline and approximately 33 Ma the climate of Antarctica changed from mild and temperate to glacial conditions, with the accumulation of a thick ice sheet. The cooling is primarily attributed to a decrease in the atmospheric CO_2, but it is also possible that the opening of Drake Passage south of Tierra del Fuego allowed the formation of the Antarctic Circumpolar Current that excluded warm water from the seas around the continent. Since that time the climate of Antarctica has been essentially the same, although the extent of ice on the continental shelves fluctuated, particularly over the last five million years or so. The West Antarctic Ice Sheet developed approximately 6 Ma, but subsequently disappeared at times of warming and then reformed.

The Arctic Ocean was formed in the Middle Eocene approximately 45 Ma. Climatic conditions in the Arctic were quite different from those in the Antarctic; the climate remained temperate until near the end of the Cenozoic era. Hence, the Earth had unipolar (Southern Hemisphere) glaciation from approximately 33 Ma until perhaps 14 Ma when the Arctic Ocean formed a perennial sea ice cover. However, the Greenland ice sheet did not form until the Pliocene period approximately 3 Ma when CO_2 levels had decreased sufficiently. The closure of the Panama Isthmus approximately 2.9 Ma that affected ocean circulation in the North Atlantic may have been a contributory factor. Ice accumulated in northern Eurasia and North America approximately 2.7 Ma, but Greenland was probably ice free during a warm interval approximately 2.4 Ma.

The start of the Quaternary era that was marked by repeated glacial and interglacial intervals is dated to 2.6 Ma. Until 0.8 Ma, global climate was characterized by cycles of 41,000-year duration that were driven by the axial tilt of the Earth (the obliquity) varying between 22.1° and 24.5°. The obliquity effect on solar radiation receipts is most marked in high latitudes giving a 10% range between obliquity extremes at latitudes 80–90°. Then, for poorly understood reasons, the cyclicity switched to ~100,000 years. This periodicity roughly coincides with the timing of the eccentricity of the Earth's orbit around the Sun, but the effect of this on solar radiation is too small by itself to cause glacial/interglacial cycles. Other processes, particularly ice-albedo feedback, must amplify the small fluctuations in solar energy. This positive feedback involves the cooling effect of an ice sheet, as a result of its high reflectivity (albedo), leading to further ice growth and temperature decrease.

Between 0.8 and 0.43 Ma, interglacial episodes have been shown by the isotopic record in the European Project for Ice Coring in Antarctica (EPICA) ice core from Dome C in Antarctica to have been longer and cooler than subsequent interglacials. The interglacial from 428 ka to 397 ka known as Marine Isotope Stage (MIS)-11 was the longest and gave rise to the disappearance of the Greenland ice sheet and the West Antarctic ice sheet. Subsequent interglacials each lasted only approximately 10% of the 100,000 years occupied by the glacial cycles. These glacial cycles, lasting between 10,000 and 100,000 years, were basically in phase in the two polar regions. Ablation on the Northern Hemisphere ice sheets responded mainly to solar variations in summer and ice-albedo feedback effects as a result of the main ice sheet margins being on land. In Antarctica, in contrast, the ice terminates in the ocean and most ice loss is by calving. Hence, the main control on temperature here is the length of the summer season.

The penultimate interglacial interval, known as the Eemian, peaked at approximately 125,000 years ago (125 ka), with temperatures approximately 1–2°C above those in the twentieth century. Trees grew as far north as northern Norway in what is now arctic tundra. However, the Greenland ice sheet persisted, although it shrank slightly.

During the last glacial cycle (115 to 15 ka), the climates of the two polar regions were slightly out of phase. Antarctic temperatures led Arctic temperatures by a quarter of a period, leading to a bipolar seesaw on a millennial timescale. The Antarctic warms while Greenland is cold and then the Antarctic cools as Greenland warms rapidly. This linkage of the two polar regions is attributed to the interhemispheric ocean circulation—the Meridional Overturning Circulation. Greenland ice cores reveal that there were 25 major climatic oscillations during the last glacial cycle (and these continued into the postglacial Holocene period). The oscillations were characterized by rapid warming followed by a gradual cooling with each event over the last 50,000 years averaging approximately 1470 years. They were probably caused by influxes of fresh water into the North Atlantic related to "binge–purge" cycles in the mass of the Laurentide ice sheet over North America.

During the Last Glacial Maximum (LGM) approximately 25–18 ka vast ice sheets covered much of North America and Fennoscandia, extending to the British Isles. In North America, the Canadian Arctic Archipelago was occupied by the Innuitian ice sheet, the western cordilleras by the Cordilleran ice sheet, and most of central and eastern Canada and the northern United States by the 3 km thick Laurentide ice sheet. The locking up of water in the ice sheets led to a drop in the sea level of approximately 130 m. Consequently, continental shelves were exposed, particularly in the Arctic, and a Bering land bridge formed, enabling humans to enter Alaska from eastern Siberia. Permafrost developed in the sediments of the exposed shelves and this persists extensively in the Laptev and East Siberia seas at present. Global mean temperature during the LGM was approximately 6°C below that in the twentieth century. In Beringia, July temperatures were approximately 4°C lower than late twentieth century.

The ice began to retreat approximately 16 ka, but then between 12,800 and 11,500 years Before Present (BP) there was a pronounced cool event known as the Younger Dryas when postglacial warming was replaced by a return to glacial conditions in high latitudes of the Northern Hemisphere. Temperatures in Greenland and Europe dropped by 15°C. Accumulation on the Greenland ice sheet was halved and there were high dust loadings in the atmosphere.

There is no evidence of the event in Antarctica, where a thousand years earlier an Antarctic Cold Reversal has been documented. The mechanism of the Younger Dryas is attributed to a surge of melt water from Glacial Lake Agassiz in North America into the Arctic Ocean and North Atlantic. This influx of freshwater shut down the subtropical ocean conveyor belt by putting a lid of low-density water at the surface. The onset and termination of the Younger Dryas each took place in only about a decade.

Holocene

The start of the post-glacial Holocene period is dated from 11,500 years ago when conditions approached those of today. In the Arctic region, the warmest conditions appear to have occurred in the early Holocene, coincident with the precessional timing of the boreal summer solstice, occurring when the Earth was nearest to the Sun (perihelion), making northern summers relatively warmer. The precession of the equinoxes has a period of approximately 23,000 years; currently, summer solstice occurs at aphelion, when the Earth is most distant from the Sun, making summers relatively cooler. In the early Holocene the distribution of bowhead whale bones indicates at least periodically ice-free summers along the length of the Northwest Passage and the same pattern is repeated much later from approximately A.D. 500–1250.

In the Northern Hemisphere the massive ice sheets delayed the Holocene warming. The Fennoscandian ice sheet disappeared approximately 9 ka and the Laurentide finally approximately 6.5 ka in northern Labrador-Ungava. Its last surviving remnant is the present day Barnes ice cap in Baffin Island. In northern mid-latitudes the Holocene thermal maximum occurred from approximately 8–5 ka, since which time there has been a slight cooling trend. This is driven by precessional effects decreasing solar radiation, which in summer at 65°N has declined by approximately 6 W/m² over the last two millennia This has been punctuated by several neoglacial events each lasting a few centuries. The best known and most studied of these is the Little Ice Age (LIA), dated around the North Atlantic to A.D. 15501850. Recent work suggests that it was triggered by four volcanic events in the late thirteenth century that each led to two to three cold summers as a result of sulfur dioxide aerosols injected into the stratosphere.[2] The effect of these was amplified by ice-albedo feedback resulting from sea ice expansion in the Arctic region, which initiated the onset of the LIA in Iceland and Baffin Island. Further volcanism in the mid-fifteenth century gave this cooling a boost and later, during the Maunder Minimum in sunspot activity from A.D. 1645 to 1715, a decrease of approximately 1% in solar radiation added to the cooling. The presence of sunspots paradoxically gives rise to increased solar radiation because, although the sunspot itself is cooler, the surrounding faculae have higher levels of emission.

The LIA followed an interval known as the Medieval Warm Period, approximately A.D. 950–1250. Temperatures were approximately 0.5–1.4°C above those of 1961–1990 for the northwestern and central

North Atlantic. During this interval the Vikings sailed the North Atlantic and built settlements in southwest Greenland that later succumbed to the deteriorating climate.

The Anthropocene

The Anthropocene is a term that has recently been proposed for the period when human influences on global climate have become widely apparent. The start date of this interval is uncertain, but it coincides approximately with the beginning of the Industrial Revolution approximately A.D. 1800. At that time atmospheric concentrations of CO_2, methane, and pollutants began to rise. From A.D. 1800, CO_2 concentrations increased by almost 40% from approximately 280 ppm to 390 ppm today due to fossil fuel burning and changes in land use, whereas methane concentrations rose 2.5 times from approximately 700 ppb to 1750 ppb due to agriculture (paddy rice and cattle), termites, and release from wetlands and landfill sites. Methane is a more potent greenhouse gas than CO_2. It has a lifetime in the atmosphere of only 12 years, but compared with CO_2 (value 1), methane has a greenhouse warming potential of 25 over a century. For comparative purposes, we may note that Antarctic ice cores show that, in response to fluctuations of CO_2 concentration from 180 to 280 ppm during glacial/interglacial cycles, mean air temperature fluctuated by approximately 8°C.

Temperatures in the twentieth century rose to an initial peak in the 1930s and 1940s, then declined or leveled off, and have risen continuously since the mid-1980s. The early to mid-century warming, which was most marked in the northern North Atlantic, is thought to be attributable to natural variability. Cool season temperatures on Svalbard rose by 2–4°C from 1912 to the 1930s associated with changes in air circulation and sea-surface temperatures.

The decline in Northern Hemisphere temperatures in the 1960s–1970s is attributed to higher levels of sulfate and soot (black carbon) aerosols in the atmosphere due to industrial activity. Over the period 1950s–1980s, there was a "global dimming," with a reduction in direct solar radiation by 4–6% attributed to this effect. Surface solar radiation decreased by 3–7 W/m² per decade in major land areas. A "global brightening" (+2–8 W/m² per decade) succeeded this during the 1980s to 2000, and the trend has continued in the United States and Europe as a result of controls on air pollution. The brightening has contributed to the post-1980s global warming.[3]

The Arctic

Since approximately 1980, the mean global temperatures have risen due to increasing greenhouse gas concentrations by 0.5°C, with the three warmest years on record (since 1850) being 2010, 2005, and 1998. The pattern differs from that of the high northern latitude warming in the early twentieth century in that it was a global phenomenon. In the Arctic, temperatures have increased at almost twice the global average rate in the past 100 years as a result of "polar amplification" involving the effects of the ice-albedo positive feedback of sea ice and snow cover. This effect, together with the advection (horizontal transport) of warm water and warm air from middle latitudes, has given rise to the amplified Arctic warming. In the central Arctic Ocean temperatures from 1979 to 1995 rose by 0.5°C or more in April–June and January and fell slightly during October–December. For locations north of 64°N there have been Arctic-wide warm conditions since the 1990s in spring and, less strongly, in summer that are unique in the instrumental record. Before this there was generally interdecadal negative covariability between northern Europe and Baffin Bay in winter. This pattern of temperature anomalies is associated with the North Atlantic Oscillation (NAO) in mean sea-level pressure between the Azores anticyclone and the Icelandic low. The Arctic-wide warming pattern of the 1990s indicates a change in atmospheric circulation.

The Arctic Ocean sea ice cover has undergone major shrinkage and thinning since the beginning of all-weather satellite mapping with passive microwave sensors in 1979. The trend in ice area from 2001 to 2012 is approximately –3% per decade in the winter months and –12.4% per decade in September when the ice reaches its annual minimum. This can be compared with a decrease of only 2.2% per decade through 1996. The annual minimum area has decreased from approximately 6.5 million km² in 1979 to 3.4 million km² in 2012.[4] The five years since 2007 have shown the lowest ice areas in the 33-year record. The ice thickness has also decreased based on submarine sonar measurements. The mean ice draft decreased from 3.1 m in 1958–1976 to 1.8 m in the 1990s. Further analysis for 1988–2003 shows a thinning of 1.31 m or 43%. Between 2004 and 2008, satellite LiDAR data show a thinning of 67 cm and a 42% decrease in the area of multiyear ice. These changes are attributed to a complex set of factors. Since 2000, air temperatures over the Arctic Ocean and absorbed solar radiation at the surface have increased, and the Pacific surface water entering via Bering Strait has warmed. Earlier, changes in air circulation led to increased export of multiyear ice via Fram Strait associated with a cyclonic circulation. This last factor was due to the positive phase of the Arctic Oscillation (low atmospheric pressure over the Arctic Ocean and high pressure in mid-latitudes) during winters of the late 1980s to early 1990s.

The shrinkage and thinning of sea ice have resulted in the Northwest Passage through the Canadian Arctic Archipelago becoming open for shipping for the first time in summer 2007 and this situation has been repeated in successive summers. The Northern Sea Route north of Siberia also became open to shipping in summer 2008 and subsequent summers. This is a major commercial benefit of global warming.

Increased winter precipitation has characterized the Arctic for the past 40–50 years. However, snow cover in northern high latitudes has generally declined.[5] Visible satellite imagery shows that Arctic snow cover extent in May–June decreased by an average of 18% over the 1966–2008 period of record. The largest and most rapid decreases in snow water equivalent (SWE) and snow cover duration have been observed in the maritime regions of the Arctic, which also have the highest precipitation amounts. The North American sector has exhibited decreases in snow cover duration and depths from approximately 1950, whereas widespread decreases in snow cover were not apparent over Eurasia until after approximately 1980. However, there have been long-term increases in winter snow depth over northern Scandinavia and Eurasia.

Lakes in the high latitudes of Canada are responding to the amplified polar warming by increasingly earlier breakup and later freezeup.[6] The respective rates increased from –0.18 d/yr and +0.123 d/yr for 1950–2004 to –0.23 d/yr and +0.16 d/yr, respectively, for 1970–2004.

Glaciers and ice caps in the Arctic region have been retreating like those in most mountain ranges of the world since the 1960s. Icelandic glaciers and ice caps were in retreat during 1930–1960 and then accelerated retreat and thinning have been observed since 1990. Glaciers and ice caps in the Canadian Arctic region retreated and thinned since records began approximately in 1960. The Ward Hunt ice shelf in the northern Ellesmere Island has lost 90% of its area since the beginning of the twentieth century and has largely disintegrated. Glaciers and ice caps on the Arctic islands north of Eurasia have also retreated and thinned since the –1960s, with an acceleration of this trend since 1990.

The area of summer melt on the Greenland ice sheet, mapped from satellite passive microwave data, has shown a steady increase, with interannual fluctuations, since 1978. The large outlet glaciers in Greenland have shown enhanced calving and thinning in the 2000s as a result of bottom melting from incursions of warm ocean water. This followed a long period of terminus stability. The mass balance of the ice sheet changed from more or less in balance in the 1970s–1980s to a loss of approximately 250 Gt per year during 2005–2011.

Arctic permafrost is warming in response to higher air temperatures and a shortening of the snow cover season. On the North Slope of Alaska this warming was pronounced in the 1980s and 1990s. New record high temperatures at 20 m depth were recorded in 2011 at all permafrost observatories on the North Slope of Alaska, where measurements began in the late 1970s.[7] In northern Russia, permafrost temperatures rose by 1–2°C over the last three decades. During the last 15 years, active-layer thickness has increased in the northern European Russia, northern East Siberia, Chukotka, Svalbard, and Greenland. However, active-layer thickness on the Alaskan North Slope and in the western Canadian Arctic was relatively stable during 1995–2008.

Antarctica

Ice-core reconstructions indicate that Antarctica has warmed approximately 0.2°C since the late nineteenth century.[8] Temperatures were broadly in phase with those in the Southern Hemisphere, but temperatures on the continent and in the Antarctic Peninsula were basically anti-phase.

Temperatures over most of Antarctica showed slight changes in the twentieth century. The exception is the Antarctic Peninsula where a 2.5°C warming has occurred since the 1950s, mainly in winter and spring, when observational records began. Analysis of records, including those from automatic weather stations, since 1957 show warming of 0.10°C per decade in East Antarctica and 0.17°C per decade in West Antarctica. A postulated bipolar seesaw in Arctic and Antarctic air temperatures in the twentieth century has been shown to be absent using ice core data and a reconstruction of the Southern Annular Mode index.[9] Neither of these indices is correlated with climate records from the Arctic or North Atlantic, nor with the Atlantic Multidecadal Oscillation. Instead, these indices are correlated with tropical Pacific sea surface temperatures. However, they point out the complexity of Antarctic climate variability.

Major changes in ice shelves along the Antarctic Peninsula have been observed beginning in the 1960s. Most losses occurred from the 1990s onward. Larsen-B on the northeast coast largely disintegrated in 2002. The Wordie and Wilkins ice shelves on the west coast underwent several large calving events in the late 1990s and 2000s. The loss of ice shelf buttressing of glaciers entering the Larsen-B ice shelf led to their acceleration and rapid thinning.

Whaling ship observations and early ice atlases indicate that Antarctic sea ice in austral summer declined from the 1930s–1950s to the satellite record of the 1970s–1980s. Over the last three decades there has been a slight increase of approximately 3% in Antarctic sea ice area overall, with a maximum in March, and with large regional variations. This increase has been attributed to a slight cooling in summer and autumn.[10] Another suggested mechanism involves increased snowfall raising the surface albedo, but further work is needed.

Records of snow accumulation from ice cores, snow pits, stake networks, and weather stations for both West and East Antarctica indicate that there have been no changes in annual mean values since the late 1950s.

In Antarctica, there is rapid thinning of glaciers in the Amundsen Sea sector, especially on Pine Island Glacier where it is up to 9 m per year. However, it has proved very difficult to estimate the overall mass balance of the ice sheet. Satellite radar altimetry covering 72% of the grounded ice sheet during 1992-2003 indicated a small mass growth. West Antarctica appears to be losing mass and East Antarctica gaining slightly.[11] Permafrost in the Antarctic Peninsula shows warming over the last few decades but there are few data points.

Concluding Remarks

On geological time scales the two polar regions have shown generally synchronous changes, as well as on the 10,000–100,000 year scale. The major exception was from 33 Ma to 14 Ma when the Antarctic had a major ice sheet and the Arctic did not. The last glacial cycle saw major Northern Hemisphere ice sheets in North America and Fennoscandia that only disappeared in the Holocene. The post-glacial warming was interrupted by the 1300-year-long Younger Dryas cold event, particularly around the North Atlantic sector. In the last millennium the same sector experienced the LIA. In the early twentieth century the northern North Atlantic warmed strongly. From the 1980s, there is a marked global warming signature, strongly amplified in the Arctic. Arctic glaciers, the Greenland ice sheet, and particularly Arctic sea ice all are showing a marked response to this warming. In contrast, in southern high latitudes only the Antarctic Peninsula has displayed substantial warming over the last five decades and this is reflected in the retreat or collapse of ice shelves. Antarctic sea ice has shown minor changes.

References

1. DeConto, R.M.; Galeotti, S.; Pagani. M.; Tracy, D.; Schaefer, K.; Zhang, T.; Pollard, D.; Beering, D.J. Past extreme warming events linked to massive carbon release from thawing permafrost. Nature **2012**, *484,* 87–91.
2. Miller, G.H.; Geirsdóttir, A.; Zhong, Y.; Larsen, D.J.; Otto-Bliesner, B.L.; Holland, M.M.; Bailey, D.A.; Refsnider, K.A.; Lehman, S.J.; Southon, J.R.; Anderson, C.; Björnsson, H.; Thordarson, T. Abrupt onset of the Little Ice Age triggered by volcanism and sustained by seaice/ocean feedbacks. Geophys. Res. Lett. **2012**, *39,* L02708
3. Wild, M. Enlightening global dimming and brightening. Bull. Am. Met. Soc. **2012**, *93,* 27–37.
4. Stroeve, J.C.; Serreze, M.C.; Holland, M.P.; Kay, J.E.; Malanik, J.; Barrett, A.P. The Arctic's rapidly shrinking sea ice cover: a research synthesis. Clim. change **2012**, *110,* 105–127.
5. Callaghan, T.V.; Johansson, M.; Brown, R.D.; Groisman, P.Y.; Labba, N.; Radionov, V.; Barry, R.G.; Bulygina, O.N.; Essery, R.L.H.; Frolov, D.M.; Golubev, V.N.; Grenfell, T.C.; Petrushina, M.N.; Razuvaev, V.N.; Robinson, D.A.; Romanov, P.; Shinedell, D.; Shmakin, A.B.; Sokratov, S.A.; Warren, S.; Yang, D. The changing face of Arctic snow cover: a synthesis of observed and projected change. Ambio **2012**, *40,* 17–31.
6. Latifovic, R.; Pouliot, D. Analysis of climate change impacts on lake ice phenology in Canada using the historical satellite data record. Remote Sens. Environ. **2007**, *106,* 492–507.
7. Romanovsky, V.E.; Gruber, S.; Instanes, A.; Jin, H.; Marchenko, S.S.; Smith, S.L.; Trombotto, D.; Walter, K.M. Frozen Ground, Chapter 7, In *Global Outlook for Ice and Snow;* Earthprint, UNEP/GRID: Arendal, Norway, 2007; 181–200.
8. Schneider, D.P: Steig, E.J; van Ommen, T.D.; Dixon, D.A.; Mayewski, P.A.; Jones, J.M.; Bitz, C.M. Antarctic temperatures over the past two centuries from ice cores. Geophys. Res. Lett. **2006**, *33,* L16707.
9. Schneider, D.P; Noone, D.C. Is a bipolar seesaw consistent with observed Antarctic climate variability and trends? Geophys. Res. Lett. **2012**, *39,* L06704.
10. Shu, Q.; Qiao, F.; Song, Z.; Wang, C. Sea ice trends in the Antarctic and their relationship to surface air temperature during 1979-2009. Clim. Dynam. **2012**, *38,* 2355-2563.
11. Wingham, D.J; Shepherd, A.; Muir, A.; Marshall, G.J. Mass balance of the Antarctic ice sheet. Phil. Trans. Roy. Soc. **2006**, *364,* 1627–1635.

Bibliography

Barry, R.G; Gan, T.Y. *The Global Cryosphere: Past, Present and Future;* Cambridge University Press: Cambridge, 2011; 472.
Serreze, M.C; Barry, R.G. *The Arctic Climate System;* Cambridge University Press: Cambridge, 2005; 385.
Turner, J; Marshall, G.J. *Climate Change in the Polar Regions;* Cambridge University Press: Cambridge, 2011; 434.

37

Climate Change: Effects on Habitat Suitability of Tree of Heaven along the Appalachian Trail

John Clark
*Woods Hole
Research Center*

Yeqiao Wang
University of Rhode Island

Introduction

The Appalachian National Scenic Trail (A.T.) is a footpath stretching from Springer Mountain in Georgia to Mount Katahdin in Maine and spanning over 3,500 km of peaks, valleys, and ridges along the Appalachian Mountains. It intersects 14 states; 8 National Forests; 6 units of the National Park System; more than 70 State Park, Forest, and Game Management units; and 287 local jurisdictions. The A.T. passes through some of the largest and least fragmented forest blocks remaining in the eastern United States [1]. The A.T.'s gradients in elevation, latitude, and moisture and north–south alignment represent a continental-scale cross-section mega-transect of eastern U.S. forest and alpine areas for monitoring the health of ecosystems and species that inhabit them.

Tree of heaven (*Ailanthus altissima*) is a deciduous member of the Simaroubaceae family native to the temperate regions of Central China. Tree of heaven is an exotic tree species pervasive throughout the United States due to its rapid growth, high fecundity, hardy tolerance, and strong competitive ability. Mapping the distribution of suitable tree of heaven habitats by integrating field-based observations, geospatial variables defining the biophysical environments of habitats, and a modeling approach capable of predicting habitat change under different projected scenarios, such as climate change effects, helps improve the understanding of ecological processes and management decision support.

Tree of heaven has successfully colonized every continent with the exception of Antarctica. Suitable climates range from temperate to subtropical and humid to arid [2]. It is particularly common in temperate regions with typical conditions consisting of long, warm growing seasons, regular winter frost, and annual precipitation >500 mm [3]. Populations have been recorded within 42 states across the United States, from Florida to Oregon and from New Mexico to Maine. It is susceptible to frost damage, particularly juveniles, restricting it from higher latitudes and elevations. It is relatively drought hearty, though extended dry periods exclude the species from extremely arid regions, and has a root system vulnerable to flooding. The strong competitive ability enables the species to severely impact native communities within its introduced range.

Climate change is a widely recognized phenomenon [4] with significant implications for the spread, impact, and management of invasive species [5]. Climate change alters temperature and precipitation patterns, and resource availability (CO_2, N), and affects management decisions and practices in land cover and land use [6]. Hellman et al. [7] identify five groups of potential interactions between climate change and biological invasion, namely, altered pathways of introduction, likelihood of new invasions, distribution of existing invasions, impacts of invasion, and effectiveness of management strategies. It is challenging to incorporate the full extent of complex factors driving tree of heaven invasion, especially potential interspecific interactions. However, the broad geographic range, high dispersal, and rapid growth suggest that it will adapt to changing conditions more readily than native species, giving them a decisive competitive advantage as the frequency of disturbances within the landscape increases [8].

This chapter summarizes a study for the prediction of the direct effects of climate change, that is, temperature and precipitation trends, on the change of tree of heaven habitat suitability. Warming trends have been predicted to correspond with horizontal migrations as rapid as 0.43 km/yr within boreal forests, globally [9]. In particular, for the A.T. corridor, temperatures are predicted to increases by 2°C–6°C by the end of the 21st century [10]. Modeling effects of climate change on habitats of tree of heaven provided insight on the future distribution of suitable habitat and potential ecological impacts [11]. Integration of seamless geospatial data and climate models along with ground-based observations can provide the capacity in habitat suitability modeling for assessment of the current and potential distributions, as well as the ecological impacts.

Method

Study Site

The spatial extent of this study is adopted from a boundary defined by the National Park Service (NPS) and U.S. Geological Survey (USGS). It was established by selecting all 10-digit Hydrological Unit Code (HUC-10) watersheds within five statute miles of the A.T. land base, termed the A.T. HUC-10 shell [12]. The A.T.-shell provides an ecologically relevant boundary around the trail to distinguish the data most relevant to the corridor area (Figure 37.1).

Data Sources

Ground-Based Observations

Ground-based observations were provided by the Forest Inventory and Analysis (FIA) program of the USDA Forest Service. FIA measurements include the species, size, and condition of trees within the plot, as well as physiographic site attributes [13]. To protect the privacy of private forest landowners, a portion of the FIA plot locations are altered before the records are made publically available. Within the selected A.T.-shell area, 3,926 FIA plots were visited and measured between 2002 and 2010, and observations of tree of heaven were recorded at 136 locations (Figure 37.1). In addition to the plot coordinates, several attributes were retained from the FIA records to examine the characteristics of sites colonized by tree of heaven and compare them to the overall study area. Plot attributes included elevation, aspect, slope, distance to improved road, land ownership, water on plot, physiographic class, stand age, stand size, and basal area of live trees.

FIGURE 37.1 **(See color insert.)** Appalachian Trail HUC-10 shell, FIA plot distribution, and the plots with tree of heaven observed.

Geospatial Data

Topographic information within the A.T.-shell was supplied by the National Elevation Dataset (NED), a 30-m resolution digital elevation model (DEM) produced by the USGS [14]. Individual tiles spanning the study area were acquired from seamless.usgs.gov, mosaicked, and clipped to the boundary of the A.T.-shell. Additional variables, for example, slope/aspects, the compound topographic index (CTI), topographic radiation aspect index (TRASP), were derived from the DEM and provided in TopoMetrics.

Tree of heaven is strongly associated with agricultural and developed land cover, less likely to thrive in saturated wetland conditions, and inherently excluded from open water. Potentially significant land-cover classes were extracted from the USGS National Land Cover Database (NLCD) 2006 [15]. Additional layers were generated by measuring the distance from each pixel to the nearest agricultural and developed feature, respectively, to reflect their strong association with tree of heaven dispersal.

Climate Data

This study adopted the climate data from NASA's Terrestrial Observation and Prediction System (TOPS) [16]. The TOPS provided baseline and projected climate data from the IPCC Fifth Assessment Report (AR5) Coupled Model Intercomparison Project Phase 5 (CMIP5) [17]. The CMIP5 is an ensemble of 16 individual general circulation models (GCMs) that predict future conditions under a set of alternative scenarios defined by representative concentration pathways (RCPs). RCPs represent the atmospheric concentration of greenhouse gasses, or radiative forcing values, in the year 2100 resulting from future scenarios with varying levels of global emissions and mitigation. RCP6.0 was selected for tree of heaven modeling, as it represents a moderate increase in radiative forcing that stabilizes by 2100 due to technologies and strategies for reducing greenhouse gas emissions. An ensemble of CMIP5 data was downscaled to 250 m and subset to the A.T. shell for two time periods, a 1950–2005 baseline and projections for 2090–2095. Multidimensional raster data layers for average monthly maximum temperature, minimum temperature, and precipitation were created, with an individual layer for each month. A set of 19 bioclimatic variables were derived to reflect annual trends, seasonality, and extreme or limiting environmental factors. The bioclimatic variables were derived from each set of climate data using the "biovars" function of the R package "dismo" [18].

Data Preparation

All environmental layers were preprocessed to conform to a uniform spatial extent, resolution, and geographic projection, and converted to ESRI ASCII grid format prior to Maxent modeling. The layers were clipped using the shapefile of the A.T.-shell boundary with a one-mile buffer. Retaining data within a buffer around the shell circumvents distortions caused by edge effects near the boundary of the study area. The layers were then resampled to 300 m using a snap raster template to ensure cell (pixel) alignment agreed perfectly between layers. Bilinear interpolation was used for downscaling continuous variables with a resolution coarser than 300 m.

Habitat Modeling

Variable Selection

A shapefile of all FIA plots within the A.T.-shell was imported into ArcGIS to append values from the collection of ecogeographical variables (EGVs) to the point data attributes. Histograms, boxplots, and t-tests were constructed to compare the distributions of the total FIA population and the *Ailanthus* presence points. Values from the EGVs were also extracted and appended at each FIA plot location, and similar statistics were calculated. Several metrics for evaluating EGV contributions are provided with the Maxent model output. Marginal variable response curves plot the change in logistic prediction from varying the value of one EGV while holding all other EGVs constant at their average sample value [19].

Maxent Modeling Parameters

The default settings for Maxent are adapted from a study that tuned parameters based on datasets for 226 species across 6 regions, and have been shown to deliver good performance across a wide range of applications [20]. Each model iteration was itself a 10-fold (replicate) cross-validation, with 122 of the *Ailanthus* presence points used for model training and the remaining 12 for testing. To assess overfitting, Maxent runs were replicated using both the full set of feature types (linear, quadratic, product, threshold, and hinge transformations) and only basic features (linear and quadratic). Model suitability scores were output in "raw" format, as required for use with ENMTools [19].

Habitat Projection

An array of models was generated by incorporating alternative sets of EGVs and Maxent parameters. Once a model was selected, TOPS AR5 GCM data were substituted for current climate variables and suitability was recalculated. The projections used clamping to ensure the values of all substituted variables were restricted to the range of values encountered while training the model under current conditions. In addition to the projected distribution, Maxent provides outputs for evaluating divergence of current and projected variables, as well as the influence of variable clamping.

Result

Analysis of FIA/EGVs

Comparing the FIA attributes throughout the study area with the subset of tree of heaven presence points revealed patterns that reflected the habitat preferences. Topographically, tree of heaven was generally observed at sites with lower elevations and moderate slope. Culturally, tree of heaven has been found at sites closer to roadways and located on private lands more often than publically held. Biologically, it was more prevalent in younger forest stands [19].

Model Projection Results

Projected climate variables, derived from the AR5 CMIP5 RCP 6.0 ensemble for 2090–2095, were substituted for current climate variables (Figure 37.2). The Maxent modeling-projected distribution for the current and future tree of heaven habitats and the change are illustrated in Figure 37.3. The Maxent

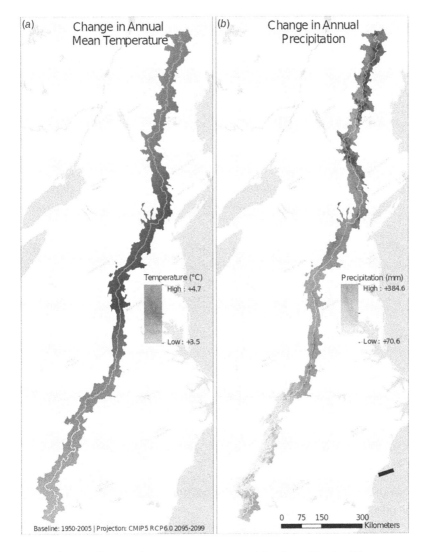

FIGURE 37.2 Distribution of projected changes in temperature (a) and precipitation (b).

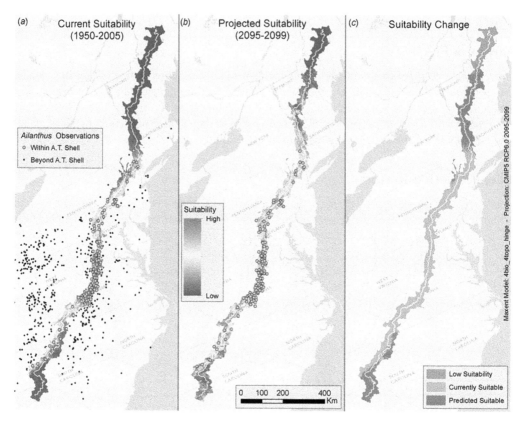

FIGURE 37.3   (See color insert.) Maxent modeling-projected suitability of tree of heaven habitats for the current (1950–2005) (a), the future (2095–2099) (b), and the change (c).

model of current conditions predicts that 57,944 km², or 53.55%, of the A.T. shell is potentially suitable for tree of heaven colonization. By 2095, the suitable areas are projected to expand to 76,940 km² (71.11%), an increase of 32.78%. The mean elevation of suitable areas is projected to increase by 12.27 m (from 420.53 to 432.79 m) and the mean latitude to shift north by 67.72 km².

Discussion

The performance of various EGVs assessed throughout Maxent modeling indicates that the distribution of suitable tree of heaven habitats is primarily constrained by climate conditions at a regional scale. The mean temperature of the coldest quarter and temperature seasonality were particularly significant.

Topographic variables were included in the final distribution model. Soil drainage made the largest contribution, likely due to its influence on flood mortality. Similarly, suitability increased with slope and slope position. While CTI was theoretically well suited to identify wet and dry positions in the landscape, it was inhibited by the coarse scale and blurring effect. The variable response for TRASP shows that tree of heaven prefers the sun exposure and increased temperatures of the south–southwest-facing slopes. Elevation (DEM) contributed to the model, although its role was likely diminished due to correlation with many other variables.

Land-cover variables, while potentially highly significant, proved difficult to incorporate into the model. The association between tree of heaven and urban areas is very prevalent throughout the literature and apparent from the analysis of FIA plot data. However, several obstacles arise when applying this knowledge to a coarse, regional-scale analysis. A 30-m mask of all urban areas was extracted from the NLCD, the distance from each cell in the study area to the nearest urban cell calculated, the layer aggregated to 300 m, and a 800 m radius moving average taken to reflect the uncertainty of FIA locations. The result is a highly distorted representation of the underlying regional patterns of development. Considering that the dispersal distance has been reported to be as little as 100–456 m [21], the variable inevitably loses much of its significance. The regions within the A.T. shell most distant from development also contain some of its most extreme conditions. Maine to the north and the Smoky Mountains to the south contain remote areas but are also at very high elevations with low temperatures and high rainfall, respectively. These novel environments may obscure the underlying relationships between tree of heaven and land-cover types.

The distribution of suitable habitats estimated by the Maxent model largely coincides with the existing knowledge of tree of heaven distribution within the eastern United States. The majority of locations with high suitability fall within the Virginian and mid-Atlantic sections of the A.T, which is unsurprising given the distribution of FIA presence points. Conditions in these regions are ideal, with moderate to low rainfall, low elevations, mild winters, and abundant development. Suitability decreases as the trail moves south into the Smoky Mountains and elevation and precipitation increase, and development thins. To the north, suitability again decreases as elevation and precipitation rise. The northeast is predicted to contain the least suitable areas along the A.T. While *Ailanthus* invasions are reported throughout the northeast, and historically abundant in New England, these records predominately occur within the low elevations and dense population centers along the Atlantic seaboard, rather than the remote, mountainous regions the A.T. passes through. The increase in suitability predicted as the A.T. leaves the Kittatinny Mountain in New Jersey and approaches the New York City metro area seems to support this conclusion. Projecting the model produces several interesting trends. Overall, there is a 32.78% increase in a suitable area, representing a dramatic increase in the potential extent of tree of heaven invasion. Additional methods and independent test data are needed to further validate the model.

Conclusion

The model projection indicates that the potential extent of tree of heaven invasion will increase significantly as the climate changes. Mapping the distribution of suitable habitats facilitates a quantitative assessment of the potential impacts on biodiversity and ecosystem services within the A.T. corridor. In particular, further investigation is needed to determine how biological communities of sensitive high-elevation areas and northern forests will be affected by the introduction of novel competitors.

This habitat suitability model successfully integrated a set of environmental variables that define habitat suitability, map the estimated current distribution of suitable habitats, and examine the potential effects of climate change on biological invasion. The FIA database provided accurate, abundant, and detailed ground observations of tree of heaven populations. Although the relatively coarse grain of the model may have obscured some relationships, geospatial data proved to be a valuable tool for determining the environmental factors. In particular, the seamless climate data products provided by TOPS were a powerful and accessible resource. This study demonstrates the utility of coupling *in situ* and geospatial data with innovative statistical techniques to investigate important ecological processes within the landscape. This modeling approach establishes a framework that can be effectively adopted to examine the distribution of additional important species in the region and inform efforts to conserve natural resources within the study area.

References

1. Dufour, C., and E. Crisfield (Eds.). 2008. *The Appalachian Trail MEGA-Transect*. Appalachian Trail Conservancy: Harpers Ferry, WV.
2. Miller, J. H. 1990. Ailanthus altissima (Mill.) Swingle ailanthus. *Silvics of North America* 2:101–104.

3. Kowarik, I., and I. Saumel. 2007. Biological flora of Central Europe: Ailanthus altissima (Mill.) Swingle. *Perspectives in Plant Ecology, Evolution and Systematics* 8:207–237.

4. IPCC. 2007. The Physical Science Basis. Intergovernmental Panel on Climate Change, *Fourth Assessment Report*, Working Group: 4.

5. Mooney, H. A., and R. J. Hobbs. 2000. *Invasive Species in a Changing World*. Island Press: Washington, DC.

6. Bradley, B. A., D. M. Blumenthal, D. S. Wilcove, and L. H. Ziska. 2010. Predicting plant invasions in an era of global change. *Trends in Ecology & Evolution* 25:310–318.

7. Hellmann, J. J., J. E. Byers, B. G. Bierwagen, and J. S. Dukes. 2008. Five potential consequences of climate change for invasive species. *Conservation Biology* 22:534–543.

8. Dale, V. H., L. A. Joyce, S. McNulty, R. P. Neilson, M. P. Ayres, M. D. Flannigan, P. J. Hanson, L. C. Irland, A. E. Lugo, C. J. Peterson, D. Simberloff, F. J. Swanson, B. J. Stocks, and B. M. Wotton. 2001. Climate change and forest disturbances. *BioScience* 51:723–734.

9. Loarie, S. R., P. B. Duffy, H. Hamilton, G. P. Asner, C. B. Field, and D. D. Ackerly. 2009. The velocity of climate change. *Nature* 462:1052–1055.

10. Hashimoto, H., S. H. Hiatt, C. Milesi, F. S. Melton, A. R. Michaelis, P. Votava, W. Wang, and R. R. Nemani. 2011. Ch. 22, Monitoring and Forecasting Ecosystem Climate Impacts on Ecosystem Dynamics Using the Terrestrial Observation and Prediction System. In Wang, Y., Editor. *Remote Sensing of Protected Lands*. CRC Press: Boca Raton, FL.

11. Jeschke, J. M., and D. L. Strayer. 2008. Usefulness of bioclimatic models for studying climate change and invasive species. *Annals of the New York Academy of Sciences* 1134:1–24.

12. Dieffenbach. 2011. "Appalachian National Scenic Trail vital signs monitoring plan. Natural Resource Technical Report NPS/NETN/NRR—2011/389.", Woodstock, VT: National Park Service, Northeast Temperate Network.

13. Woudenberg, S. W., B. L. Conkling, B. M. O'Connell, E. B. LaPoint, J. A. Turner, and K. L. Waddell. 2010. *The Forest Inventory and Analysis Database: Database Description and Users Manual Version 4.0 for Phase 2*. US Department of Agriculture, Forest Service, Rocky Mountain Research Station: Newtown Square, PA.

14. Gesch, D., M. Oimoen, S. Greenlee, C. Nelson, M. Steuck, and D. Tyler. 2002. The national elevation dataset. *Photogrammetric Engineering & Remote Sensing* 68:5–11.

15. Fry, J.A., G.X. Xian, S. Jin, and J. Dewitz. 2011. Completion of the 2006 national land cover database for the conterminous United States. *Photogrammetric Engineering & Remote Sensing*, 77(9):858–864.

16. Nemani, R., H. Hashimoto, P. Votava, F. Melton, W. Wang, A. Michaelis, L. Mutch, C. Milesi, S. Hiatt, and M. White. 2009. Monitoring and forecasting ecosystem dynamics using the terrestrial observation and prediction system (TOPS). *Remote Sensing of Environment*, 113:1497–1509.

17. Taylor, K. E., R. J. Stouffer, and G. A. Meehl. 2012. An overview of CMIP5 and the experiment design. *Bulletin of the American Meteorological Society* 93:485–498.

18. Hijmans, R. J., S. Phillips, J. Leathwick, and J. Elith. 2012. Package "dismo."

19. Clark, J. 2013. Effects of Climate Change on the Habitats of the Invasive Species Ailanthus altissima along the Appalachian Trail. *M.S. Thesis*, University of Rhode Island.

20. Phillips, S. J., and M. Dudik. 2008. Modeling of species distributions with Maxent: New extensions and a comprehensive evaluation. *Ecography* 31:161–175.

21. Landenberger, R. E., N. L. Kota, and J. B. McGraw. 2006. Seed dispersal of the non-native invasive tree Ailanthus altissima into contrasting environments. *Plant Ecology* 192:55–70.

38

Climate Change: Ecosystem Dynamics along the Appalachian Trail

Yeqiao Wang
University of Rhode Island

Introduction

Appalachian Mountains, also known as the Appalachians, are a system of mountains in eastern North America. The Appalachians run from Central Alabama to the New England region of the northeast United States and extend into sections of Quebec, New Brunswick, and Newfoundland of Canada. The mountain range forms a natural barrier between the eastern coastal regions and the interior lowlands of northeast America. The forests of the Appalachians sustain a variety of ecosystems and native biological diversity. The environmental effects and ecosystem functions and services associated with the Appalachians have long been the focus of scientific research in understanding the mountain range as a unique geographic entity and in comparison with different spatial contexts for providing a regional reference and validation.

The Appalachian Trail traverses most of the high-elevation ridges of the eastern United States, extending 3,676 km across 14 states, from Springer Mountain in northern Georgia to Mount Katahdin in Central Maine. The Appalachian Trail intersects 8 National Forests and 6 National Park units; crosses more than 70 State Park, Forest, and Game Management units; and passes through 287 local jurisdictions. Its gradients in elevation, latitude, and climate sustain a rich biological assemblage of temperate-zone forest species. The Appalachian Trail and its surrounding protected lands harbor forests with some of the greatest biological diversity in the United States, including rare, threatened, and endangered species, and diverse bird and wildlife habitats. They are also the source of the headwaters of important water resources of millions of people. The north–south alignment of the trail represents a cross-sectional mega-transect of the eastern United States forests and alpine areas, and offers a setting for collecting data on the health of the ecosystems and the species that inhabit them. The high-elevation setting and its protected corridor provide a barometer for early detection of undesirable changes in the natural resources, such as development encroachment, acid precipitation, invasions of exotic species, and climate change impacts (Dufour and Crisfield, 2008; Zhao et al., 2012). Within the corridor, the decrease in mean temperature toward the north becomes evident around the 39°N, illustrating the latitudinal variation of climate (Figure 38.1). The northern section of the Appalachian Trail corridor provides an important reference for monitoring the effects of climate change in the northeastern United States.

Climate change studies reveal the effects of global warming on the growing season of terrestrial vegetation at mid- and high latitudes (Karl et al., 1996; Myneni et al., 1997; Zhang et al., 2004; Burakowski, 2008). Tracking variation of landscape dynamics provides an understanding of the changing environment, the impacts and threats caused by changes, and the likely trends in the future for natural resources and associated ecosystems (Wang et al., 2009).

Adaptive management requires data and indicators to assess the state of ecosystems and natural as well as human resources managed, including change over time to verify whether or not management strategies are effective. This is especially important in view of a changing climate, with direct and indirect consequences on the resources managed. Nevertheless, monitoring programs to systematically

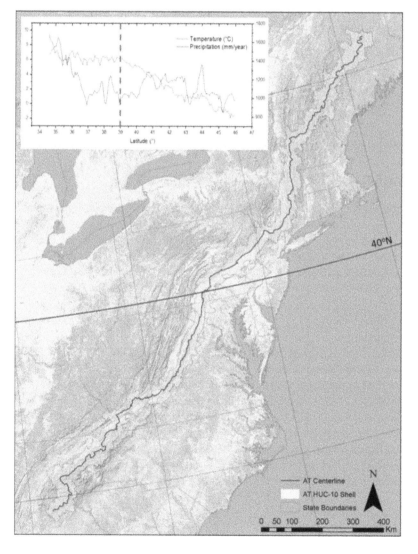

FIGURE 38.1 Spatial extent of the Appalachian Trail corridor and the patterns of decreasing mean temperature at around 39°N with variations in precipitations toward northern latitudes (see inset) that reflect latitudinal effects of the climate.

gather and analyze such data are rare, often opportunistic, and not necessarily driven by the needs of adaptive management efforts. Monitoring efforts are also expensive and difficult to maintain over time; therefore, there is a need for strategic and targeted efforts with sustained sources of dedicated funding. Finally, ecosystems and natural resources do not recognize political barriers, and it is a significant challenge for states and governmental entities to work together to plan and execute truly regional efforts. This chapter presents an example of monitoring climate change impacts on ecosystem dynamics along the Appalachian Trail corridor as a cross-region mega-transect perspective along the Appalachian Mountain regions. It is noteworthy that, despite the differences in the nature of those efforts, successful monitoring benefits from well-defined indicators and requires extensive collaborations across agencies and state boundaries.

The current patterns in climate and vegetation along the Appalachian Trail corridor indicate that

1. Mean temperature increases from north to south with a difference of over 10°C between the northern terminus in Maine and the southern terminus in Georgia;
2. Most of the study area receives more than 900 mm/year of precipitation. Precipitation decreases from 35°N to 37°N by 500 mm/year, increases from 37°N to 41°N, and then decreases again north of latitude 41°N;
3. Annual peak leaf area index (LAI) shows a large decline from 39°N to 41°N due to the amount of cropland and urban area in the landscape;
4. Net primary production (NPP) decreases from south to north, and the decrease can be explained by the temperature gradient with the exception of a decline induced by cropland and urban area around 40°N (Hashimoto et al., 2011). A comparison between climate variables and NPP shows that the latitudinal gradient of NPP is mostly controlled by temperature through its effect on modulating growing season length (Jenkins et al., 2002).

Methods and Results

Hashimoto et al. (2011) analyzed time-series data from Moderate Resolution Imaging Spectroradiometer (MODIS) and other remote sensing data products, Global Inventory Modeling and Mapping Studies (GIMMS), and Surface Observation and Gridding System (SOGS), using the Terrestrial Observation and Prediction System (TOPS). The study projected the regional impacts of climate change along the Appalachian Trail corridor area by downscaling general circulation model (GCM) scenarios and using the scenarios to drive dynamic ecosystem models to assess the vegetation response to the projected climate scenarios. The study used climate scenarios derived from the World Climate Research Program (WCRP) Coupled Model Intercomparison Project (CMIP3) multimodel datasets, which are based on the climate scenarios produced for the Fourth Assessment Report (AR4) of the Intergovernmental Panel on Climate Change (IPCC, 2007). The study used the outputs from 11 models for the Special Report on Emission Scenarios A1B (SRES A1B), which assumes a future with high economic growth, a well-balanced energy resource portfolio, and new technology development, with atmospheric CO_2 concentration stabilizing at 720 ppm.

The study found that all the climate models project a steady temperature increase, ranging from 2°C to 6°C by the end of the 21st century. The ensemble mean temperature increased from 11°C to 14.5°C, while precipitation did not show a clear trend (Figure 38.2). To evaluate the regional impacts of climate scenarios on ecosystems and protected areas, the study ran a dynamic ecosystem model to simulate the ecosystem response to changes in climate from 1980 to the end of the 21st century. The simulated NPP is projected to increase. Measured in grams of carbon bound into organic material per square meter per year (gC/m²/year), the ensemble mean NPP increases from 60 to 80 gC/m²/year. The increase can be attributed to the effect of increasing atmospheric CO_2 concentrations and CO_2 fertilization on the vegetation in the region. However, the net ecosystem exchange (NEE) is predicted to be constant at a rate of approximately –10 gC/m²/year (Figure 38.3). Negative NEE indicates a flux of carbon from

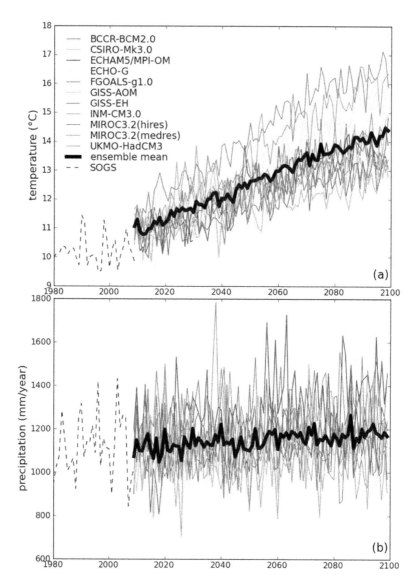

FIGURE 38.2 (See color insert.) Projected mean temperature (a) and precipitation (b) through the end of the 21st century downscaled from CMIP3 multimodel dataset of SRES A1B scenario for 11 global climate models (GCMs) (Hashimoto et al., 2011).

terrestrial ecosystems to the atmosphere. This means that the predicted increase in respiration due to rising temperature exceeds the predicted increase in NPP. The results suggest that the current carbon sink in the eastern United States could turn into a carbon source in the future under the SRES A1B scenario. If we were to use the projections that follow the trajectory of the highest emission scenarios, the forests along the Appalachian Trail corridor would be expected to start releasing even more carbon and eventually decline in growth.

Changes in the seasonal distribution of runoff have a great impact on downstream water availability and potentially affect water allocation planning. Although the projected annual runoff does not show a clear trend (Figure 38.3), the peak in runoff is projected to take place earlier in the year, advancing from April to March by the end of the 21st century for the region as a whole. Also, the cumulative winter runoff is projected to increase, while peak runoff is projected to decrease (Hashimoto et al., 2011).

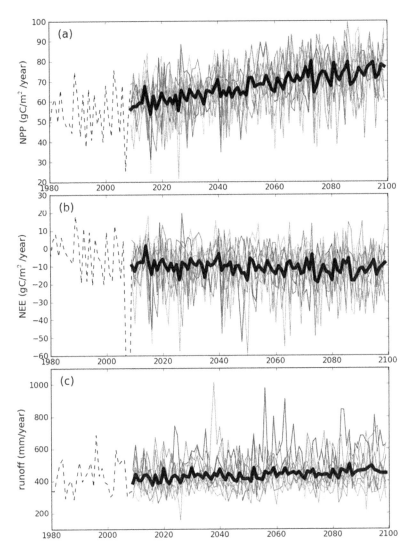

FIGURE 38.3 **(See color insert.)** Projected NPP (a), NEE (b), and runoff (c) through the end of the 21st century downscaled from CMIP3 multimodel dataset of SRES A1B scenario for 11 GCMs (Hashimoto et al., 2011).

This projected change in runoff can be attributed to increased winter snowmelt caused by higher temperatures and increased winter rainfall.

Land surface phenology (LSP) is one of the measures of landscape dynamics. As an indicator, LSP reflects the response of vegetated surfaces to seasonal and annual changes in climate, including the hydrologic cycle. An increasing number of studies are reported on phenology shifts in spatial pattern and timing of the growing season (e.g., de Beurs and Henebry, 2004; Wolfe et al., 2005; Jeong et al., 2011; Wang et al., 2012). Using data from time-series satellite remote sensing, LSP metrics typically retrieve the time-of-onset greenness as the start of the season (SOS), onset of senescence or time-of-end greenness as the end of the season (EOS), time of maximum of the growing season by peak vegetation indices, and growing season length or duration of greenness (LOS). Studies of LSP in the Appalachian Trail corridor for the 24 years between 1982 and 2006 reveal trends of 2.6-day delay for SOS and 9.7-day delay for EOS, respectively. The trend of delayed EOS was more evident for the sections within the northern provinces of ecosystem regions, for example, in the Adirondack–New England

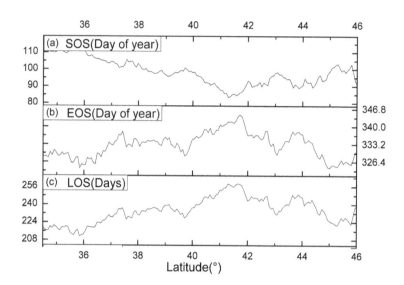

FIGURE 38.4 The latitude and longitude trends of LSP metrics of the SOS (a), EOS (b), and growing season length or duration of greenness (LOS) (c) for the entire Appalachian Trail corridor area between 1982 and 2006.

Mixed Forest–Coniferous Forest–Alpine Meadow Province than that within the Central Appalachian Broadleaf Forest–Coniferous Forest–Meadow Province in the south (Figure 38.4). For the corridor area from 39°N north, the SOS shows a trend of advances over 3 days in the period between 1982 and 1998 and delayed over 2 days in the later period between 1999 and 2006, respectively. The trend of EOS was delayed over 7 days between 1982 and 1998 and delayed more for the 24 years between 1982 and 2006.

Conclusion

This chapter summarizes the case study in which TOPS was used to analyze recent trends in vegetation condition and predict the response of vegetation to climate change along the Appalachian Trail corridor. Climate processing, satellite data ingestion, and ecosystem modeling are the main capabilities that are integrated into TOPS. TOPS produced the climatology of recent temperature, precipitation, and vegetation patterns, and found a north–south temperature-dependent gradient of NPP along the Appalachian Trail corridor. Time series of climate and NDVI derived from TOPS revealed a forest cover decline during the past three decades. TOPS was used to downscale future climate change projections to a spatial resolution of 8 km, and run a biophysical ecosystem model to understand the impacts on the Appalachian Trail ecosystems. This analysis illustrated a projected increase in both NPP and carbon release, with the overall result that ecosystems along the Appalachian Trail were projected to serve as a net source of CO_2.

Measurements of LSP metrics provided information about patterns in interannual variation and long-term trends over the time period. Across the entire corridor and for the 24 years between 1982 and 2006, the SOS had a trend of 4-day delay and EOS had a trend of 10-day delay. Since 2000, the SOS was later than the long-term average, which was the reason that SOS showed a delayed trend in this time period. Different trends were observed for the time period from 1982 to1999. The SOS showed an advanced trend (−0.38 days/year) from 1982 to 1999. Affected by delayed EOS, the LOS was extended accordingly (Zhao et al., 2012).

The outcome of this study provided information about the effects of climate change along the ridges of Appalachian Mountains in the eastern United States. The data and scientific conclusions were imbedded into an Internet-based decision support system (DSS) for monitoring, reporting, and forecasting ecological conditions of the Appalachian Trail (Clark et al., 2014).

References

Burakowski, E.A. Trends in wintertime climate in the northeastern United States: 1965–2005. *Journal of Geophysical Research. Biogeosciences*, **2008**, 113:D20114.

Clark, J., Wang, Y., August, P. Assessing current and projected suitable habitats for tree-of-heaven along the Appalachian Trail. *Philosophical Transactions of the Royal Society B*, **2014**, 369:20130192.

de Beurs, K.M., Henebry, G.M. Land surface phenology, climatic variation, and institutional change: Analyzing agricultural land cover change in Kazakhstan. *Remote Sensing of Environment*, **2004**, 89 (4):497–509.

Dufour, C., Crisfield, E. *The Appalachian Trail MEGA-Transect*, **2008**, Harpers Ferry, WV: Appalachian Trail Conservancy.

Hashimoto, H., Hiatt, S.H., Milesi, C., Melton, F.S., Michaelis, A.R., Votava, P., Wang, W., Nemani, R.R. Monitoring and Forecasting Climate Impacts on Ecosystem Dynamics in Protected Areas Using the Terrestrial Observation and Prediction System. In Wang, Y. (Ed.), *Remote Sensing of Protected Lands*, **2011**, Boca Raton, FL: CRC Press.

IPCC, 2007: *Climate Change 2007: The Physical Science Basis. Contribution of Working Group I to the Fourth Assessment Report of the Intergovernmental Panel on Climate Change*, S. Solomon, D. Qin, M. Manning, Z. Chen,M.Marquis, K.B.Averyt,M. Tignor and H.L.Miller, Eds., Cambridge University Press, Cambridge, 996 pp.

Jenkins, J.P., Braswell, B.H., Frolking, S.E., Aber, J.D. Detecting and predicting spatial and interannual patterns of temperate forest springtime phenology in the eastern US. *Geophysical Research Letters*, **2002**, 29 (24):54.

Jeong, S.J., Ho, C.H., Gim, H.J., Brown, M.E. Phenology shifts at start vs. end of growing season in temperate vegetation over the Northern Hemisphere for the period 1982–2008. *Global Change Biology*, **2011**, 17 (7):2385–2399.

Karl, T.R., Knight, R.W., Easterling, D.R., Quayle, R.J. Indices of climate change for the United States. *Bulletin of the American Meteorological Society*, **1996**, 77 (2):279–292.

Wang, Y., Mitchell, B.R., Nugranad-Marzilli, J., Bonynge, G., Zhou, Y., Shriver, G. Remote sensing of land-cover change and landscape context of the National parks: A case study of the Northeast temperate network. *Remote Sensing of Environment*, **2009**, 113:1453–1461.

Wang, Y., Zhao, J., Zhou, Y., Zhang, H. Variation and trends of landscape dynamics, land surface phenology and net primary production of the Appalachian mountains. *Journal of Applied Remote Sensing*, **2012**, 6 (061708):1–15.

Wolfe, D.W., Schwartz, M.D., Lakso, A.N., Otsuki, Y., Pool, R.M., Shaulis, N.J. Climate change and shifts in spring phenology of three horticultural woody perennials in northeastern USA. *International Journal of Biometeorology*, **2005**, 49 (5):303–309.

Zhang, X., Friedl, M.A., Schaaf, C.B., Strahler, A.H., Schneider, A. The footprint of urban climates on vegetation phenology. *Geophysical Research Letters*, **2004**, 31 (12):L12209.

Zhao, J., Wang, Y., Hashimoto, H., Melton, F.S., Hiatt, S.H., Zhang, H., Nemani, R.R. The variation of land surface phenology from 1982 to 2006 along the Appalachian Trail. *IEEE Transactions on Geoscience and Remote Sensing*, **2012**, 51(4):2087–2095.

39

Spatial and Temporal Variations in Global Land Surface Phenology

Jianjun Zhao,
Xiaoyi Guo, and
Hongyan Zhang
*Northeast Normal
University*

Introduction

The past century has been affected by global warming. From 1880 to 2012, the global temperature increased by an average of 0.85°C. In addition, the temperature has increased more in the past 30 years than in any other period since 1850 [1]. Climate change has had tremendous impacts on ecosystems worldwide by constantly changing their structure, function, and production and directly affecting the sustainable development of the environment and society [2,3]. Moreover, these impacts will certainly become aggravated in the future [1].

Vegetation, a natural "link" between soil, atmosphere, and water, acts as a comprehensive indicator in climate change research. Accordingly, vegetation dynamics represent the most distinct and comprehensive reflection of the impacts of climate change, which can noticeably affect the production, carbon budget, and phenology of land vegetation and even the composition of species. Furthermore, the feedback of vegetation to the climate may either accelerate or slow the rate of global climate change. For example, reported studies suggest that the advance of forest phenology slows the rate of global warming [4], while an increase in vegetation coverage leads to a rise in precipitation [5], and changes in various vegetation types affect regional climate [6]. Land surface phenology (LSP), which is closely related to climatic factors [7], has been widely used in research on the responses of ecosystems to climate change [8]. It is applied in monitoring and understanding of vegetation in the life cycle [9].

Traditional phenological studies have mainly focused on field investigations to record the characteristics of phenological change by field-based observation. Alternatively, the development of remote sensing technology has provided a means for large-scale phenological observations, compensating for the shortcomings of traditional observation methods [10]. Remote sensing provides the possibility for further understanding of the relationships between regional, continental, and global vegetation phenological dynamics and climate change [11,12]. Remote sensing inversion method for vegetation phenology has been developing using time-series remote sensing data [13]. At present, major methods employed to extract phenological information from remote sensing data include the following: (i) The threshold method [14,15] uses the percentage of the amplitude of a given time-series curve to determine phenological information. Its advantage lies in its ease of use, while its disadvantage is that the noise from sensors and the atmosphere cannot be completely eliminated. (ii) The midpoint method extracts phenological information based on the midpoint of the annual range of the vegetation index [16,17], that is, the average value of the difference between the maximum and minimum vegetation indexes. This approach suffers from the same disadvantage. (iii) The maximal slope method estimates the growing season according to the change rate of vegetation index; the dates of the fastest rising point on the left side and the fastest falling point on the right side of the vegetation index curve are calculated. This method is also affected by noise in measured data [18]. (iv) The spectral characteristics analysis method [19] eliminates the influence of clouds on changes in the vegetation index. (v) The fitting method obtains phenological parameters by fitting time-series changes in the vegetation index [20]. This method can eliminate the influence of pixel noise from the fitting curve. (vi) The cumulative frequency method obtains phenological parameters using cumulative growth days in time series for the vegetation index [21]. This method requires additional climate data to estimate the rate of change in the date per degree.

Over the past 30 years, many studies have thoroughly investigated vegetation phenological dynamics and the relationship between phenology and climate in the global scale [22,23], throughout the Northern Hemisphere [14,24], and at the continental [25,26] and regional scales [27,28]. The findings of the studies revealed that the growth stage of vegetation in spring was advanced and that the withering period in autumn was delayed [11,29]. This study integrates three different time-series reconstruction methods and two phenological parameter extraction methods to extract global land surface phenological parameters, which provide a reference for understanding the spatial and temporal variations caused by global climate change.

Data and Methods

Advanced Very High Resolution Radiometer Global Inventory Modeling and Mapping Studies (AVHRR GIMMS) Dataset

The main normalized difference vegetation index (NDVI) product is the latest global vegetation index data product (NDVI3g) released by the National Aeronautics and Space Administration (NASA) developed by GIMMS (https://ecocast.arc.nasa.gov/data/pub/gimms/3g.v1/). GIMMS NDVI3g series products are developed on the basis of the seven National Oceanic and Atmospheric Administration (NOAA) AVHRR series, namely, the satellites NOAA-7, NOAA-8, NOAA-11, NOAA-14, NOAA-16, NOAA-17, and NOAA-18. The spatial resolution of the data is 8×8 km, the temporal resolution is 15 days, and the temporal coverage ranges from January 1982 to December 2015. GIMMS NDVI3g data are preprocessed by applying a radiation correction and atmospheric correction and a coordinate transformation, after which daily and track-by-track images are further processed through an accurate geometric correction and the removal of clouds and bad lines. Finally, the maximum value composite (MVC) is determined to obtain the final NDVI dataset.

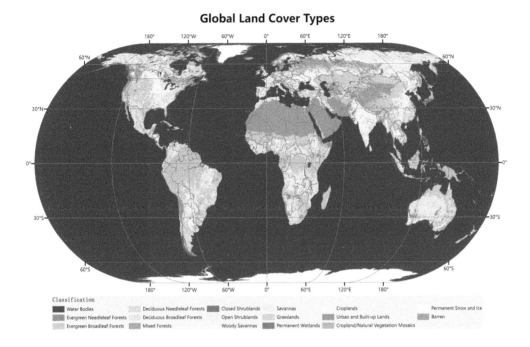

FIGURE 39.1 (**See color insert.**) Global land-cover types.

Land-Cover Data

The Moderate Resolution Imaging Spectroradiometer (MODIS) Land cover Type Product (MCD12Q1) provides a suite of science datasets (SDSs) that map global land cover at a 500 m spatial resolution with an annual time step for six land-cover legends. The maps were created from the classifications of spectrotemporal features derived from MODIS data. This chapter used the International Geosphere-Biosphere Programme (IGBP) classification system in 2015 with a 500 m spatial resolution [30].

In terms of the global vegetation distribution, deciduous needleleaf forests, evergreen needleleaf forests, and mixed forests are distributed mainly at the middle and high latitudes of the Northern Hemisphere. Evergreen broadleaf forests are distributed mainly near the equator of South America, Africa, and Oceania, while deciduous broadleaf forests are mainly distributed in eastern North America and East Asia. Grasslands and cropland/natural vegetation mosaics are distributed throughout the Northern and Southern Hemispheres (Figure 39.1). Table 39.1 shows the values and descriptions of each land-cover type.

Methods

To avoid uncertainty caused by using a single phenology extraction method, the NDVI time-series reconstruction methods used in this chapter include linear interpolation cubic spline function method (spline) and double logistic fitting [31]. In addition, two extraction methods are utilized: the threshold method [14,15,16] and the derivative method [18]. These three fitting methods and two extraction methods are combined to obtain six phenological parameters at the start of the growing season (SOS) and the end of the growing season (EOS). The averages of these six results were used for the analysis presented in this chapter.

TABLE 39.1 MCD12C1 IGBP Legend and Class Descriptions

Land-Cover Types	Name	Description
Land cover 0	Water bodies	At least 60% of the area is covered by permanent water bodies.
Land cover 1	Evergreen needleleaf forests	Dominated by evergreen conifer trees (canopy >2 m). Tree cover >60%.
Land cover 2	Evergreen broadleaf forests	Dominated by evergreen broadleaf and palmate trees (canopy >2 m). Tree cover >60%.
Land cover 3	Deciduous needleleaf forests	Dominated by deciduous needleleaf (larch) trees (canopy >2 m). Tree cover >60%.
Land cover 4	Deciduous broadleaf forests	Dominated by deciduous broadleaf trees (canopy >2 m). Tree cover >60%.
Land cover 5	Mixed forests	Dominated by neither deciduous nor evergreen (40%–60% of each) tree types (canopy >2 m). Tree cover >60%.
Land cover 6	Closed shrublands	Dominated by woody perennials (1–2 m height). Tree cover >60%.
Land cover 7	Open shrublands	Dominated by woody perennials (1–2 m height). Tree cover 10%–60%.
Land cover 8	Woody savannas	Tree cover 30%–60% (canopy > 2 m).
Land cover 9	Savannas	Tree cover 10%–30% (canopy > 2 m).
Land cover 10	Grasslands	Dominated by herbaceous annuals (<2 m).
Land cover 11	Permanent wetlands	Permanently inundated lands with 30%–60% water cover and >10% vegetated cover.
Land cover 12	Croplands	At least 60% of the area is cultivated cropland.
Land cover 13	Urban and builtup lands	At least 30% is impervious surface area, including building materials, asphalt, and vehicles.
Land cover 14	Cropland/Natural vegetation mosaics	Mosaics of small-scale cultivation covering 40%–60% with natural trees, shrubs, or herbaceous vegetation.
Land cover 15	Permanent snow and ice	At least 60% of the area is covered by snow and ice for at least 10 months of the year.
Land cover 16	Barren	At least 60% of the area is non-vegetated and barren (sand, rock, or soil) areas with less than 10% vegetation.
Land cover 17	Unclassified	Has not received a map label because of missing inputs.

Results and Discussion

Spatial Distribution Patterns of Global Land Surface Phenology

The mean SOS and EOS dates from 1982 to 2015 were calculated to obtain the global mean SOS and EOS spatial variation maps over the past 34 years (Figure 39.2a, b and Table 39.2, respectively).

The LSP changes in global vegetation from 1982 to 2015 exhibit obvious spatial differences, and their distribution pattern follows certain regularity (Figure 39.2).

According to the SOS distribution frequency (Figure 39.3), the SOS dates in the Northern Hemisphere are concentrated mainly from 60 days to 200 days, while the cumulative frequency of the SOS in the Southern Hemisphere does not show a substantial change and displays two small peaks on day 130 and day 280. The SOS for global vegetation is concentrated mainly between 80 days and 160 days (Figure 39.2a). Moreover, the SOS date varies substantially along a defined latitude; the SOS is gradually and increasingly delayed with an increase in latitude. The SOS before day 80 is mainly distributed in the area near 30°N in southern North America and southern Europe. The latest SOS (later than day 160) occurs mainly at the middle and low latitudes from 10°N to 30°N and latitudes higher than 10°S. Within the same latitude area, the SOS dates in Asia occur later than those in Europe and North America, and

(a) The Start of the Growing Seasons (SOS) Distribution from 1982 to 2015

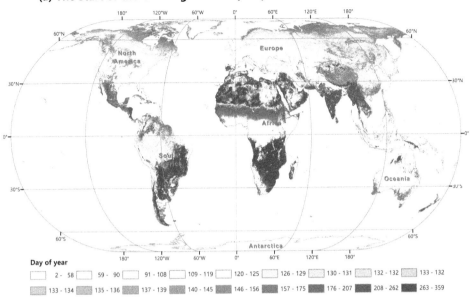

(b) The End of the Growing Seasons (EOS) Distribution from 1982 to 2015

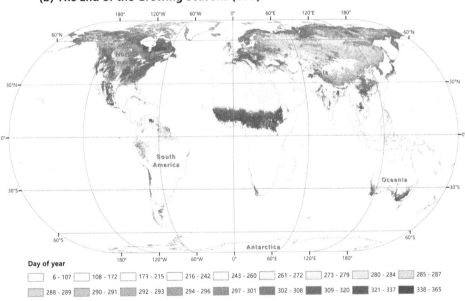

FIGURE 39.2 **(See color insert.)** Averages of the SOS (a) and the EOS (b), around the globe between 1982 and 2015.

TABLE 39.2 Mean SOS and EOS Dates of the Continental LSP

	Asia	Europe	North America	Oceania	Africa	South America
SOS	137.91	112.48	131.78	123.61	210.42	195.23
EOS	274.36	286.59	293.17	224.67	223.17	191.00

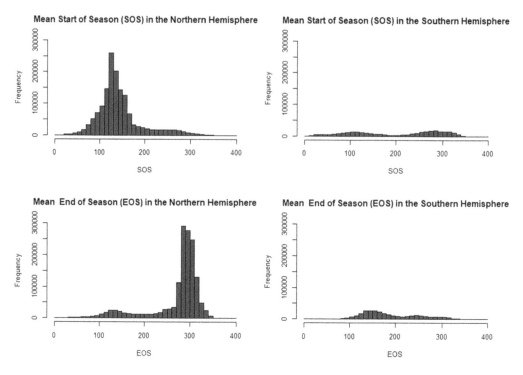

FIGURE 39.3   Frequency distributions of the global LSP.

the distributions of the SOS dates in North America and Asia are markedly different. The SOS dates throughout Europe are relatively consistent and occur mostly from 100 days to 120 days. The SOS dates in South America, Africa, and Oceania are relatively late compared with those in other continents, occurring mainly from 200 days to 350 days. The SOS in the region near the equator in South America and Africa occurs from 0 days to 200 days.

The spatial distributions of the SOS and EOS dates exhibit opposite trends (Figure 39.2b). The peak cumulative frequency of the EOS in the Northern Hemisphere appears mainly from 270 days to 330 days (Figure 39.3). The EOS is delayed with increases in latitude. The earliest EOS occurs mainly in the Arctic Circle (earlier than day 260), and the latest EOS occurs in the low latitudes over North America and Europe (later than day 320). The EOS is concentrated from 280 days to 300 days throughout most regions of Asia. The distribution frequency of the EOS in the Southern Hemisphere does not change substantially. Most of the EOS dates at the pixel scale occur from 100 days to 200 days, while a small number of EOS dates occur from 200 days to 300 days. The spatial regularity of the EOS in the Southern Hemisphere is the opposite of that in the Northern Hemisphere. The distributions of the EOS dates throughout North America, South America, and Oceania are relatively consistent, while the spatial distributions of the EOS dates in the Northern Hemisphere are substantially different.

Variations in the Land Surface Phenology with Latitude

This chapter analyzes the change characteristics of the mean dates of land surface phenological pixels (vegetation-covered areas) along the latitudes from 55°S to 77°N at the same longitude. The changes in the SOS and EOS with latitude show opposite trends; that is, the SOS increases with an increase in latitude and the EOS decreases with a northward increase in latitude.

Figure 39.4 shows the characteristics of the SOS dates along a latitude that are analyzed from north to south. In the Northern Hemisphere, the SOS is significantly advanced with a decrease in latitude from

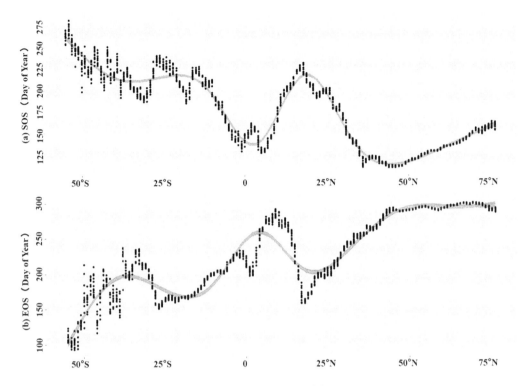

FIGURE 39.4   Variations in the LSP with the latitude (a) SOS and (b) EOS.

77°N to 45°N; the advancing rate of the SOS is 1.58 days with a 1° decrease in latitude. The SOS reaches a minimum (i.e., the beginning time of the growing season is the earliest) at approximately 45°N. From 44°N to 17°N, the SOS date becomes significantly delayed with a decrease in latitude; the delay rate of the SOS is 4.71 days with a 1° decrease in latitude. The SOS reaches a peak at approximately 17°N. With a decrease in latitude, the SOS continues to advance and reaches a new minimum at approximately 5°N. From 16°N to 5°N, the SOS advances by 9.19 days with a 1° decrease in latitude, and there is a minor fluctuation from 5°N to the equator. An upward trend is generally observed from the equator southward. From 1°S to 15°S, the SOS is delayed by 6.71 days with a 1° increase in latitude. From 16°S to 30°S, the SOS shows an advance–delay–advance fluctuation with a continuing increase in latitude. Finally, from 31°S to 55°S, the SOS is advanced by 2.52 days with a 1° increase in latitude.

The variation trend of the EOS along a latitude is basically opposite to that of the SOS. The minimum of the EOS corresponds to the peak of the SOS; that is, the EOS occurs later, whereas the SOS occurs earlier, which coincides with a long growing season. No EOS trend is evident between 77°N and 45°N. In the range from 44°N to 18°N, the EOS is advanced by 4.66 days with a 1° decrease in latitude. The EOS initially increases and then falls between 17°N and 0°. In the range from 1°S to 27°S, the EOS is advanced by 6.7 days with a 1° increase in latitude. The EOS shows a delay trend from 28°S to 34°S. Finally, in the range from 35°S to 55°S, the EOS is advanced by 4.63 days with a 1° increase in latitude.

The minimum SOS occurs at approximately 45°N instead of near the equator, which is caused mainly by differences among the different land-cover types. Moreover, the vegetation at low latitudes has periodic characteristics composed of two growing seasons in a year or three growing seasons in 2 years, resulting in two peaks for the phenology of many vegetation types. Therefore, at approximately 45°N, the SOS dates are later and the EOS dates are earlier than those in high-latitude areas. Many studies have found that phenology is closely related not only to the accumulation of high temperatures but also to the stimulation and accumulation of the lowest temperatures in winter, among other factors.

Furthermore, low- and high-temperature demands are met in succession; that is, plants initially require low temperatures and then require an accumulation of heat. If a plant is not sufficiently stimulated by low temperatures, its recovery and growth may be delayed, thus affecting the plant phenology. Precipitation is another important factor affecting phenology, and the vegetation differences among different climate zones can also affect the observed length of seasons.

Land Surface Phenology Trends at the Pixel Scale

At the pixel level, the linear change trends for the SOS and EOS dates of global vegetation are calculated at a 95% confidence level (Figure 39.5). According to the results, in 1982–2015, the SOS dates of 20.66% of all pixels are significantly advanced or delayed ($P < 0.05$), while the pixels with a negative (advanced)

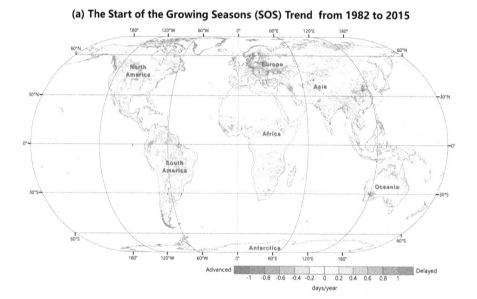

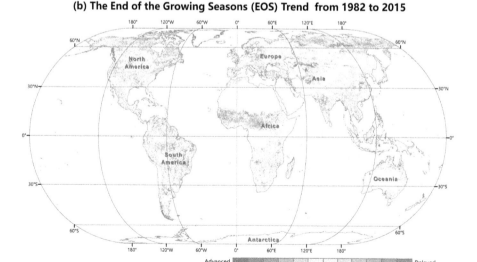

FIGURE 39.5 (See color insert.) Trends of the SOS (a) and the EOS (b), from 1982 to 2015.

SOS account for 63.90% of all pixels with a significant change in the SOS ($P < 0.05$); hence, there are more pixels with an advanced SOS than there are pixels with a delayed SOS ($P < 0.05$). The regions with significant advanced SOS dates are Europe, East Asia, and some areas in Australia, while the regions with significant delayed SOS dates are North America, South America, and Africa.

For the EOS, 26.7% of the pixels experience a significant change, and the pixels with a delayed EOS account for 58.85% of all pixels with a significantly changed EOS; hence, there are slightly more pixels with a delayed EOS than there are pixels with an advanced EOS ($P < 0.05$). The regions with delayed EOS dates are concentrated mainly between 40°N and 60°N, especially in the United States, Europe, and some regions in Asia. The EOS dates in Central Africa and Oceania are also significantly delayed, while the EOS in Eurasia above 60°N is more notably advanced.

Land Surface Phenology Changes at the Global Scale

The mean, minimum, maximum, slope, r, and p value of different land-cover types are calculated to analyze the changes in the LSP for different land-cover types at the global scale (Table 39.3).

The SOS dates are significantly advanced for deciduous broadleaf forests, mixed forests, closed shrublands, open shrublands, and urban and builtup lands ($P < 0.05$); among them, closed shrublands show the most obvious advancing trend with an average advancing rate of 0.32 days/year. The advancing

TABLE 39.3 LSP Parameters at the Global Scale

	Land-Cover Types	Mean	Min	Max	Slope	r	P
SOS	Land cover 1	122.69	113.46	128.36	−0.05	−0.14	0.43
	Land cover 2	152.89	140.12	167.47	0.12	0.21	0.23
	Land cover 3	130.07	118.15	139.56	−0.07	−0.15	0.39
	Land cover 4	140.71	133.93	145.74	−0.1	−0.43	0.01
	Land cover 5	128.07	123.62	135.5	−0.17	−0.55	0
	Land cover 6	166.51	145.43	183.1	−0.32	−0.38	0.03
	Land cover 7	145.05	138.45	153.33	−0.14	−0.38	0.03
	Land cover 8	136.09	132.05	139.93	−0.03	−0.18	0.32
	Land cover 9	161.73	158.11	167.11	−0.02	−0.14	0.43
	Land cover 10	154.96	149.93	159.49	−0.02	−0.11	0.53
	Land cover 11	139.52	134.8	144.39	0	−0.02	0.92
	Land cover 12	139.13	134.33	142.66	−0.01	−0.03	0.85
	Land cover 13	122.61	116.44	130.8	−0.12	−0.36	0.03
	Land cover 14	129.7	123.54	135.12	−0.05	−0.17	0.32
EOS	Land cover 1	286.26	279.07	293.46	0.17	0.51	0
	Land cover 2	213.96	194.75	225.27	0.06	0.09	0.62
	Land cover 3	282.6	274.02	290.04	−0.13	−0.28	0.11
	Land cover 4	272.84	266.17	278.08	0.17	0.63	0
	Land cover 5	278.96	272.27	286.55	0.18	0.48	0
	Land cover 6	235.8	224.57	243.88	−0.26	−0.48	0
	Land cover 7	276.42	265.11	282.17	−0.28	−0.65	0
	Land cover 8	274.88	269.39	279.2	−0.01	−0.05	0.77
	Land cover 9	258.71	252.3	262.68	−0.1	−0.42	0.01
	Land cover 10	260.28	256.96	263.52	−0.07	−0.39	0.02
	Land cover 11	292.17	281.46	298.73	−0.27	−0.62	0
	Land cover 12	263.83	260.41	266.94	−0.02	−0.13	0.46
	Land cover 13	268.91	262.39	274.71	0.09	0.36	0.04
	Land cover 14	273.96	265.73	281.66	0.11	0.32	0.07

rates of deciduous broadleaf forests, mixed forests, open shrublands, and urban and builtup lands are 0.1, 0.17, 0.14, and 0.12 days/year, respectively, while the change trends for the other vegetation types are not significant. The mean SOS is concentrated mainly from 123 days to 167 days and occurs mostly at approximately day 141. Among these land-cover types, urban and builtup lands have the earliest SOS on day 123 and closed shrublands have the latest SOS on day 167.

The EOS dates for evergreen needleleaf forests, deciduous broadleaf forests, mixed forests, closed shrublands, open shrublands, savannas, grasslands, permanent wetlands, and urban and builtup lands are significantly advanced or delayed ($P < 0.05$). The mean EOS is concentrated mainly from 214 days to 293 days and occurs mostly on day 271. The EOS dates for evergreen needleleaf forests, deciduous broadleaf forests, mixed forests, and urban and builtup lands are positively delayed with an average delay rate of 0.15 days/year. The EOS is concentrated mainly from 225 days to 299 days and occurs mostly on day 265. The EOS date occurs the earliest for urban and builtup lands, while the EOS date occurs the latest for evergreen needleleaf forests. The EOS date for urban and builtup lands is consistent with the SOS date, which may be the result of intense human activities. Moreover, there is a negative correlation between the EOS advancement for closed shrublands, open shrublands, savannas, grasslands, and permanent wetlands with an average rate of 0.20 days/year; the EOS is concentrated mainly from 263 days to 294 days and occurs mostly at approximately day 277. The EOS occurs the earliest for closed shrublands and the latest for permanent wetlands. The EOS trend for closed shrublands is opposite to the SOS trend, which may be related to the relatively few disturbances to its growth environment corresponding to human activities.

The changes in the SOS and EOS for evergreen broadleaf forests, woody savannas, croplands, and cropland/natural vegetation mosaics are not statistically significant, while the SOS and EOS for deciduous broadleaf forests, mixed forests, closed shrublands, open shrublands, and urban and builtup lands are either significantly advanced or significantly delayed. The SOS and EOS for closed shrublands and open shrublands are significantly advanced ($P < 0.05$), which may be due to the low requirements that shrubs have for their growth environment and their strong adaptability to changes in the natural environment.

Land Surface Phenology Trends for Different Land-Cover Types at the Intercontinental Scale

At the intercontinental scale, here, the change trends of the SOS and EOS for different land-cover types on different continents are calculated at the intercontinental scale ($P < 0.05$, Figure 39.6).

The SOS dates for different land-cover types on different continents show different characteristics. In Africa, the SOS dates for evergreen needleleaf forests and deciduous broadleaf forests are significantly advanced by 0.77 days/year and delayed by 0.25 days/year. In Asia, closed shrublands and open shrublands show advancing trends of 0.2 days/year and 0.19 days/year, respectively. Deciduous broadleaf forests, closed shrublands, open shrublands, woody savannas, savannas, and urban and builtup lands in Oceania all exhibit strong advancing trends of 1.19, 0.92, 0.60, 0.81, 0.62, and 0.68 days/year, respectively. In Europe, the SOS for closed shrublands is significantly delayed, whereas advancing trends are observed for most of the other vegetation types, such as evergreen needleleaf forests, mixed forests, and urban and builtup lands, whose advancing rates are 0.42, 0.32, and 0.36 days/year, respectively. The SOS dates for closed shrublands and croplands in North America are delayed, while the change trends for all other vegetation types are not evident. In South America, evergreen needleleaf forests, deciduous broadleaf forests, and permanent wetlands show delay trends, while the change trends for all other vegetation types are not significant. Overall, the largest advance in the SOS is observed in Oceania, followed by Europe and Asia, while the SOS change trends in Africa, North America, and South America are not significant.

The EOS dates of most vegetation types in Africa are delayed, of which cropland/natural vegetation mosaics exhibit the strongest trend with a delay rate of 0.41 days/year, followed by deciduous broadleaf

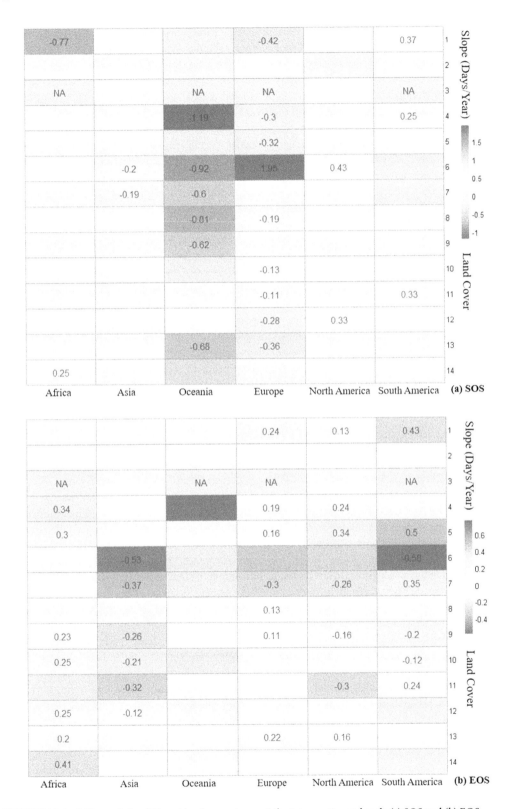

FIGURE 39.6 LSP trends for different land-cover types at the intercontinental scale (a) SOS and (b) EOS.

forests with a delay rate of 0.34 days/year. In Asia, most vegetation types exhibit advancing trends, while Oceania has no significant change trend. In Europe, most vegetation types show a significant delay trend except for open shrublands; evergreen needleleaf forests show a maximum delay trend of 0.24 days/year, followed by urban and builtup lands with a delay trend of 0.22 days/year. In North America, evergreen needleleaf forests, deciduous broadleaf forests, mixed forests, and urban and builtup lands display delay trends, while open shrublands, savannas, and permanent wetlands present advancing trends. In South America, the change trends vary for different types of vegetation; evergreen needleleaf forests, mixed forests, open shrublands, and permanent wetlands display significant delay trends with mixed forests showing the strongest delay trend of 0.50 days/year, while closed shrublands, savannas, and grasslands show significant advancing trends.

Overall, the SOS change trend in Africa is not significant, but the EOS delay trend is notable. In contrast, both the SOS and the EOS in Asia show advancing trends. The SOS in Oceania exhibits a significant advancing trend, but the change trend of the EOS is not evident. Moreover, Europe has a strong SOS advancing trend and an EOS delay trend. The SOS advancing trends in North America and South America are not significant, but the EOS delay trends are significant. In summary, the SOS and EOS change trends vary for different vegetation types on different continents, and SOS advances and EOS delays do not occur simultaneously.

The SOS change trend in North America is not significant, but the EOS delay trend is notable, which is consistent with the conclusions of previous studies that the extended length of the growing season (LOS) in North America is mainly attributable to the delay of the EOS [32]. In this chapter, the phenological change trend evident within the Northern Hemisphere is greater in Europe than in Asia and North America, which is consistent with the findings of previous studies indicating that the change trend in Eurasia is greater than that in North America and that Europe shows the strongest change [21,23,26].

Land Surface Phenology Trends for Different Land-Cover Types in the 10° Latitude Zone

To analyze the LSP changes for different land-cover types at different latitudes, land covers 1–14 (Table 39.1) are selected to calculate the mean change trend for each vegetation type in the 10° latitude zone. As shown in Figure 39.7, NA indicates that no vegetation exists within the specified area, while numerical labels represent significant change trends, and the remaining boxes show no significant change trends.

Evidently, the regions with relatively obvious SOS advancing trends are situated mainly at the high latitudes of the Northern Hemisphere. In the range from 70°N to 80°N, all the areas (except for those without vegetation) show a significant advancing trend; among the land-cover types, deciduous needleleaf forests exhibit the highest advancing trend of 0.69 days/year. At low latitudes ranging from 0°N to 30°N in the Northern Hemisphere, the advancing and delay trends of the continents are not the same. In the Northern Hemisphere, woody savannas, savannas, grasslands, permanent wetlands, croplands, urban and builtup lands, and cropland/natural vegetation mosaics show delay trends in the range from 0°N to 10°N, while the change trends for the other vegetation types are not significant.

In the Southern Hemisphere, closed shrublands and croplands show advancing trends between 20°S and 50°S, while open shrublands and woody savannas exhibit significant advancing trends between 30°S and 40°S, and urban and builtup lands present an advancing trend between 30°S and 40°S. The other vegetation types at the corresponding latitudes in the Southern Hemisphere have no significant change trends or otherwise exhibit delay trends. For example, open shrublands show a strong delay trend between 70°S and 80°S.

From a global perspective, the change trends are different in land-cover types. Not all land-cover types have advancing or delay trends, and the same land-cover type can have different change trends at different latitudes. Overall, the SOS advancing trends in the Northern Hemisphere are more evident than those in the Southern Hemisphere. At low latitudes, the SOS advancing trends in the Northern

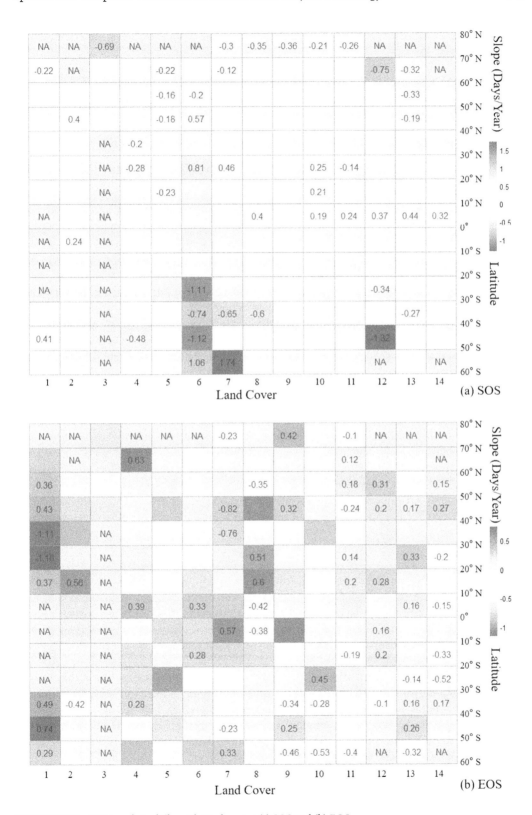

FIGURE 39.7 LSP trends in different latitude zones (a) SOS and (b) EOS.

Hemisphere and the Southern Hemisphere are not significant, and at some latitudes, sometimes delay trends are observed.

At the global scale, there are more regions with significant EOS change trends than those with significant SOS change trends. The EOS for evergreen needleleaf forests exhibits different change trends at different latitudes; this land-cover type shows a significant advancing trend between 20°N and 40°N in the Northern Hemisphere with a maximum advancing rate of 1.16 days/year, while at other latitudes, it displays delay trends, the most obvious of which lies between 40°S and 50°S in the Southern Hemisphere at 0.74 days/year. Evergreen broadleaf forests have a delay trend of 0.56 days/year between 10°N and 20°N and an advancing trend of 0.42 days/year between 30°S and 40°S, while the change trends at other latitudes are not evident. Deciduous needleleaf forests do not exhibit significant change trends between 40°N and 80°N. Deciduous broadleaf forests show a strong delay trend of 0.63 days/year between 60°N and 70°N; delay trends are also observed between 0° and 10°N and between 10°S and 20°S. Mixed forests do not show significant change trends between 60°S and 70°N. Closed shrublands have a significant delay trend between 0°N and 10°N and between 10°S and 20°S. Open shrublands have a complex change trend with an advancing trend in the Northern Hemisphere and a delay trend in the Southern Hemisphere. Woody savannas show a delay trend at latitudes between 10°N and 30°N with a maximum of 0.60 days/year, but an advancing trend is evident between 50°N and 60°N and between 10°S and 10°N. Savannas show a delay trend between 40°N and 50°N and between 70°N and 80°N in the Northern Hemisphere and between 40°S and 50°S in the Southern Hemisphere, and they show an advancing trend between 30°S and 40°S and between 50°S and 60°S in the Southern Hemisphere. Grasslands have no obvious change trends in the Northern Hemisphere, but they show a delay trend between 20°S and 30°S in the Southern Hemisphere and an advancing trend at middle and high latitudes between 30°S and 40°S and between 50°S and 60°S. Permanent wetlands mainly show a delay trend in the Northern Hemisphere but display an advancing trend between 40°N and 50°N, between 10°S and 20°S, and between 50°S and 60°S. Croplands show a delay trend between 40°N and 60°N, between 10°N and 20°N, and between 0° and 20°S; the most obvious delay trend reaches 0.31 days/year at latitudes from 50°N to 60°N, and an advancing trend appears between 30°S and 40°S. Urban and builtup lands mainly show a delay trend in the Northern Hemisphere, where the most obvious delay trend reaches 0.33 days/year at latitudes from 20°N to 30°N; additionally, a delay trend is observed between 30°S and 50°S. Cropland/natural vegetation mosaics exhibit a delay trend in the Northern Hemisphere above 30°N and an advancing trend between 10°S and 30°S in the Southern Hemisphere.

Conclusions

The phenological changes in global vegetation from 1982 to 2015 display notable spatial differences. The distribution pattern follows a defined regularity. The distributions of the SOS and EOS in the vegetation pixel frequency map of the Northern Hemisphere for the past 34 years show obvious peaks, while the distributions of the SOS and EOS in the Southern Hemisphere exhibit no obvious peaks, indicating that the phenological differences in the Northern Hemisphere are greater than those in the Southern Hemisphere.

On the basis of the global distribution of the mean change in the phenological index, the change trend of the SOS in the Northern Hemisphere is most evident in Asia, followed by North America and Europe, while that in the Southern Hemisphere is most evident in Africa, followed by South America and Oceania. Similarly, the EOS change trend in the Northern Hemisphere is most evident in North America, followed by Europe and Asia, while that in the Southern Hemisphere is most evident in Oceania, followed by Africa and South America.

According to the LSP distribution between 55°S and 77°N, the overall SOS shows an advancing trend, while the overall EOS indicates a delay trend, but both exhibit many fluctuations. The fluctuations of the SOS and EOS within this latitude zone are opposite to each other; that is, the peaks and valleys of the SOS correspond to the valleys and peaks, respectively, of the EOS. The range of fluctuations in the Northern Hemisphere is noticeably larger than that in the Southern Hemisphere. Several extreme points are found at 30°S, 16°S, 5°N, 17°N, and 45°N from south to north. These fluctuations in the phenological index are attributable to many factors, including the vegetation type, stimulation by low temperatures, accumulation of temperatures, and precipitation.

In terms of the pixel-scale change trends, 20.66% of the SOS pixels between 1982 and 2015 are significantly advanced or delayed ($P < 0.05$), and pixels with a negative (advanced) SOS account for 63.90% of all pixels with a significant change (P<0.05). The pixels with an SOS advancing trend are located mainly in Europe, Asia, North America, and Oceania. For the EOS, 26.7% of the pixels show a significant change trend, while the pixels with a delay trend account for 58.85% of all pixels with a significant change during the 34-year study period; hence, there are more pixels with a delay trend than there are pixels with an advancing trend ($P < 0.05$). The pixels with EOS delay trends are located mainly in North America, Europe, and Africa.

At the global scale, the change trends of the phenological indexes for different vegetation types with time are noticeably different. The SOS dates for deciduous broadleaf forests, mixed forests, closed shrublands, open shrublands, and urban and builtup lands show significant advancing trends; among them, closed shrublands exhibit the most obvious advancing trend with an average rate of 0.32 days/year. The change trends for the other vegetation types are not significant. The EOS dates for evergreen needleleaf forests, deciduous broadleaf forests, mixed forests, and urban and builtup lands show delay trends with an average rate of 0.15 days/year; in contrast, closed shrublands, open shrublands, savannas, grasslands, and permanent wetlands exhibit advancing trends with an average rate of 0.20 days/year. Furthermore, the SOS and EOS dates for closed shrublands and open shrublands show significant advancing trends.

In terms of the phenological index change trends for different vegetation types at the intercontinental scale, the SOS dates for evergreen needleleaf forests and cropland/natural vegetation mosaics in Africa show more notable advancing trends, while the change trends for the other vegetation types are not significant. In addition, the EOS dates for most vegetation types show significant delay trends. The SOS and EOS in Asia present advancing trends, but the change values are small. The SOS in Oceania shows a significant advancing trend; among the land-cover types, deciduous broadleaf forests exhibit the most significant advancing trend with an average rate of 1.19 days/year, but the change trend for the EOS is not obvious. The SOS for closed shrublands in Europe shows a significant delay trend, while the other vegetation types exhibit mostly advancing trends. The EOS dates of most vegetation types except open shrublands show significant delay trends. In North and South America, the SOS advancing trend for each vegetation type is not significant, and the EOS dates for most vegetation types show significant delay trends.

According to the change trends for different vegetation types in the 10° latitude zone, it is evident that the change trends vary for different vegetation types in different latitude zones, and not all vegetation types exhibit advancing or delay trends. Generally, the SOS advancing trends in the Northern Hemisphere are greater than those in the Southern Hemisphere. At low latitudes, the SOS advancing trends in the Northern Hemisphere and the Southern Hemisphere are not significant, and sometimes delay trends are observed at some latitudes.

Over the past 34 years, human activities such as deforestation for land reclamation have changed the vegetation types in some areas. Moreover, some natural factors, such as the disappearance of a forest due to wildfire, result in the uncertainty of phenological changes over a long period of time. In this chapter, the global LSP of different vegetation types is investigated at the pixel scale, in different latitude zones, and at the intercontinental and global scales to reduce the uncertainties caused by changes in the vegetation types at different scales to reflect the change trend of LSP over the past 34 years. The results provide the references for analyzing phenological changes at the global scale and surface phenology changes at regional scales as well as for studies on the response of vegetation to global climate change.

References

1. Stocker, T.F.; Qin, D.; Plattner, G.K.; Alexander, L.V.; Allen, S.K.; Bindoff, N.L.; Bréon, F.M.; Church, J.A.; Cubasch, U.; Emori, S. IPCC, 2013: Technical Summary. In: *Climate Change 2013: The Physical Science Basis. Contribution of Working Group I to the Fifth Assessment Report of the Intergovernmental Panel on Climate Change* eds., Stocker, T.F., Qin, D., Plattner, G.-K., Tignor, M., Bindoff, S.K. Computational Geometry **2013**, *2007*, 1–21.

2. Walther, G.-R.; Post, E.; Convey, P.; Menzel, A.; Parmesan, C.; Beebee, T.J.C.; Fromentin, J.-M.; Hoegh-Guldberg, O.; Bairlein, F. Ecological responses to recent climate change. *Nature* **2002**, *416*, 389–395.

3. Parmesan, C.; Yohe, G. A globally coherent fingerprint of climate change impacts across natural systems. *Nature* **2003**, *421*, 37–42.

4. Fu, Y.H.; Zhao, H.; Piao, S.; Peaucelle, M.; Peng, S.; Zhou, G.; Ciais, P.; Huang, M.; Menzel, A.; Peñuelas, J. Declining global warming effects on the phenology of spring leaf unfolding. *Nature* **2015**, *526*, 104.

5. Zhang, J.; Walsh, J.E. Thermodynamic and hydrological impacts of increasing greenness in Northern high latitudes. *Journal of Hydrometeorology* **2006**, *7*, 1147.

6. Lim, Y.K.; Cai, M.; Kalnay, E.; Zhou, L. Impact of vegetation types on surface temperature change. *Journal of Applied Meteorology & Climatology* **2008**, *47*, 411–424.

7. Heumann, B.W.; Seaquist, J.W.; Eklundh, L.; Jönsson, P. AVHRR derived phenological change in the Sahel and Soudan, Africa, 1982–2005. *Remote Sensing of Environment* **2007**, *108*, 385–392.

8. Cleland, E.E.; Chuine, I.; Menzel, A.; Mooney, H.A.; Schwartz, M.D. Shifting plant phenology in response to global change. *Trends in Ecology & Evolution* **2007**, *22*, 357–365.

9. Zhao, J.; Zhang, H.; Zhang, Z.; Guo, X.; Li, X.; Chen, C. Spatial and temporal changes in vegetation phenology at middle and high latitudes of the Northern hemisphere over the past three decades. *Remote Sensing* **2015**, *7*, 10973–10995.

10. Moulin, S.; Kergoat, L.; Viovy, N.; Dedieu, G. Global-scale assessment of vegetation phenology using NOAA/AVHRR Satel. *Journal of Climate* **2010**, *10*, 1154–1170.

11. Jeong, S.J.; Ho, C.H.; Gim, H.J.; Brown, M.E. Phenology shifts at start vs. end of growing season in temperate vegetation over the Northern hemisphere for the period 1982–2008. *Global Change Biology* **2011**, *17*, 2385–2399.

12. Zeng, H.; Jia, G.; Epstein, H. Recent changes in phenology over the northern high latitudes detected from multi-satellite data. *Environmental Research Letters* 2011, *6*, 045508.

13. Justice, C.O.; Townshend, J.R.G.; Holben, B.N.; Tucker, C.J. Analysis of the phenology of global vegetation using meteorological satellite data. *International Journal of Remote Sensing* **1985**, *6*, 1271–1318.

14. Myneni, R.B.; Keeling, C.D.; Tucker, C.J.; Asrar, G.; Nemani, R.R. Increased plant growth in the northern high latitudes from 1981 to 1991. *Nature* **1997**, *386*, 698–702.

15. Fischer, A. A model for the seasonal variations of vegetation indices in coarse resolution data and its inversion to extract crop parameters. *Remote Sensing of Environment* **1994**, *48*, 220–230.

16. White, M.; Thornton, P.; Running, S. A continental phenology model for monitoring vegetation responses to interannual climatic variability. *Global Biogeochemical Cycles* **1997**, *11*, 217–234.

17. Delbart, N.; Kergoat, L.; Le Toan, T.; Lhermitte, J.; Picard, G. Determination of phenological dates in boreal regions using normalized difference water index. *Remote Sensing of Environment* **2005**, *97*, 26–38.

18. Piao, S.; Fang, J.; Zhou, L.; Ciais, P.; Zhu, B. Variations in satellite-derived phenology in China's temperate vegetation. *Global Change Biology* **2006**, *12*, 672–685.

19. Moody, A.; Johnson, D.M. Land-surface phenologies from AVHRR using the discrete Fourier transform. *Remote Sensing of Environment* **2001**, *75*, 305–323.

20. Badhwar, G.D. Automatic corn-soybean classification using Landsat MSS data. I. Near-harvest crop proportion estimation. *Remote Sensing of Environment* **1984**, *14*, 15–29.

21. De Beurs, K.M.; Henebry, G.M. Land surface phenology and temperature variation in the international geosphere–biosphere program high-latitude transects. *Global Change Biology* **2005**, *11*, 779–790.

22. Julien, Y.; Sobrino, J.A. Global land surface phenology trends from GIMMS database. *International Journal of Remote Sensing* **2009**, *30*, 3495–3513.

23. White, M.A.; Running, S.W.; Thornton, P.E. The impact of growing-season length variability on carbon assimilation and evapotranspiration over 88 years in the eastern US deciduous forest. *International Journal of Biometeorology* **1999**, *42*, 139–145.

24. Jeong, S.-J.; Ho, C.-H.; Gim, H.-J.; Brown, M.E. Phenology shifts at start vs. end of growing season in temperate vegetation over the Northern hemisphere for the period 1982–2008. *Global Change Biology* **2011**, *17*, 2385–2399.

25. Zhang, X.; Friedl, M.A.; Schaaf, C.B.; Strahler, A.H.; Liu, Z. Monitoring the response of vegetation phenology to precipitation in Africa by coupling MODIS and TRMM instruments. *Journal of Geophysical Research* **2005**, *110*, D12103.

26. Robbirt, K.M.; Davy, A.J.; Hutchings, M.J.; Roberts, D.L. Validation of biological collections as a source of phenological data for use in climate change studies: A case study with the orchid Ophrys sphegodes. 2011, *99*, 235–241.

27. Fitter, A.H.; Fitter, R.S.R.; Harris, I.T.B.; Williamson, M.H. Relationships between first flowering date and temperature in the flora of a locality in central England. *Functional Ecology* **1995**, *9*, 55–60.

28. Dash, J.; Jeganathan, C.; Atkinson, P.M. The use of MERIS terrestrial chlorophyll index to study spatio-temporal variation in vegetation phenology over India. *Remote Sensing of Environment* **2010**, *114*, 1388–1402.

29. Zhu, W.; Tian, H.; Xu, X.; Pan, Y.; Chen, G.; Lin, W. Extension of the growing season due to delayed autumn over mid and high latitudes in North America during 1982–2006. *Global Ecology & Biogeography* **2012**, *21*, 260–271.

30. Hansen, M.C.; DeFries, R.S.; Townshend, J.R.; Sohlberg, R. Global land cover classification at 1 km spatial resolution using a classification tree approach. *International Journal of Remote Sensing* **2000**, *21*(6–7), 1331–1364. Available online: www.tandfonline.com/doi/citedby/10.1080/01431160021 0209?scroll=top&needAccess=true (accessed on May 16, 2019).

31. Elmore, A.J.; Guinn, S.M.; Minsley, B.J.; Richardson, A.D. Landscape controls on the timing of spring, autumn, and growing season length in mid-Atlantic forests. *Global Change Biology* **2012**, *18*, 656–674.

32. Zhu, W.; Tian, H.; Xu, X.; Pan, Y.; Chen, G.; Lin, W. Extension of the growing season due to delayed autumn over mid and high latitudes in North America during 1982–2006. *Global Ecology and Biogeography* **2012**, *21*, 260–271.

40

Spatiotemporal Variations in Precipitation and Temperature over Northeastern Eurasia, 1961–2010

Ying Zhang,
Haibo Du, and
Zhengfang Wu
Northeast Normal University

Introduction

Climate is currently undergoing rapid changes on both regional and global scales [1–3], thus increasing the risk of natural disasters [4], resource scarcity [5], food crises [6], diseases [7], and extreme events [8–13]. However, due to different regional characteristics, the responses to global climate change vary from region to region [14]. Studies of the changes in climate over sensitive regions responded to global warming will improve our understanding of the regional effects of global climate change.

Many studies have investigated the characteristics of regional climate change over the past few decades [15–17]. For instance, the spring temperature indices in Europe exhibited opposite trends between mid-high and mid-low latitudes [18]. Similarly, during the winter, temperature increased in northern Europe, whereas the opposite condition occurred in Greenland and Baffin Bay, and vice versa [19]. Comparisons of Northeast China with surrounding areas (such as Japan and South Korea) indicated that the temperature in Northeast China recorded a higher rate of increase than it did in the other regions on an annual scale, but that these differences were especially significant on the seasonal scale [16,20–22]. The total annual precipitation (PRCP) decreased in Spain, but increased in the region spanning from France across the British Isles to Scandinavia, which was found to be related to the North Atlantic Oscillation (NAO) or Arctic Oscillation (AO) patterns. Nevertheless, most studies have only focused on changes in mean temperature, although variations in the maximum temperature (T_{max}), minimum temperature (T_{min}), and daily temperature range (DTR) are considered to be more important for the environment and agriculture [23].

Northeast Eurasia is part of the largest continent and adjacent to the largest ocean in the world, which is also one of the typical vulnerable areas and the most sensitive areas affected by climate warming. Therefore, we focused on the climate changes in northeastern Eurasian continent at middle and high latitudes in this study. Northeastern Eurasia is located in the middle and high latitudes of the Northern Hemisphere with a temperate monsoon climate and subpolar continental climate. By analyzing the temporal and spatial variations of temperature and precipitation in northeastern Eurasia, we can further understand the characteristics of climate change in this region and the regional responses to global warming. This study aims to answer the following questions: (i) What are the temporal and spatial variations of precipitation in northeastern Eurasia from 1961 to 2010? (ii) What are the temporal and spatial variations of temperature in northeastern Eurasia from 1961 to 2010?

Study Region and Data

The study area is the northeastern part of the Eurasian continent at middle and high latitudes (northeastern Eurasia), which is located at 120°E–180°E, 40°N–80°N and includes Northeast China, northern Japan, and Russian Far East (Figure 40.1).

The values of observed daily PRCP, T_{max}, and T_{min} from 1961 to 2010 used in this study were derived from the China Meteorological Data Sharing Service System (http://data.cma.cn/), Russian Research Institute of Hydrometeorological Information – World Data Centre [24], and Japan Meteorological Agency (www.data.jma.go.jp/gmd/risk/obsdl/index.php). DTR series were computed using T_{max} minus

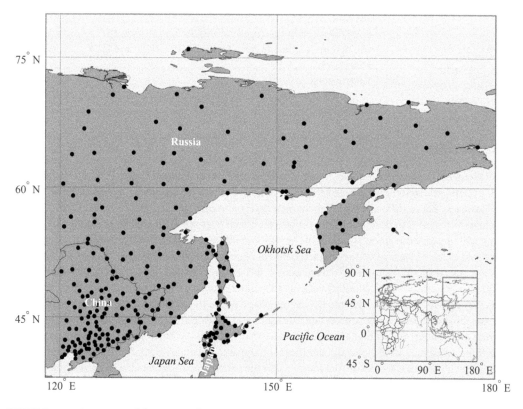

FIGURE 40.1 Locations of the 257 weather stations in northeastern Eurasia at middle and high latitudes that were utilized in this study, for which precipitation and temperature data collected during the period from 1961 to 2010 were analyzed.

T_{min} series. Annual and seasonal averaged PRCP, T_{max}, T_{min}, and DTR series were calculated from their corresponding daily series. Conventionally, seasons were divided into spring (March, April, and May), summer (June, July, and August), autumn (September, October, and November), and winter (December, January, and February).

Although almost 500 stations were obtained in this study, many of them could not be used due to their missing or insufficient data. We defined a missing record year as a year missing data for more than 36 days (~10% of a year). Stations with short data series, obvious wide gaps, or more than five missing record years were excluded. The quality control and homogeneity test strongly affected the accuracy of the data analysis. Therefore, we used RClimDex software to control the quality of the data and RHtestsV3 software to assess the homogeneity of the data (http://etccdi.pacificclimate.org/software.shtml). The RClimDex software can recognize logical errors of data series, such as $T_{min} > T_{max}$ or PRCP < 0. We removed and corrected the outliers that were off by over four standard deviations [25]. The RHtestsV3 software was able to detect and adjust change points (shifts) within a data series by utilizing a two-phase regression model [26]. This eliminated the effects of changes in instrumentation, location, exposure, and observation practices. Eventually, a high-quality data series, which was distributed relatively uniformly across 257 weather stations in northeastern Eurasia, was utilized in this study. We analyzed the temporal and spatial variability of climate variables.

Method

To analyze the changes in climatic variables (PRCP, T_{max}, T_{min}, and DTR), we utilized the ordinary least-squares method to compute the linear tendency [27]. The formula was as follows:

$$y_i = a + bt_i \qquad (40.1)$$

where y_i represents the seasonal and annual mean values of PRCP, T_{max}, T_{min}, and DTR; t_i is the time series from 1961 to 2010, b represents the linear tendency, and a is a constant. We used this method to calculate the climate changes of individual stations and the regionally mean time series. The significance of the trend was estimated by the nonparametric Mann–Kendall test.

Spatial Pattern of the Trends of PRCP, T_{max}, T_{min}, and DTR

The trends of annual PRCP (mm per decade), T_{max} (°C per decade), T_{min} (°C per decade), and DTR (°C per decade) from 1961 to 2010 in northeastern Eurasia showed obvious regional characteristics (Figure 40.2). The trends in precipitation ranged from −68 to 41 mm per decade (Figure 40.2a). A total of 99 stations (38.5%) recorded rising trends, in which 14 stations (14%) showed significant increases ($p < 0.05$). Conversely, 158 stations (61.5%) experienced decreasing trends, with 19 stations (12%) recording significant decreases ($p < 0.05$). The increasing trends were mainly distributed in north part of northeastern Eurasia and Sakhalin Oblast. The decreasing trends spread throughout south part of northeastern Eurasia, especially for Northeast China and most parts of north Japan. The highest increasing trends were distributed in northwest part of northeastern Eurasia and Sakhalin Oblast, whereas the strongest decreasing trends were spread throughout northern Japan and south part of Northeast China. Generally, the trend of precipitation in northeastern Eurasia shows a zonal spatial pattern of increase–decrease from high to low latitudes.

Almost all stations (99%) experienced increasing trends in annual T_{max} for the past 50 years, ranging from −0.04°C to 0.55°C per decade for northeastern Eurasia. For these changes, 220 stations (86.3%) exhibited a significant rising trend ($p < 0.05$) and the stations that recorded stronger warming were located in the Russian parts of northeastern Eurasia. Overall, stations in higher latitudes recorded strong warming trends, whereas those in most northern Japan and Northeast China exhibited slower increasing trends (Figure 40.2b).

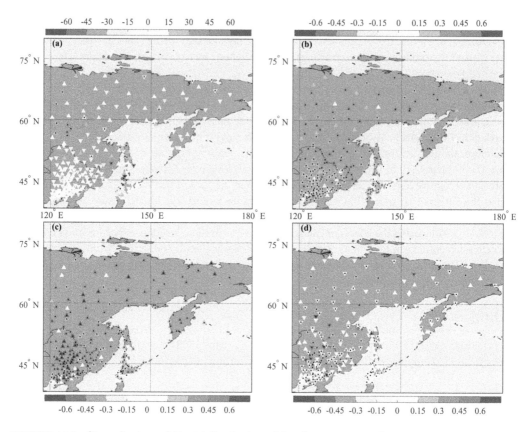

FIGURE 40.2 **(See color insert.)** Spatial distribution of the changes in annual PRCP, T_{max}, T_{min}, and DTR in northeastern Eurasia (a, b, c, and d, respectively) from 1961 to 2010. The upward-pointing and downward-pointing triangles represent increasing and decreasing trends, respectively. The change magnitudes are shown in different colors. The units are mm per decade for precipitation and °C per decade for temperature. The overlying dots on triangles indicate significant trends ($p < 0.05$) estimated by a Mann–Kendall test.

Regarding the changes in T_{min}, a majority of stations (97.3%) in the study area recorded warming trends, except for seven stations in south part of northeastern Eurasia with a nonsignificant weak decrease. Over northeastern Eurasia, the trends of T_{min} ranged from −0.47 to 1.05°C per decade. Approximately 92% of stations recorded significant increasing trends ($p < 0.05$) in northeastern Eurasia, and the stronger warming trends were observed primarily over Northeast China and parts of northeast Russia (Figure 40.2c).

In terms of the changes in annual DTR, the majority of stations (72.0%) in northeastern Eurasia showed decreasing trends in DTR, with 132 stations (71.4%) of these trends being statistically significant ($p < 0.05$). The changes in DTR ranged from −0.72 to 0.77°C per decade, and strong downtrends were observed mainly over Northeast China (Figure 40.2d).

Changes in Regionally Average PRCP, T_{max}, T_{min}, and DTR

Based on regionally average time series of PRCP and temperature, we calculated the seasonal and annual trends of PRCP, T_{max}, T_{min}, and DTR over the entire northeastern Eurasia. Only spring PRCP showed an increasing trend in northeastern Eurasia with a change of 1.2 mm per decade, while PRCP was decreasing in summer, autumn, and winter, with the trends of −2.9, −1.6, and −0.8, respectively. At an annual scale, PRCP showed a decreasing trend in northeastern Eurasia with −2.4 mm per decade for the entire northeastern Eurasia. Nevertheless, all trends of seasonal and annual PRCP in the study areas were not significant at a 95% confidence level (Figure 40.3a–e).

For temperature indices, both the annual and seasonal T_{max} and T_{min} values in northeastern Eurasia increased significantly ($p < 0.01$). Compared to the changes in T_{max} (0.25°C, 0.21°C, 0.26°C, 0.31°C, and 0.24°C per decade for spring, summer, autumn, winter, and annual time series, respectively), the trends of T_{min} in northeastern Eurasia were found to be increasing faster, with values of 0.39°C, 0.27°C, 0.36°C, 0.48°C, and 0.37°C per decade for spring, summer, autumn, winter, and annual, respectively (Figure 40.3f–j). All of these changes were significant ($p < 0.01$). In general, the changes in T_{max} and T_{min} represented relatively coincident variability, with the largest warming in winter and the smallest warming in summer. A faster increase in annual T_{min} (0.37°C per decade) than in T_{max} (0.24°C per decade) contributed to the decreasing annual DTR (−0.12°C per decade) that occurred in northeastern Eurasia ($p < 0.01$). Seasonally, the trends of DTR were consistent with the annual variations, with the strongest trends of −0.16°C per decade in winter and the smallest change in summer with a value of −0.06°C per decade.

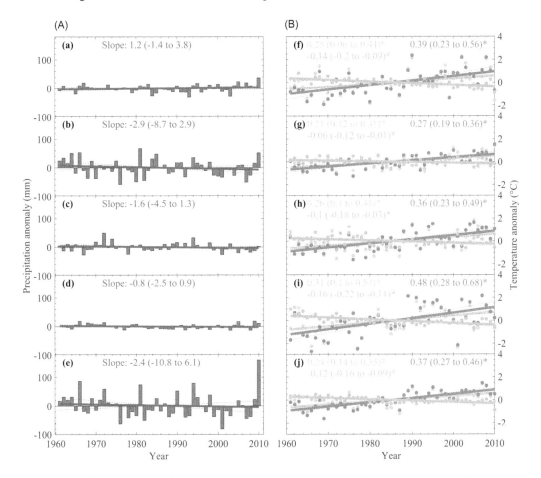

FIGURE 40.3 **(See color insert.)** Regionally averaged seasonal and annual precipitation (A) and temperature (B) time series for northeastern Eurasia during the period from 1961 to 2010. Subfigures from up to down represent spring, summer, autumn, winter, and annual time series, respectively. Anomalies were relative to the average during the period from 1961 to 2010. Slope indicates the linear trend of time series, and values in the parentheses represent the 5%–95% uncertainty range of the linear regression slope. The units of slope were mm per decade for precipitation and °C per decade for temperature. No trends were significant at a 95% confidence level for precipitation. For temperature, T_{max}, T_{min}, and DTR are shown in orange, carmine, and green, respectively. The symbol "*" indicates that the linear trend was significant at the 0.05 level.

Discussion

This study revealed the characteristics of climate change in northeastern Eurasia. Compared to temperature changes, the spatiotemporal variations of PRCP were more complex, weaker, and spatially coherent. For example, increasing PRCP was mainly observed in higher latitudes in northeastern Eurasia, especially for approximately north of 50°N, whereas decreasing trends in PRCP were primarily located in lower latitudes. This indicated that the variation in PRCP was not just a local phenomenon but part of a large-scale change [28]. The Intergovernmental Panel on Climate Change (IPCC) reported that global warming resulted in increased PRCP in the regions north of 30°N during the past century [29]. Following the increases in PRCP over the temperate areas in the Northern Hemisphere since the early 20th century [30], the annual PRCP has recorded increasing trends in northeastern Eurasia in recent decades.

In northeastern Eurasia, significant upward trends were found in all temperature indices, but T_{min} exhibited higher rates of increase than T_{max} on both the seasonal and annual scales, suggesting that the warming tendency in northeastern Eurasia was mainly caused by T_{min} warming [23]. As a result, DTR showed a significant decrease in northeastern Eurasia, especially in Northeast China [31,32], where human influences (e.g., human activities and land-use and land-cover change) on regional climate change have been proven [33]. Some studies reported that the declining DTR was associated with increasing cloud amount, which could have backscattered the incoming solar radiation and suppressed the T_{max}, thus heating the land surface by enhancing the downward long-wave radiation [34]. Nevertheless, other studies also demonstrated that the decrease in sunshine duration played a more important role in decreasing DTR compared to cloud cover, which could affect the changes in T_{max} [32].

Conclusions

Utilizing daily precipitation and temperature data collected from 257 stations from 1961 to 2010, we comparatively analyzed the spatiotemporal changes in different climate variables (PRCP, T_{max}, T_{min}, and DTR) in northeastern Eurasia on both the annual and seasonal scales. We found that PRCP showed zonal variations with increase–decrease trends from high to low latitudes. In terms of temperature indices, both T_{max} and T_{min} recorded significant warming ($p < 0.01$) in northeastern Eurasia on the seasonal and annual scales. For T_{max}, the high latitudes showed a strong warming trend, while the stations in most northern Japan and Northeast China exhibited slower increasing trends. Regarding the changes in T_{min}, strong warming trend was observed mainly in northeastern China and parts of northeastern Russia. Higher increases in minimum temperature than in maximum temperature result in a decreased DTR at both the seasonal and annual scales.

Acknowledgments

This study was financially supported by the National Key Research and Development Project (No. 2016YFA0602301) and Science and Technology Development Plan of Jilin Province (20190201291JC and 20180520098JH).

References

1. Mariotti, A.; Pan, Y.; Zeng, N.; Alessandri, A. Long-term climate change in the Mediterranean region in the midst of decadal variability. *Climate Dynamics* 2015, 44, 1437–1456.
2. Marshall, G.J.; Vignols, R.M.; Rees, W.G. Climate change in the Kola Peninsula, Arctic Russia, during the last 50 years from meteorological observations. *Journal of Climate* 2016, 29, 6823–6840.
3. Paeth, H.; Müller, M.; Mannig, B. Global versus local effects on climate change in Asia. *Climate Dynamics* 2015, 45, 2151–2164.

4. Hirabayashi, Y.; Mahendran, R.; Koirala, S.; Konoshima, L.; Yamazaki, D.; Watanabe, S.; Kim, H.; Kanae, S. Global flood risk under climate change. *Nature Climate Change* 2013, 3, 816.

5. Schewe, J.; Heinke, J.; Gerten, D.; Haddeland, I.; Arnell, N.W.; Clark, D.B.; Dankers, R.; Eisner, S.; Fekete, B.M.; Colón-González, F.J.; Gosling, S.N. Multimodel assessment of water scarcity under climate change. *Proceedings of the National Academy of Sciences of the United States of America* 2014, 111, 3245–3250.

6. Wheeler, T.; Von Braun, J. Climate change impacts on global food security. *Science* 2013, 341, 508–513.

7. Altizer, S.; Ostfeld, R.S.; Johnson, P.T.J.; Kutz, S.; Harvell, C.D. Climate change and infectious diseases: From evidence to a predictive framework. *Science* 2013, 341, 514–519.

8. Ingram, W. Extreme precipitation increases all round. *Nature Climate Change* 2016, 6, 443–444.

9. Sachindra, D.A.; Ng, A.W.M.; Muthukumaran, S.; Perera, B.J.C. Impact of climate change on urban heat island effect and extreme temperatures: A case-study. *Quarterly Journal of the Royal Meteorological Society* 2016, 142, 172–186.

10. Yabi, I.; Afouda, F. Extreme rainfall years in Benin (West Africa). *Quaternary International* 2012, 262, 39–43.

11. Zhang, Q.; Zhang, J.; Yan, D.; Wang, Y. Extreme precipitation events identified using detrended fluctuation analysis (DFA) in Anhui, China. *Theoretical and Applied Climatology* 2014, 117, 169–174.

12. Han, J.; Du, H.; Wu, Z.; He, H.S. Changes in extreme precipitation over dry and wet regions of China during 1961–2014. *Journal of Geophysical Research: Atmospheres* 2019, 124, 5847–5859.

13. Du, H.; Alexander, L.V.; Donat, M.G.; Lippmann, T.; Srivastava, A.; Salinger, J.; Kruger, A.; Choi, G.; He, H.S.; Fujibe, F.; Rusticucci, M. Precipitation from persistent extremes is increasing in most regions and globally. *Geophysical Research Letters* 2019, 46, 6041–6049.

14. Zhao, W.; Du, H.; Wang, L.; He, H.S.; Wu, Z.; Liu, K.; Guo, X.; Yang, Y. A comparison of recent trends in precipitation and temperature over Western and Eastern Eurasia. *Quarterly Journal of the Royal Meteorological Society* 2018, 144, 604–613.

15. Huntingford, C.; Marsh, T.; Scaife, A.A.; Kendon, E.J.; Hannaford, J.; Kay, A.L.; Lockwood, M.; Prudhomme, C.; Reynard, N.S.; Parry, S.; Lowe, J.A. Potential influences on the United Kingdom's floods of winter 2013/14. *Nature Climate Change* 2014, 4, 769–777.

16. Wang, L.; Wu, Z.; Wang, F.; Du, H.; Zong, S. Comparative analysis of the extreme temperature event change over Northeast China and Hokkaido, Japan from 1951 to 2011. *Theoretical and Applied Climatology* 2016, 124, 375–384.

17. Yan, Z.; Jones, P.D.; Davies, T.D.; Moberg, A.; Bergström, H.; Camuffo, D.; Cocheo, C.; Maugeri, M.; Demarée, G.R.; Verhoeve, T.; Thoen, E.; Barriendos, M.; Rodríguez, R.; Martín-Vide, J.; Yang, C. Trends of Extreme Temperatures in Europe and China Based on Daily Observations. In: Improved Understanding of Past Climatic Variability from Early Daily European Instrumental Sources. (eds Camuffo, D.; Jones, P.) Springer: Dordrecht, Netherlands, 2002.

18. Rogers, J.C.; Loon, H.V. The see-saw in winter temperatures between Greenland and Northern Europe, Part II: Some oceanic and atmospheric effects in middle and high latitudes. *Monthly Weather Review* 1979, 107, 509–519.

19. Overland, J.E.; Spillane, M.C.; Percival, D.B.; Wang, M.; Mofjeld, H.O. Seasonal and regional variation of pan-Arctic surface air temperature over the instrumental record. *Journal of Climate* 2004, 17, 3263–3282.

20. Fujibe, F. Urban warming in Japanese cities and its relation to climate change monitoring. *International Journal of Climatology* 2011, 31, 162–173.

21. Gong, D.-Y.; Guo, D.; Ho, C.-H. Weekend effect in diurnal temperature range in China: Opposite signals between winter and summer. *Journal of Geophysical Research-Atmospheres* 2006, 111, D18113.

22. Li, H.; Zhou, T.; Nam, J.-C. Comparison of daily extreme temperatures over Eastern China and South Korea between 1996–2005. *Advances in Atmospheric Sciences* 2009, 26, 253–264.

23. Wang, L.; Xie, X.; Su, W.; Guo, X. Changes of maximum and minimum temperature and their impacts in Northern China over the second half of the 20th century. *Journal of Natural Resources* 2004, 19, 337–343.

24. Bulygina, O.N.; Razuvaev, V.N. 2012. Daily Temperature and Precipitation Data for 518 Russian Meteorological Stations (1881 - 2010). United States.

25. Wang, L.; Wu, Z.; He, H.; Wang, F.; Dua, H.; Zong, S. Changes in summer extreme precipitation in Northeast Asia and their relationships with the East Asian summer monsoon during 1961-2009. *International Journal of Climatology* 2017, 37, 25–35.

26. Wang, X.L. Comments on "detection of undocumented changepoints: A revision of the two-phase Regression Model". *Journal of Climate* 2003, 16, 3383–3385.

27. Ma, S.; Zhou, T.; Dai, A.; Han, Z. Observed changes in the distributions of daily precipitation frequency and amount over China from 1960 to 2013. *Journal of Climate* 2015, 28, 6960–6978.

28. Vautard, R.; Gobiet, A.; Sobolowski, S.; Kjellström, E.; Stegehuis, A.; Watkiss, P.; Mendlik, T.; Landgren, O.; Nikulin, G.; Teichmann, C.; Jacob, D. The European climate under a 2°C global warming. *Environmental Research Letters* 2014, 9, 034006.

29. Ipcc. *Climate Change 2013: The Physical Science Basis: Working Group I Contribution to the Fifth Assessment Report of the Intergovernmental Panel on Climate Change.* Cambridge University Press: Cambridge, United Kingdom and New York, 2014.

30. Bradley, R.S.; Diaz, H.F.; Eischeid, J.K.; Jones, P.D.; Kelly, P.M.; Goodess, C.M. Precipitation fluctuations over Northern hemisphere land areas since the mid-19th century. *Science* 1987, 237, 171–175.

31. Easterling, R.D. Maximum and minimum temperature trends for the Globe. *Science* 1997, 277, 364–367.

32. Shen, X.; Liu, B.; Li, G.; Wu, Z.; Jin, Y.; Yu, P.; Zhou, D. Spatiotemporal change of diurnal temperature range and its relationship with sunshine duration and precipitation in China. *Journal of Geophysical Research Atmospheres* 2015, 119, 13163–13179.

33. Du, H.; He, H.S.; Wu, Z.; Wang, L.; Zong, S.; Liu, J. Human influences on regional temperature change – comparing adjacent plains of China and Russia. *International Journal of Climatology* 2017, 37, 2913–2922.

34. Lindvall, J.; Svensson, G. The diurnal temperature range in the CMIP5 models. *Climate Dynamics* 2015, 44, 405–421.

Index